Coding Theory,
Cryptography
and
Related Areas

Springer
Berlin
Heidelberg
New York
Barcelona
Hong Kong
London
Milan
Paris
Singapore
Tokyo

Johannes Buchmann
Tom Høholdt
Henning Stichtenoth
Horacio Tapia-Recillas
Editors

Coding Theory, Cryptography and Related Areas

Proceedings of an International Conference
on Coding Theory, Cryptography and Related Areas,
held in Guanajuato, Mexico, in April 1998

 Springer

Johannes Buchmann
Fachbereich Informatik
Technische Universität Darmstadt
Alexanderstrasse 10
64283 Darmstadt, Germany

Tom Høholdt
Department of Mathematics, Bldg. 303
Technical University of Denmark
2800 Lyngby, Denmark

Henning Stichtenoth
Fachbereich 6, Mathematik und Informatik
Universität Gesamthochschule Essen
45117 Essen, Germany

Horacio Tapia-Recillas
Departamento de Matemáticas
Universidad Autónoma Metropolitana-Iztapalapa
Apartado Postal 55-532, C.P.
09340 México, D. F., México

Library of Congress Cataloging-in-Publication Data applied for

Die Deutsche Bibliothek - CIP-Einheitsaufnahme
Coding theory, cryptography and related areas : proceedings of an
International Conference on Coding Theory, Cryptography and
Related Areas, held in Guanajuato, Mexico, in April 1998 / Johannes
Buchmann ... (ed.). - Berlin ; Heidelberg ; New York ; Barcelona ;
Hong Kong ; London ; Milan ; Paris ; Singapore ; Tokyo : Springer,
2000
  ISBN 3-540-66248-0

Mathematics Subject Classification (1991): 11T71, 11Y16, 14C40, 94A60, 68P25, 12Fxx

ISBN 3-540-66248-0 Springer-Verlag Berlin Heidelberg New York

© Springer-Verlag Berlin Heidelberg 2000
Printed in Germany

Cover design: *design & production* GmbH, Heidelberg

Typeset by the authors. Reformatted by Kurt Mattes, Heidelberg

Printed on acid-free paper   SPIN 10716213   46/3143/LK – 5 4 3 2 1 0

Preface

The corruption by noise of information transmitted over a particular channel is addressed by error-correcting codes which systematically use the redundance inherent in the messages to allow recovery of the original information. Many new developments in this field, both theoretical and applied, have appeared since the seminal work of Shannon over a half century ago. Coding theory has become an integral component of many routine procedures, ranging from dynamic memories (Hamming codes) to compact discs (Reed-Solomon codes) and the transmission of information from satellites to ground stations (convolutional codes).

Noise is not the only form of interference that occurs in the transmission of information, however. The rapid worldwide growth of electronic communications that has led to our digital society implies enormous risks, including catastrophic failures or break-ins, with potential damage to those who depend upon such transmissions. The security of these systems is crucial for the smooth functioning of our world. Today, efficient and satisfactory protection mechanisms are being developed to provide diverse information security services, mostly based on cryptographic techniques. Research in cryptography is inherently an interdisciplinary endeavor in which areas from pure mathematics (number theory, algebraic geometry), computer sciences (design and analysis of algorithms, protocols), and electrical engineering (software and hardware implementations) converge.

Following a meeting on coding theory that took place in 1996 at the Université des Antilles et de la Guyane, Guadeloupe, it was suggested that a future meeting be held in Mexico. The relationship between coding theory and cryptography, and the need for further applications of these areas in modern society, led us to organize the **International Conference on Coding Theory, Cryptography and Related Areas** (ICCC) which took place in the city of Guanajuato, Mexico from 20–24 April, 1998. Several well-known researchers in both fields participated in the conference, including colleagues from Latin America, the Caribbean, Europe, and the USA. Invited speakers included: J. Buchmann (Tech. U. Darmstadt, Germany), R. Calderbank (AT&T Lab. Research, USA), T. Høholdt (Tech. Univ. of Denmark, Denmark), G. Lachaud (CNRS, Luminy, France), A.K. Lenstra (Citibank, USA), A. Odlyzko (AT&T Lab. Research, USA), R. Pellikaan (Eindhoven U. of Tech., The Netherlands), H. Stichtenoth (U. Essen, Germany). This conference also provided an opportunity to link the development of these areas in Mexico with the international community and to establish contacts between the new generations of students and researchers working on the front line. As a part of the conference program, minicourses on cryptography (N. Koblitz) and algebraic coding theory (C. Moreno), as well as a workshop on combinatorial cryptography for high school students (also conducted by N. Koblitz) were included.

It was a policy of the Editorial Committee of the Proceedings to maintain a high scientific standard, comparable to that of a journal, thanks to the cooperation of numerous referees, who willingly contributed to this effort.

The conference was organized by the Universidad Autónoma Metropolitana-Iztapalapa and the Instituto Politécnico Nacional. Sponsors include the above institutions as well as the Consejo Nacional de Ciencia y Tecnología (CONACyT); Sociedad Matemática Mexicana; International Centre for Theoretical Physics (ICTP), Trieste, Italy; Oficina Regional de Ciencia y Tecnología para América Latina, UNESCO; Banco de México; Citibank de México; Silicon Graphics de México S.A. de C.V.; SeguriData Privada S.A. de C.V., Mexico; and Infosel S.A. de C.V., Mexico.

We express our thanks to the staff and to all of those who helped in the organization of the Conference, particularly to Beatriz Arce and Emily McClung, as well as to the Centro de Investigación en Matemáticas (CIMAT, Guanajuato, Mexico) and the Tourism Office of the State of Guanajuato for their help in the local arrangements of the conference. We would also like to thank the Springer-Verlag staff for the preparation of these Proceedings.

Conference Committee

J.P. Cherdieu (U. Antilles et Guyane, Guadeloupe), T. Høholdt (Tech. Univ.of Denmark), N. Koblitz (U. of Washington, USA), G. Lachaud (CNRS, Luminy, France), D. LeBrigand (Paris VI, France), A. Menezes (U. Waterloo, Canada), O. Moreno (U. Puerto Rico, P.R.), C. Rentería (ESFM-IPN, Mexico), R. Rolland (CNRS, Luminy, France), J. Stern (ENS, France), H. Tapia-Recillas (UAM-I, Mexico), S. Vanstone (U. Waterloo, Canada).

Editorial Committee

J. Buchmann (Germany), T. Høholdt (Denmark), H. Stichtenoth (Germany), H. Tapia-Recillas (Mexico)

Mexico, D.F., April, 1999

If the error vectors are predetermined for each syndrome, we do not need an error correcting code which is practically attractive in the traditional sense. Most classes of error correcting codes used in practice are popular because they have a large minimum distance and because there exist efficient decoding algorithms for them, that is, a decoding algorithm that efficiently determines the *closest* codeword to a given received vector. However, these properties are not at all important in the application with the family of RN schemes.

3.1 NEW-1: Error Vectors from Inverses

¿From the discussion above, we might just select a parity check matrix $\mathbf{H}$ for a random $[n, k]$ code. On the other hand, what Alice and Bob do require in this application is an efficient way to find an error vector for a given syndrome. Given a syndrome $\mathbf{s}$, $\mathbf{e}(\mathbf{s}) = \mathbf{s} \cdot (\mathbf{H}^{-1})^T$ is always an error vector whose syndrome is $\mathbf{s}$, where $\mathbf{H}^{-1}$ is a right inverse of $\mathbf{H}$, that is a matrix such that $\mathbf{H}\mathbf{H}^{-1}$ is the identity matrix. Both users can compute the same error vector $\mathbf{e}(\mathbf{s})$. Alice will do it after choosing $\mathbf{s}$ randomly, and Bob will retrieve the same $\mathbf{e}(\mathbf{s})$ after computing the syndrome $\mathbf{s}$ in the usual way.

This right inverse is not unique, but if both Alice and Bob have access to a common deterministic procedure for obtaining $\mathbf{H}^{-1}$ from $\mathbf{H}$, then they no longer need to exchange the secret tables of error vectors as part of the key. The size of the exchanged key by now is limited to $n(n - k)$ q-ary symbols.

Hence, in the new scheme, which we will denote by the NEW-1 scheme, Alice will encrypt a message by

- selecting a random syndrome vector $\mathbf{s} \in \mathbb{F}_q^{n-k}$
- obtaining an error vector $\mathbf{e} = \mathbf{e}(\mathbf{s}) = \mathbf{s} \cdot (\mathbf{H}^{-1})^T$
- encrypting according to (2).

In order to retrieve the message $\mathbf{m}$, Bob will

- calculate the syndrome $\mathbf{s} = \mathbf{c}\mathbf{H}^T$,
- obtain $f_\mathbf{s}(\mathbf{m})\mathbf{G}$ by subtracting from $\mathbf{c}$ the error vector $\mathbf{e} = \mathbf{e}(\mathbf{s}) = \mathbf{s} \cdot (\mathbf{H}^{-1})^T$
- obtain $f_\mathbf{s}(\mathbf{m})$,
- apply the inverse permutation $f_\mathbf{s}^{-1}$ to obtain the message $\mathbf{m}$.

3.2 NEW-2: Permutation as Key

In order to further reduce the size of the exchanged key, we could use a public, standard $[n, k]$ code defined by (random) matrices $\mathbf{G_{PUBL}}$ and/or $\mathbf{H_{PUBL}}$. We will denote this by the NEW-2 scheme. The secret key is then just a permutation $\mathbf{P}$ of the n coordinates. Clearly the number of keys in this case is $n!$, so the key size (depending on the representation) is lower bounded by $\lceil \log_2(n!) \rceil$ bits. However, when n is large it could be a complex task to map an arbitrary permutation into $\lceil \log_2(n!) \rceil$ bits and back. A straightforward encoding of $\mathbf{P}$, where each coordinate's position in the permutation is given by a $\lceil \log_2(n) \rceil$-bit number, yields a key of size $n \lceil \log_2(n) \rceil$ bits. It is however easy to obtain a more efficient representation of the key, for instance in the following way.

Let $n' = 2^{\lfloor \log_2(n) \rfloor}$ be the largest power of two smaller than or equal to n. The $n - n'$ first positions in the permutation are represented by $\log_2 n' + 1$ bits each. When these positions have been listed, the remaining n' positions can be renumbered, and the next $n'/2$ positions can be represented by a $\log_2 n'$ bit-number each, and so on. In this way the required number of bits to represent the permutation is

$$
\begin{aligned}
(n - n') \, (\log_2 n' + 1) + \sum_{i=1}^{\log_2 n'} i 2^{i-1} \\
= (n - n') \, (\log_2 n' + 1) + n'(\log_2 n' - 1) + 1 \\
= n \, (\log_2 n' + 1) - 2n' + 1,
\end{aligned}
\tag{4}
$$

which is usually much smaller than $n(n - k)$. Figure 1 shows the number of bits needed to represent the permutation for the different representation schemes.

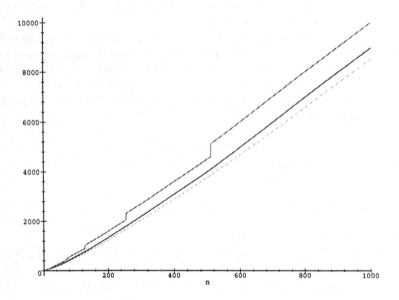

Fig. 1. Key size of different representation schemes for the permutation. Upper line: $n(\log_2 n' + 1)$. Middle line: $n(\log_2 n' + 1) - 2n' + 1$. Lower line: $\lceil \log_2(n!) \rceil$.

With respect to encryption and decryption, NEW-2 works like NEW-1, with generator matrix $\mathbf{G} = \mathbf{G_{PUBL}P}$ and parity check matrix $\mathbf{H} = \mathbf{H_{PUBL}P}$, where the permutation is represented as a matrix product. In order to increase the security of the system, the non-linear function f acting on the message in eq. (2) should be indexed by both $\mathbf{s}$ and $\mathbf{P}$.

Thus, the scheme NEW-2 will work as follows:

- Public: Parity check matrix $\mathbf{H_{PUBL}}$.
- Secret key: A binary vector $\mathbf{P}$, denoting a permutation. This will also be used as an index into the nonlinear functions
- Preprocessing: Compute $\mathbf{H} = \mathbf{H_{PUBL}P}$, and a generator matrix $\mathbf{G}$ and a right inverse $\mathbf{H^{-1}}$ both derived from $\mathbf{H}$ by deterministic algorithms.

Encryption:

- select a random syndrome vector $\mathbf{s} \in \mathbb{F}_q^{n-k}$,
- obtain an error vector $\mathbf{e} = \mathbf{e(s)} = \mathbf{s} \cdot [\mathbf{H^{-1}}]^T$. Thus the syndrome of this vector will always be $\mathbf{s}$.
- encrypt according to (2).

In order to retrieve the message $\mathbf{m}$, Bob will

- calculate the syndrome $\mathbf{s} = \mathbf{cH}^T$.
- obtain $f_{\mathbf{s},\mathbf{P}}(\mathbf{m})\mathbf{G}$ by subtracting from $\mathbf{c}$ the error vector $\mathbf{e} = \mathbf{e(s)} = \mathbf{s} \cdot (\mathbf{H^{-1}})^T$
- obtain $f_{\mathbf{s},\mathbf{P}}(\mathbf{m})$,
- apply the inverse function $f_{\mathbf{s},\mathbf{P}}^{-1}$ to obtain the message $\mathbf{m}$.

4   Security

The security of the system relies critically on the implementation of the nonlinearizing function $f_{\mathbf{s},\mathbf{P}}()$. In general, both the syndrome $\mathbf{s}$ and the secret key $\mathbf{P}$ should influence that function, and it could be implemented by putting together various substitution and permutation components, each indexed by $\mathbf{s}$ and/or $\mathbf{P}$. A simple example is shown in Section 5. In the current section we first discuss how the nonlinearizing function $f_{\mathbf{s},\mathbf{P}}()$ should be implemented. Then some potential attacks are presented and analyzed.

4.1   Nonlinearizing Functions: Design Principles

The design of the nonlinearizing functions of (2) is not a straightforward task, and there are a number of pitfalls that should be avoided. In this section we discuss some principles for the design.

- The function $f_{\mathbf{s},\mathbf{P}}()$ should depend on the secret key $\mathbf{P}$. Otherwise, for each message $\mathbf{m}$, the sets $C = \{f_{\mathbf{s}}(\mathbf{m})\mathbf{G_{PUBL}} + \mathbf{e(s)}, \mathbf{s} \in F_{q^{n-k}}\}$ and $C_P = \{f_{\mathbf{s}}(\mathbf{m})\mathbf{G_{PUBL}P} + \mathbf{e(s)P}, \mathbf{s} \in F_{q^{n-k}}\}$ will differ in only the permutation $\mathbf{P}$. Assuming that Eve can study chosen plaintext-ciphertext pairs, both of these sets are accessible to her, and this gives Eve a possibility of attacking the system. One way to accomplish this is by grouping the words $\mathbf{x}$ of C and C_P by their *weight vector* $w(\mathbf{x}) = (a_0, a_1, ...a_{q-1})$, where $a_j = |\{i : x_i = j\}|$ for $j = 0, ..., q-1$, in each set. Unless the sets are too large, Eve can attempt to align them by use of the weight vectors, and then she can learn a lot about the permutation $\mathbf{P}$ by comparing each pair of those sets.

– Another design decision regards what should be the effect of the syndrome
  s on the permutations $f_{\mathbf{s},\mathbf{P}}()$. The syndrome is an $(n-k)$-symbol vector,
  while the function $f_{s,\mathbf{P}}()$ is a permutation of the space of k-symbol vectors.

In view of these two remarks, as a first attempt we would suggest to include
(for instance) the first ($\mathbf{e_F}$) and last ($\mathbf{e_L}$) k-bit subvector of $\mathbf{e(s)}$ as additive
elements of $f_{\mathbf{s},\mathbf{P}}()$. That is, as a building block for the $f_{\mathbf{s},\mathbf{P}}()$ function we propose
using the function $\mathbf{m} \mapsto \mathbf{m} + \mathbf{e_F}(\mathbf{s}) + \mathbf{e_L}(\mathbf{s}) = \mathbf{m} + \mathbf{e(s)}\mathbf{P}(\mathbf{A}_1 + \mathbf{A}_2)$ where
$\mathbf{A}_1 = (\mathbf{I}_k|0)^T$, and $\mathbf{A}_2 = (0|\mathbf{I}_k)^T$ (both $\mathbf{A}_1$ and $\mathbf{A}_2$ of size $k \times n$).

However, note that in this building block of $f_{\mathbf{s},\mathbf{P}}$, the dependency on **s** is
linear and also the dependency on **P** disappears when $\mathbf{s} = \mathbf{0}$. There should also
be a nonlinear dependency of $f_{\mathbf{s},\mathbf{P}}()$ on **s** and a stronger and also nonlinear
dependency on **P**.

– To achieve a nonlinear dependency on **s**, we propose the use of a function
  $P_\mathbf{s}$ consisting of a composition of permutations both on the k message coordinates and on the whole message space, indexed by **s** and with an easy
  description in terms of that **s**. For instance, simultaneous swaps of successive
  subdivisions of the message coordinates or of the whole message space, rotational shifts, and/or arithmetical operations modulo some radix. An example
  of this can be seen in Section 5.
– A natural way of constructing a non-linear function depending on the key
  permutation **P** and acting on the q^k elements of the whole message space
  would be, for instance, as follows.
  Let $\mathbf{P} = [p(1), p(2), ...p(n)]$ be the secret key in list format. To construct
  a permutation $P_\mathbf{P}$ that acts on the message space with q^k messages, let
  $P_\mathbf{P} = [p(1), p(2)....., p(n), n + p(1), n + p(2), ...n + p(n), 2n + p(1), ...]$. For
  convenient implementation and a balanced output of $P_\mathbf{P}$, n should divide
  q^k, that is, n should be a power of p where $q = p^r$ for some r.

Thus the nonlinearizing function could be on the form $f_{\mathbf{s},\mathbf{P}}(\mathbf{m}) = P_\mathbf{P}(P_\mathbf{s}(\mathbf{m} + \mathbf{e_F}(\mathbf{s}) + \mathbf{e_L}(\mathbf{s})))$.

– It can be argued that the function $f_{\mathbf{s},\mathbf{P}}(\mathbf{m})$ should satisfy the condition
  that, for all (or most?) permutations **P** and for each message **m**, the sum
  $\sum_\mathbf{s} f_{\mathbf{s},\mathbf{P}}(\mathbf{m})$ should be zero, to defend against attack number 3 in the next
  subsection. This does however impose a restriction on the function that may
  be unwanted since at the same time we want it to be as "unpredictable" as
  possible.

4.2 Attacks

Here is a list of potential attacks on the system:

1. Computing and storing all cryptograms of chosen plaintexts, in order to
   determine the key. Workload and storage requirements: $O(q^n)$.

2. Attacks in the particular and degenerate case of $k = n$: In this case the system becomes a transposition cipher, $\mathbf{c} = f_{\mathbf{P}}(\mathbf{m})\mathbf{G}$. Then in a chosen plaintext attack, Eve can easily obtain a basis for the vector space of cryptograms. Even if it is still a complicated task to align this permuted vector space with the equivalent space defined by $\mathbf{H}_{\mathbf{PUBL}}$, Eve will obtain a method to systematically characterize and perhaps eventually determine $\mathbf{P}$, so we want to avoid this. If $n - k > 0$, then the error vector corresponding to the random syndrome serves to make it much more difficult to attempt a similar attack.

3. Note that if Eve can generate all the random cryptograms corresponding to a fixed message $\mathbf{m}$, she can calculate

$$\sum_{\mathbf{s}} \left[f_{\mathbf{s},\mathbf{P}}(\mathbf{m})\mathbf{G} + \mathbf{s} \cdot [\mathbf{H}^{-1}]^T \right] = \sum_{\mathbf{s}} f_{\mathbf{s},\mathbf{P}}(\mathbf{m})\mathbf{G} + \sum_{\mathbf{s}} \mathbf{s} \cdot [\mathbf{H}^{-1}]^T \quad (5)$$

$$= \sum_{\mathbf{s}} f_{\mathbf{s},\mathbf{P}}(\mathbf{m})\mathbf{G},$$

which is a codeword. By comparing with the public parity check matrix $\mathbf{H}_{\mathbf{PUBL}}$, this yields much information about the secret key $\mathbf{P}$. To defend against this attack, we can require that $\sum_{\mathbf{s}} f_{\mathbf{s},\mathbf{P}}(\mathbf{m})$ always is equal to zero. Alternatively, we can make the redundancy $n - k$ of the system so large that it is infeasible to calculate all the q^{n-k} syndrome vectors. A third solution to the problem will be presented in a future paper.

4. Trying all permutations. Workload and storage requirements: $O(n!)$. Recall that for fixed q and n sufficiently large, $n! >> q^n$ so attack 1 is more efficient.

5. Attacks based on the specific choice of public parity check matrix $\mathbf{H}_{\mathbf{PUBL}}$, for instance by exploiting peculiarities of the weight distribution. However, almost all codes have a "binomial-like" weight distribution, that is, the number of codewords of weight w is approximately

$$\binom{n}{w} q^{k-n}.$$

There are in general no such peculiarities (or at least they are easy to avoid when setting up the system.)

4.3   Suggested Parameters

Based on the attacks listed in the previous subsection, we should select the parameters so that (attack 1) q^n is a large number. Similarly (attack 2) $n - k$ should be nonzero, although from our discussion above it is not obvious how large this redundancy should be. In fact, if attack 3 is not dealt with in other ways, q^{n-k} should be so large that it will be infeasible to calculate all the syndrome vectors. For example, for the NEW-2 scheme with $q = 2^8$, $k = 20$, and $n = 30$, each of the attacks (1,3,4) above will require a work factor of at least $q = 2^{80}$, for a keysize of (see (4)) 119 bits.

5   A Small Example

This section contains a toy example of the NEW-2 scheme. In this example $q = 2$.

Let

$$\mathbf{H_{PUBL}} = \begin{pmatrix} 1 1 1 0 \\ 1 0 0 1 \end{pmatrix} \tag{6}$$

Since $n = 4$, it takes $5 = 2 + 2 + 1$ bits to specify a permutation. For instance, if we assume that the coordinates are labelled 0,1,2,3, then the first two key bits specify the new position of the zeroth coordinate, the next two specify the new position of the first coordinate, and the final key bit determine which of the two remaining positions will be the new position of the second coordinate. For instance, if $\mathbf{P} = (1, 3, 0) = (01, 11, 0) = (p_0, p_1, p_2, p_3, p_4)$, then column 0 of $\mathbf{H_{PUBL}}$ becomes the column in position 1 in $\mathbf{H}$, and the column in position 1 of $\mathbf{H_{PUBL}}$ becomes the column in position 3 in $\mathbf{H}$. Now the column in position 2 of $\mathbf{H_{PUBL}}$ can end up in two possible positions, 0 or 2; the final key bit 0 specifies the first one of these. Thus

$$\mathbf{H} = \begin{pmatrix} 1 1 0 1 \\ 0 1 1 0 \end{pmatrix}. \tag{7}$$

Suppose that some deterministic algorithm gives us the generator matrix

$$\mathbf{G} = \begin{pmatrix} 0 1 1 1 \\ 1 0 0 1 \end{pmatrix}. \tag{8}$$

and the inverse matrix

$$(\mathbf{H^{-1}})^T = \begin{pmatrix} 1 1 1 1 \\ 0 0 1 0 \end{pmatrix}. \tag{9}$$

Now suppose that the syndrome $\mathbf{s} = (s_0, s_1)$ and that the nonlinear function $f_{\mathbf{s}, \mathbf{P}}()$ is given in the following way:

- first $\mathbf{m} \mapsto \mathbf{m}' = \mathbf{m} + e_\mathbf{F}(\mathbf{s}) + e_\mathbf{L}(\mathbf{s})$
- now the syndrome $\mathbf{s}$ can represent a number from 0 to 3, so we can use the permutation $\mathbf{P}$ acting on $\mathbf{s}$. We implement this in two steps, so that the first step is a function of $\mathbf{s}, \mathbf{P}$ and $\mathbf{m}$ into the space $\mathbb{F}_2^2$, highly nonlinear but not evenly distributed over $\mathbb{F}_2^2$, and the second step is a simple function from $\mathbb{F}_2^2 \times \mathbb{F}_2^2$ into $\mathbb{F}_2^2$ that guarantees an even distribution. For example (1) $\mathbf{m}'' := \mathbf{m}' +$ (the Hamming weight of $\mathbf{s}$ plus the Hamming weight of $\mathbf{P}$) mod 4, (2) $\mathbf{m}''' := \mathbf{P}((\mathbf{m}'' + 2s_0 + s_1) \mathrm{mod} 4)$, return $\mathbf{m}'''$. (For the sake of the example, please observe that here the additions are interpreted as normal mod 4 additions.)

For larger parameters we would prefer this function to be composed of various operations like swapping of left and right halves, rotational shifts, and arithmetic operations modulo some radix, as we had already remarked.

Now both the sender and the receiver can proceed to carry out encryption and decryption as outlined in Subsection 3.2.

For example, assume that the message $\mathbf{m} = (1,0)$. Then Alice obtains a random syndrome $\mathbf{s}$, say $\mathbf{s} = 2 = (1,0)$. This gives $\mathbf{e}(\mathbf{s}) = \mathbf{s} \cdot [\mathbf{H}^{-1}]^T = (1,1,1,1)$, Hence,

$$\mathbf{m}' = (1,0) + (1,1) + (1,1) = (1,0) = 2 \tag{10}$$

$$\mathbf{m}'' = 2 + 1 + 3 \bmod 4 = 2 \tag{11}$$

$$\mathbf{m}''' = \mathbf{P}(2 + 2 + 0) = \mathbf{P}(0) = 1 = (0,1) \tag{12}$$

Now $f_{\mathbf{s},\mathbf{P}}(\mathbf{m}) \cdot \mathbf{G} = (1,0,0,1)$, $\mathbf{e}(\mathbf{s}) = \mathbf{s} \cdot [\mathbf{H}^{-1}]^T = (1,1,1,1)$, so the cryptogram is $\mathbf{c} = (0,1,1,0)$.

Bob will calculate the syndrome, $\mathbf{s} = \mathbf{c} \cdot \mathbf{H}^T = (1,0)$ (unless we made a mistake), and can then proceed to calculate and eliminate $\mathbf{e}$. He can also determine $f_{\mathbf{s},\mathbf{P}}(\mathbf{m})$ and eventually $\mathbf{m} = (1,0)$.

For this particular key $\mathbf{P}$, the possible cryptograms are

$\mathbf{m} \downarrow \mathbf{s} \rightarrow$	00	01	10	11
00	0111	0101	1111	1101
01	1001	1011	1000	1010
10	1110	1100	0110	0100
11	0000	0010	0001	0011

We chose this example because it is simple and illustrates some of the ideas we have discussed. However, the table above shows some apparent weaknesses, note for instance that the set of cryptograms associated with $\mathbf{m} = 11$ make up a vector space. This will always happen for $q = k = n - k = 2$ with this type of f function (which acts, for a fixed $\mathbf{m}$ and $\mathbf{P}$, as a permutation of the space of syndrome vectors), but it will be a rare event for larger parameters.

6   Summary

We have described a way to simplify the key management in the Rao-Nam cryptosystem. The size of the exchanged and stored keys is dramatically reduced, compared to the RNST scheme. This reduction in key size does not in an obvious way decrease the security of the system. However, the actual security depends on the implementation of the nonlinear functions involved.

References

1. E. R. Berlekamp, R. J. McEliece, and H. van Tilborg, "On the inherent intractability of certain coding problems", *IEEE Transactions on Information Theory*, 24 (1978), pp. 384-386.
2. T. Berson, "Failure of the McEliece public-key cryptosystem under message-resend and related-message attack", in: *Lecture Notes in Computer Science 330; Advances in Cryptology - Proceedings of CRYPTO'97*, Springer, 1998.
3. E. F. Brickell and A. M. Odlyzko, "Cryptanalysis - A Survey of Recent Results", in: *Contemporary Cryptology: The Science of Information Integrity*, G. Simmons (Ed.), Ch. 10, IEEE Press, 1992.
4. P. J. M. Hin, "Channel-error-correcting privacy cryptosystem", Ph. D. Thesis, Delft University of Technology (1986, in Dutch).
5. P. J. Lee and E. F. Brickell, "An observation on the security of McEliece's public-key cryptosystem", in: *Lecture Notes in Computer Science 330; Advances in Cryptology - Proceedings of CRYPTO'88*, pp. 275-280, Springer, 1988.
6. R. J. McEliece, "A public-key cryptosystem based on algebraic coding theory", DSN progress Report 42-44, Jet Propulsion Laboratory, Pasadena, 1978.
7. T. R. N. Rao and K. H. Nam, "Private-Key Algebraic-Coded Cryptosystem", in: *Lecture Notes in Computer Science 263; Advances in Cryptology - Proceedings of CRYPTO'86*, A. M. Odlyzko (Ed.), pp. 35-48, Springer, 1987.
8. R. Struik and J. van Tilburg, "The Rao-Nam scheme is insecure against a chosen-plaintext attack", in: *Lecture Notes in Computer Science 293; Advances in Cryptology - Proceedings of CRYPTO'87*, C. Pomerance (Ed.), pp. 445-457, Springer, 1988.

Efficient Reduction on the Jacobian Variety
of Picard Curves

Ernesto Reinaldo Barreiro[1], Jorge Estrada Sarlabous[1],
and Jean-Pierre Cherdieu[2]

[1] Department of Geometry and Combinatorics. CEMAFIT/ICIMAF, Calle E No.
309, esquina 15, Vedado, La Habana, Cuba. matdis@cidet.icmf.inf.cu
[2] Département de Mathématiques et Informatique. Université des Antilles et de la
Guyane. Campus de Fouillole, F97159 Pointe-à-Pitre CEDEX.
jean-pierre.cherdieu@univ-ag.fr

Abstract. In this paper, a system of coordinates for the elements on the
Jacobian Variety of Picard curves is presented. These coordinates possess
a nice geometric interpretation and provide us with an unifying environment to obtain an efficient algorithm for the reduction and addition of
divisors. Exploiting the geometry of the Picard curves, a completely effective reduction algorithm is developed, which works for curves defined
over any ground field k, with $char(k) \neq 3$.

In the generic case, the algorithm works recursively with the system
of coordinates representing the divisors, instead of solving for points in
their support. Hence, only one factorization is needed (at the end of the
algorithm) and the processing of the system of coordinates involves only
linear algebra and evaluation of polynomials in the definition field of
the divisor D to be reduced. The complexity of this deterministic reduction algorithm is $O(deg(D))$. The addition of divisors may be performed
iterating the reduction algorithm.

1 Introduction

In the present paper we present a fast and completelly effective algorithm for the
reduction of divisors on the Jacobian Variety of Picard curves. Our algorithm
works correctly for any Picard curve $C(k)$ defined over a field k with $char(k) \neq$
3. This algorithm is an improvement the seudo-algorithm presented in [5]: the
modifications introduced are based on a refinement of the system of coordinates
guiven in [5]. The complexity of this new algorithm is linear in the degree of the
effective divisor D to be reduced. Additionaly it has certain features permiting us
to diminish the cost of the computation of a large multiple of a point, as well as
obtaining explicit formulas wich in fact are useful to reduce the complexity. We
have tested the algorithm using the symbolic computation language MAPLE V,
furthermore, some non-trivial examples are shown.

The Picard Curves are genus *three* plane projective curve which has been
intensively studied due to their conection with certain Hilbert's problems (c.f.
[7–10]) as well as to the study of some linear error correcting codes (c.f. [6]).

2   Notations and Terminology

Let k be an arbitrary field and $\overline{k}$ its algebraic closure. Let $X(\overline{k})$ be a k-defined plane projective curve in $\mathbb{P}^2_{\overline{k}}$ (here k-defined means that the polynomial defining $X(k)$ has all its coefficients in k) and $K_{X(\overline{k})}$ be the field of rational functions on $X(\overline{k})$. Let also $X(k)$ be the subset of k-rational points of $X(\overline{k})$ and $K_{X(k)}$ be the subfield of k-rational functions on $X(k)$.

A divisor D on $X(\overline{k})$ is a formal sum

$$D = \sum_{P \in X(\overline{k})} m_P P, \ m_P \in \mathbb{Z},$$

where all but a finite set of the m_P are zero (i.e. D is an element of the free abelian group $Div(X(\overline{k}))$ generated by the elements of $X(\overline{k})$). Given D we associate to it the number

$$\deg(D) = \sum_{P \in X(\overline{k})} m_P.$$

The map $D \to deg(D)$ is an homorphism from $Div(X(\overline{k}))$ onto $\mathbb{Z}$. A divisor D is said to be k-rational iff all its points have coordinates in k (i.e $m_P \neq 0 \Rightarrow P \in X(k)$) and k-defined iff $D = D^\sigma$ for all σ in the absolute Galois group $G = Gal(\overline{k}/k)$.

To any element f in $K_{X(\overline{k})}$ we associate the divisors $(f)_0$ and $(f)_\infty$ of zeros and poles of f, respectivelly. Denote also by $(f) = (f)_0 - (f)_\infty$ the divisor of f.

A divisor D is said to be principal iff there exists a rational function f such that $D = (f)$. The fact $\deg((f)) = 0$ joined to $(f \cdot g) = (f) + (g)$ shows that the set $\mathcal{P}(X(\overline{k}))$ of principal divisors forms a subgroup of the group $Div_0(X(\overline{k}))$ of divisor of degree zero. Then, the quotient group

$$J(X(\overline{k})) = Div_0(X(\overline{k}))/\mathbb{P}(X(\overline{k}))$$

is called the Jacobian variety of $X(\overline{k})$. We may consider also the subgroup $J_k(X(\overline{k}))$ of k-rational points of the Jacobian, i.e. the set of classes $[x]$ wich are invariant by the action of the absolute Galois group G ($[x] = [x]^\sigma$ for all $\sigma \in G$).

If k is a finite field it is known that $J_k(X(\overline{k}))$ is a finite abelian group.

3   Some Geometric Facts About Picard Curves

Let k be an arbitrary field of $char(k) \neq 3$, and let $\overline{k}$ denote its algebraic closure.

Definition 1. *A Picard curve $C_{p_4}(k)$ is a genus three plane projective curve with model:*

$$C_{p_4}(k) : WY^3 - W^4 p_4(\frac{X}{W}) = 0 \tag{1}$$

where $p_4(x) = x^4 + a_3 x^3 + a_2 x^2 + a_1 x + a_0$ is a polynomial in $k[x]$.

For $char(k) \neq 3$ it is not difficult to prove that $C_{p_4}(k)$ will be non-singular if and only if the discriminant of p_4 is different from *zero* (i.e. p_4 has no multiple roots in $\bar{k}$). Moreover, every curve $C_{p_4}(k)$ is birationally equivalent to a Picard curve $C_{\hat{p}_4}(k)$, with $\hat{p}_4(x) = x^4 + \hat{a}_2 x^2 + \hat{a}_1 x + \hat{a}_0$ (c.f. [4]), hence, without lost of generality, we may suppose in (1) $a_3 = 0$.

If the field k is algebraically closed every Picard curve $C_{p_4}(k)$ has five total ramification points $R_1, \ldots, R_5$ with respect to the covering morphism

$$\pi_x : C_{p_4}(k) \xrightarrow{(x:y:w) \longmapsto x} \mathbb{P}^1_k$$

The points $R_i = (r_i : 0 : 1)$, $i = 1, \ldots, 4$, where r_i are the roots of $p_4(x)$ and $R_5 = P_\infty = (0 : 1 : 0)$, the point at infinity on $C_{p_4}(k)$. Moreover, if ξ represents a primitive cubic root of unity in k (i.e. $\xi^2 + \xi + 1 = 0$) the mapping

$$\sigma : C_{p_4}(k) \xrightarrow{(x:y:w) \longmapsto (x:\xi y:w)} C_{p_4}(k)$$

is an automorphism of $C_{p_4}(k)$ satisfying:

$$\pi_x \circ \sigma = id_{\mathbb{P}^1_k} \text{ and } \sigma^3 = id_{C_{p_4}(k)}.$$

Given two points P_1 and P_2 we call them *conjugate* if $P_1 = \sigma(P_2)$ or $P_2 = \sigma(P_1)$ (from here on, we will denote $\sigma(P)$ simply by σP).

Lemma 1. *Let $C_{p_4}(k)$ be a non-singular Picard curve. Then the effective divisors of the canonical class K, of $C_{p_4}(k)$, are those which are the intersection of lines with $C_{p_4}(k)$.*

Proof. It is an easy consequence of the fact that:

$$\omega_1 = \frac{1}{y}dx, \ \omega_2 = \frac{x}{y^2}dx, \ \omega_3 = \frac{1}{y^2}dx \tag{2}$$

where

$$x = \frac{X}{W} \text{ and } y = \frac{Y}{W} \tag{3}$$

constitute a basis of $\Omega(0)$ (c.f. [2]). $\square$

4   Assigning Coordinates to Divisors

In this section $C_{p_4}(k)$ will be a fixed Picard curve, hence we will denote it simply by C.

Definition 2. *Given an affine effective divisor D, i.e. $D \succeq 0$ and $P_\infty \notin supp(D)$, we call it* semireduced *if there exists no P_1 such that $D \succeq P_1 + \sigma P_1 + \sigma^2 P_1$. We set also*

$$Div^{+,i}(C) := \{D \in Div(C) \mid D \text{ k-rational semireduced of degree } i\}$$

for $i \geq 0$, and $\mathcal{D}(r,s) = \bigcup_{i=r}^{s} Div^{+,i}(C)$ for $0 \leq r < s$.

Given an effective affine divisor D we will consider the linear space $\mathcal{H}(D)$ of rational functions of the form $f(x, y)$, for $f \in k[X_1, X_2]$, $x = \frac{X}{W}$ and $y = \frac{Y}{W}$, satisfying $(f)_0 \succeq D$. Note that $\nu_{P_\infty}(x) = -3$ and $\nu_{P_\infty}(y) = -4$ ($\nu_{P_\infty}(*)$ denotes the valuation of $*$ at P_∞), hence, for $f \in \mathcal{H}(D)$, $f(x, y) = \sum_{i,j} a_{ij} x^i y^j$ we have:

$$\nu_{P_\infty}(f(x, y)) = \min_{i,j} \nu_{P_\infty}(a_{ij} x^i y^j).$$

Then, we call the term $a_{i_1 j_1} x^{i1} y^{j1}$ that exactly realizes the previous equality the *leading term* of $f(x, y)$ and if $a_{i_1 j_1} = 1$ then $f(x, y)$ is called *monic*. Given an affine effective divisor D, we will assign to it the unique monic function $f(x, y)$ in $\mathcal{H}(D)$ with greatest valuation at P_∞.

For $D \in \mathcal{D}(2, 4)$ this "minimal" polynomial rational function has the general form

$$v_D(x, y) = a_{20} x^2 + a_{10} x + a_{01} y + a_{11} xy + a_{20} y^2 + a_{00},$$

hence, we call v_D the *interpolating conic* of D. Note that in certain cases the conic may degenerate in a line and that $v_D(x, y)$ satisfies

$$\nu_{P_\infty}(v_D(x, y)) \le (5 - deg(D)) - 8. \tag{4}$$

From here on, we will identify $f(x, y) \in \mathcal{H}(D)$ with the coresponding polynomial in $k[X_1, X_2]$.

Lemma 2. *Let* $v(x, y) = a_{20} x^2 + a_{10} x + a_{01} y + a_{11} xy + a_{20} y^2 + a_{00}$. *Then the following equivalences hold:*

1. $\nu_{P_\infty}(v(x, y)) = -7 \Leftrightarrow a_{02} = 0$ *and* $a_{11} \ne 0$.
2. $\nu_{P_\infty}(v(x, y)) = -6 \Leftrightarrow a_{02} = 0$, $a_{11} = 0$ *and* $a_{20} \ne 0$.
3. $\nu_{P_\infty}(v(x, y)) = -4 \Leftrightarrow a_{02} = 0$, $a_{11} = 0$, $a_{20} = 0$ *and* $a_{01} \ne 0$.

Proof. Consider the local parameter $t = \frac{X}{Y}$ at P_∞, and impose the required vanishing conditions on $v_D(x, y)$. $\square$

Definition 3. *Given a divisor* D *of degree* ≥ 3 *we call* D *collinear if there exist three points* P_1, P_2, P_3 *in* supp(D) *and a line* r_0 *such that* $(r_0)_0 \succeq P_1 + P_2 + P_3$. *Otherwise,* D *is called* generic.

Lemma 3. *Given a divisor* D *in* $\mathcal{D}(3, 4)$, *the following propositions are equivalent:*

a) $v_D(x, y)$ *is* linear *or factorizes in* linear *factors.*
b) $D + (5 - deg(D))P_\infty$ *is collinear.*
c) $v_D(x, y) = a_{20} x^2 + a_{10} x + a_{01} y + a_{11} xy + a_{00}$ *with* $a_{11}^2 a_{00} + a_{01}^2 a_{20} - a_{11} a_{01} a_{10} = 0$.

Note: recall that after (4) and lemma 2 we may assume $a_{02} = 0$.

Proof. a) $\Rightarrow$ c). If $v_D(x, y)$ is a line then $a_{11} = a_{20} = 0$. Hence $a_{11}^2 a_{00} + a_{01}^2 a_{20} - a_{11}a_{01}a_{10} = 0$ holds. If $v_D(x, y)$ is of *degree two* then, after lemma 2, $a_{02} = 0$. Furthermore, $v_D(x, y)$ factorizes in linear factors if and only if

$$4 \times \det \begin{pmatrix} a_{20} & \frac{a_{11}}{2} & \frac{a_{10}}{2} \\ \frac{a_{11}}{2} & 0 & \frac{a_{01}}{2} \\ \frac{a_{10}}{2} & \frac{a_{01}}{2} & a_{00} \end{pmatrix} = a_{11}^2 a_{00} + a_{01}^2 a_{20} - a_{11}a_{01}a_{10} = 0.$$

c) $\Rightarrow$ b). If $a_{11}^2 a_{00} + a_{01}^2 a_{20} - a_{11}a_{01}a_{10} = 0$ then if $a_{11} = a_{20} = 0$ then obviously holds b). Else, depending on whether $a_{11} = a_{01} = 0$ or not we get either

$$v_D(x, y) = (a_{20}x^2 + a_{10}x + a_{00}) = r_1 r_2$$

or

$$v_D(x, y) = (a_{20}x + a_{11}y + a_{10} - a_{20}a_{01}/a_{11})(x + a_{01}/a_{11}) = r_1 r_2.$$

In any case v_D factorizes as a product of lines. Then, if $D \in Div^{+,4}(C)$, $D + P_\infty$ contains five points and at least three of them belong to r_1 or r_2. If $D \in Div^{+,3}(C)$ then the same reasoning applies to $D + 2P_\infty$.

b) $\Rightarrow$ a). Follows directly from Bezout's theorem. $\square$

Let's denote by Φ the correspondence

$$\Phi : \mathcal{D}(2, 4) \longrightarrow k[x] \times k[x, y] \times k[y]$$

which assigns to a divisor D the 3-uple of polynomials $(u_D(x), v_D(x, y), w_D(y))$ where:

$$u_D(x) = \prod_{P_i \in supp(D)} (x - x_i) \tag{5}$$

$$w_D(y) = \prod_{P_i \in supp(D)} (y - y_i) \tag{6}$$

$$v_D(x, y) = \text{the } \textit{interpolating conic} \text{ at } D, \tag{7}$$

where, $P_i = (x_i : y_i : 1)$. Unfortunately, the mapping Φ fails to be injective on $\mathcal{D}(2, 4)$: let x_1 and x_2 be elements of k satisfying $p_4(x_1) = p_4(x_2) \neq 0$, and suppose y_0 is a root of $y^3 - p_4(x_1) = 0$, then the divisors

$$D_1 = (x_1 : y_0 : 1) + (x_1 : \xi y_0 : 1) + (x_2 : y_0 : 1) + (x_2 : \xi^2 y_0 : 1),$$
$$D_2 = (x_1 : y_0 : 1) + (x_1 : \xi^2 y_0 : 1) + (x_2 : y_0 : 1) + (x_2 : \xi y_0 : 1)$$

have the same image by Φ. Nevertheless, if we restrict Φ to the set

$$\mathcal{D}_0(2, 4) = \bigcup_{i=2}^{4} Div_0^{+,i},$$

where,

$$Div_0^{+,i}(C) = \{D \in Div^{+,i}(C) \mid D \text{ contains no pair of } conjugate \text{ points}\},$$

for $i = 2, 3$, and

$$Div_0^{+,4}(C) = \{D \in Div^{+,i}(C) \mid D \neq P_1 + \sigma P_1 + P_2 + \sigma P_2,\}$$

we obtain:

Lemma 4. *The correspondence Φ restricted to $\mathcal{D}_0(2,4)$ defines a bijection onto its image $\Phi(\mathcal{D}_0(2,4))$.*

Proof. For D in $Div_0^{+,2}$, $Div_0^{+,3}$ or D in $Div_0^{+,4}$, with $D + P_\infty$ *generic*, after lemma 3, we obtain that $v_D(x,y)$ is a conic (or a line) whose coefficient of y is a polynomial in x not vanishing in the x-coordinates of the points in D. Therefore, factoring $u_D(x)$ we can recover the x-coordinates of the points on $supp(D)$ and substituting in $v_D(x,y)$ we find the y-coordinates. The remaining cases are:

1. $D = P_1 + P_2 + P_3 + P_4$ with $P_1 + P_2 + P_3$ *collinear*, $P_4 \neq \sigma^k P_i$, $k = 1, 2$, $i = 1, 2, 3$. Then $v_D(x,y) = r_1(x,y) \cdot (x - x_4)$, where $(r_1)_0 \succeq P_1 + P_2 + P_3$, $r_1 = \alpha x + \beta y + \gamma$, $\beta \neq 0$. Factoring $u_D(x)$ and substituting in r_1 we recover P_1, P_2, P_3. The y-coordinate of P_4 is obtained as the root of the linear polynomial

$$L = \frac{w_D(y)}{(y - y_1)(y - y_2)(y - y_3)}$$

2. D is *generic* but $D + P_\infty$ is *collinear*. Then $D = P_1 + P_2 + P_3 + \sigma P_3$, with $P_1 \neq \sigma^k P_2$, $k = 1, 2$; and $v_D(x,y) = r_1(x,y) \cdot (x - x_3)$, with $(r_1)_0 \succeq P_1 + P_2$, $r_1 = \alpha x + \beta y + \gamma$, $\beta \neq 0$. Factoring $u_D(x)$ and substituting in r_1 we recover P_1, P_2. We find the y-coordinate of $\sigma^3 P_3$ as the root of the linear polynomial

$$L = \frac{R_x(v_D, C)}{g.c.d(R_x(r_1, C), R_x(v_D, C))}$$

$\square$

5　An efficient reduction algorithm in the Jacobian of a Picard curve

In the present section we will construct an efficient effective reduction algorithm in the Jacobian variety of a Picard curve. This algorithm works correctly in any field k with $char(k) \neq 3$, but our main interest (motivated by applications) will be the case when k is a finite field $\mathbb{F}_q$. Let's state, clearly, the problem we will solve:

Reduction problem: Given an effective affine divisor D find an effective affine divisor D_f, with $deg(D_f) \leq 3$, such that: $D - deg(D)P_\infty \cong D_f - deg(D_f)P_\infty$.

The reduction algorithm we present in this paper is based on the following geometric idea. Suppose given an effective affine divisor $D_0 = P_1 + P_2 + P_3 + P_4$ of degree *four*. If the points on the divisor D_0 are collinear, then, by lemma 1, D_0 is in the canonical class and $D_0 - 4P_\infty \cong 0$. Otherwise, to find the *reduction* of $D_0 - 4P_\infty$, we take the *interpolating conic* v_0 (we denote $v_i = v_{D_i}$) of the divisor D_0. Then, after the relation (4), v_0 intercepts C, counting multiplicities, in at most *three* more affine points H_1, H_2, H_3. Therefore, we obtain:

$$(v_0) = (D_0 - 4P_\infty) + (D_1 - 3P_\infty)$$
$$D_0 - 4P_\infty \cong -(D_1 - 3P_\infty) \tag{8}$$

where $D_1 = H_1 + H_2 + H_3$. Now, consider the *interpolating conic* v_1 of the divisor D_1; v_1 intercepts C in the additional points M_1, M_2, M_3, then holds:

$$D_1 - 3P_\infty \cong -(D_2 - 3P_\infty) \tag{9}$$

with $D_2 = M_1 + M_2 + M_3$. Combining (8) and (9), we get:

$$(D_0 - 4P_\infty) \cong (D_2 - 3P_\infty). \tag{10}$$

Therefore, the degree *three* divisor D_2 will be the reduced divisor of D_0.

A possible reduction algorithm for an effective affine divisor D, of arbitrary *degree*, could be the **Algorithm1** in Table 1 (c.f. pag. 19). From the computational point of view, **Algorithm1** may be very expensive, since in two of its steps it is necessary to factorize polynomials in $k[x]$.

Algorithm1 (receives D and returns D_f)
  1-**If** $deg(D) \leq 3$ **then** D is already reduced, set $D_f := D$ and go to **End**.
    else take $D_0 \preceq D$, $deg(D_0) = 4$ and set $D = D - D_0$.
  2-Compute the *interpolating conic* v_0 of D_0.
  3-Factorize $R_y(v_0, C)$ (resultant with respect to y) to obtain the x- coordinates
    of the points on $supp(D_1)$, using v_0 compute their y-coordinates.
  4-Known D_1 compute the conic v_1 interpolating C at $D_1 + 2P_\infty$.
    then from $R_y(v_1, C)/u_0$ and v_1 recover D_2.
  5-**If** $deg(D) < 4 - deg(D_2)$ **then** set $D_f = D_2 + D$ and go to **End**. **else** take
    $E_0 \preceq D$, $deg(E_0) = 4 - deg(D_2)$, set $D_3 := D_2 + E_0$, $D_0 := D_3$ and go to 2.
End Return(D_f)

Table 1. Algorithm1: the naive one

Our next objective will be to modify algorithm **Algorithm1** constructing a *factorization free* reduction algorithm with computational complexity linear in $deg(D)$. The modified algorithm we will present may be summarized as follows:

1. Suppose that the divisor D is partitioned as $D = D_0 + E_0 + E_1 + \ldots + E_{N-1}$, with E_j affine and effective, for $j = 1, \ldots, N - 1$; and the reduction

process (in algorithm **Algorithm1**) is performed by constructing a sequence of effective affine divisors

$$D_0, D_1, D_2, D_3, \ldots, D_{3j}, D_{3j+1}, D_{3j+2} \ldots, D_{3N}, D_{3N+1}, D_{3N+2} \qquad (11)$$

where $D_{3j} = D_{3(j-1)+2} + E_{(j-1)}$, for $j = 1, \ldots, N$ and

$$D_{3j} - 4P_\infty \cong -(D_{3j+1} - deg(D_{3j+1})P_\infty) \cong (D_{3j+2} - deg(D_{3j+2})P_\infty).$$

With $0 \leq deg(D_{3j+1})$, $deg(D_{3j+2}) \leq 3$, $deg(D_{3j}) = 4$ and $deg(E_{j-1}) = 4 - deg(D_{3j+2})$. Hence

$$D - deg(D)P_\infty \cong (D_{3N+2} - deg(D_{3N+2})P_\infty)$$

and D_{3N+2} is the reduction of D.

2. If the divisors D_h, $h = 0, \ldots, 3N + 2$, are in $\mathcal{D}_0(2,4)$ we will assign to D_h its coordinates $\overline{D}_h = \Phi(D_h)$. Then we obtain a "dual" sequence

$$\overline{D}_0, \overline{D}_1, \overline{D}_2, \overline{D}_3, \ldots, \overline{D}_{3j}, \overline{D}_{3j+1}, \overline{D}_{3j+2} \ldots, \overline{D}_{3N}, \overline{D}_{3N+1}, \overline{D}_{3N+2}. \qquad (12)$$

3. The basic idea is: given $\overline{D}_0$ (resp. D_0), depending on whether $D_0 \in \mathcal{D}_0(2,4)$ or not, we will compute $\overline{D}_h$ or D_h, for $h \geq 1$, recursively from the previous divisors in the sequences (11) and (12). The recursive computation of each $\overline{D}_h$ or D_h will be done, in the worst case, by solving a small k-defined linear system in each step. Finally, known $\overline{D}_{3N+2} = (u_{3N+2}, v_{3N+2}, w_{3N+2})$ we will recover the points in $supp(D_{3N+2})$ using Lemma 4.

Remark 1. Given $\overline{D}_{3j+1}, \overline{D}_{3j+2}$, we can prove (c.f. [5]) the equalities:

$$v_{3j+1} = v_{3j+2} \qquad (13)$$

and

$$u_{3j+2} = \left(\frac{R_y(v_{3j+1}, C)}{u_{3j+1}} \right)^* \qquad (14)$$

$$w_{3j+2} = \left(\frac{R_x(v_{3j+1}, C)}{w_{3j+1}} \right)^* \qquad (15)$$

where $(**)^*$ means that the polynomial $**$ is divided by the coefficient of its leading term. Note, also, that if v_{3j+1} does not depends explicitly on x then

$$w_{3j+2} = w_{3j+1}. \qquad (16)$$

Lemma 5. *Let be $D_{3j} \in Div^{+,4}$, explicitly known, then we can compute:*

1. $\overline{D}_{3j}$ provided $D_{3j} \in Div_0^{+,4}$.
2. $\overline{D}_{3j+1}$ and $\overline{D}_{3j+2}$ provided $D_{3j} \notin Div_0^{+,4}$.

Proof. 1. We compute u_{3j+1} and w_{3j+1} as in (5) and (6) and v_{3j+1} by solving linear systems of sizes at most 4×4.

2. Necessarily $D_{3j} = P_1 + \sigma P_1 + P_2 + \sigma P_2$, with $P_1 \neq \sigma^k P_2$, $k = 1, 2$. Then $D_{3j+1} = \sigma^2 P_1 + \sigma^2 P_2$, hence we compute u_{3j+1} and w_{3j+1} as in (5) and (6). The *interpolating conic* v_{3j+1} is, clearly, the line joining $\sigma^2 P_1$ with $\sigma^2 P_2$ (in case $P_1 = P_2$, the tangent line to $\sigma^2 P_1$). Known $\overline{D}_{3j+1} = (u_{3j+1}, v_{3j+1}, w_{3j+1})$ we compute $\overline{D}_{3j+2} = (u_{3j+2}, v_{3j+2}, w_{3j+2})$ using remark 1.   □

Lemma 6. *Let* $\overline{D}_{3j} = (u_{3j}, v_{3j}, w_{3j})$ *be the coordinates of a divisor* D_{3j} *in* $Div_0^{+,4}(C)$, *then, one of the following possibilities holds:*

1. *we can compute* $\overline{D}_{3j+1} = (u_{3j+1}, v_{3j+1}, w_{3j+1})$ *and* $\overline{D}_{3j+2} = (u_{3j+2}, v_{3j+2}, w_{3j+2})$, *with* v_{3j+1} *(and therefore* v_{3j+2}*) dependent on* y.
2. *we can compute* D_{3j+2} *explicitly.*

(it is not necessary to know D_{3j} *explicitly.)*

Proof. It is necessary to consider the cases:

1. Case $v_{3j}(x, y)$ is linear. Then the points in $supp(D_{3j})$ are collinear and $D_{3j} - 4P_\infty \cong 0$, hence $D_{3j+2} = 0$.
2. Case $v_{3j}(x, y)$ is a conic not factorizing in linear factors (i.e. $v_{3j}(x, y) = a_{20}x^2 + a_{10}x + a_{01}y + a_{11}xy + a_{00}$ with $a_{11}^2 a_{00} + a_{01}^2 a_{20} - a_{11}a_{01}a_{10} \neq 0$). We begin computing u_{3j+1} and w_{3j+1} using (14) and (15). To recover $v_{3j+1}(x, y) = b_{20}x^2 + b_{10}x + b_{00} + b_{01}y$, we solve the 4×4 linear system

$$R_y(v_{3j}, v_{3j+1}) = \lambda u_{3j+1} \text{ where } \lambda \text{ is aconstant} \neq 0. \tag{17}$$

This system has determinant $a_{11}^2 a_{00} + a_{01}^2 a_{20} - a_{11}a_{01}a_{10} \neq 0$, hence, it has a unique solution. Selecting λ conveniently we normalize v_{3j+1}.

3. Case $v_{3j}(x, y) = r_1(x, y)(x + a_{01}/a_{11})$ with $r_1(x, y) = (a_{20}x + a_{11}y + a_{10} - a_{20}a_{01}/a_{11}))$

3.a. If $(x + a_{01}/a_{11})^2 \mid u_{3j}$, then $D_{3j} = P_1 + P_2 + P_3 + \sigma P_3$ with P_1, P_2, P_3 unknown. In this case, as in case 5, we will try to compute $\overline{D}_{3j+1}$. First we compute u_{3j+1} and w_{3j+1}. Clearly, we can not use the system (17) to recover v_{3j+1}. Next, we find $\sigma^2 P_3$: $x_3 = -a_{01}/a_{11}$ and $\xi^2 y_3$ is the root of the linear polynomial

$$L = \frac{w_{3j+1}}{g.c.d(R_x(r_1, C), w_{3j+1})}$$

if $g.c.d(R_x(r_1, C), w_{3j+1})$ is a polynomial of *degree* 2. If it is not the case that means $\sigma^2 P_3$ is a root of r_1 and substituting x_3 in r_1 we recover $\xi^2 y_3$. Once we have $\sigma^2 P_3$, we must consider the cases:

3.a.1. If P_3 (resp. σP_3) anulates $r_1(x, y)$ then the underlying divisor D_{3j} is collinear and as the polynomial $u_{3j+1}/(x + a_{01}/a_{11})^2$ is linear we may recover the other interception point, M, of $r_1(x, y)$ with C. Clearly, $D_{3j+2} = \sigma M + \sigma^2 M + P_3$ (resp. $D_{3j+2} = \sigma M + \sigma^2 M + \sigma P_3$)

3.a.2 we try to find v_{3j+1} as the solution of the 4×4 linear system

$$v_{3j+1}(\sigma^2 P_3) = 0$$
$$R_y(v_{3j+1}, r_1) = \lambda(u_{3j+1}/(x + a_{01}/a_{11})) \tag{18}$$

This system has determinant $(a_{11})^2 \cdot r_1(\sigma^2 P_3)$ which if different from zero iff $r_1(\sigma^2 P_3) \neq 0$. If this is the case, we can recover v_{3j+1} and $\overline{D}_{3j+1}$. Otherwise, $(r_1(x,y))_0 \succeq P_1 + P_2 + \sigma^2 P_3 + M$ and we can recover M: x_M is the root of the linear polynomial

$$L = \frac{u_{3j+1}}{(x + a_{01}/a_{11})^2},$$

and evaluating r_1 in x_M recover the y_M. Now, $D_{3j+1} = 2 \cdot \sigma^2 P_3 + M$. and we may find v_{3j+1} from one of the systems:

$$v_{3j+1}(\sigma^2 P_3) = 0 \text{ of order two}$$
$$v_{3j+1}(M) = 0 \tag{19}$$

if $\sigma^2 P_3 \neq M$ or

$$v_{3j+1}(\sigma^2 P_3) = 0 \text{ of order three} \tag{20}$$

if $\sigma^2 P_3 = M$.

3.b. If $(x + a_{01}/a_{11})^2 \,/u_{3j}$ the unknown D_{3j} is necessarily equal to $D_{3j} = P_1 + P_2 + P_3 + P_4$ with, let's say, P_1, P_2, P_3 collinear (i.e. $(r_1(x,y))_0 \succeq P_1 + P_2 + P_3$), then,

$$D_{3j} - 4P_\infty \cong -(\sigma P_4 + \sigma^2 P_4 + M - 3P_\infty) \cong (\sigma M + \sigma^2 M + P_4 - 3P_\infty)$$

where M is the fourth point in which $r_1(x,y)$ intersects C. To find P_4 and M we proceed as follows: $x_4 = -a_{01}/a_{11}$, x_M is the root of the linear polynomial

$$L_M = \frac{R_y(r_1, C)(x + a_{01}/a_{11})}{u_{3j}},$$

if r_1 depends on x, then y_M is obtained evaluating r_1 in x_M and y_4 is the root of the lineal polynomial

$$L_1 = \frac{w_{3j}(y - y_M)}{R_x(r_1, C)},$$

otherwise, y_M is the solution (in y) of $r_1 = 0$ and y_4 is the root of the linear polynomial

$$L_1 = \frac{w_{3j}}{(y - y_M)^3},$$

Hence we may recover $D_{3j+2} = \sigma M + \sigma^2 M + P_4$ explicitly.

In those cases where we have computed $\overline{D}_{3j+1}$ then using remark 1, we may compute $\overline{D}_{3j+2}$. □

Lemma 7. *Given* $\overline{D}_{3j+1} = (u_{3j+1}, v_{3j+1}, w_{3j+1})$ *and* $\overline{D}_{3j+2} = (u_{3j+2}, v_{3j+2}, w_{3j+2})$ *and known the divisor* E_{j-1}, *then exactly one of the following cases hold:*

1. *we can compute* $\overline{D}_{3(j+1)} = (u_{3(j+1)}, v_{3(j+1)}, w_{3(j+1)})$ *explicitly.*
2. *we can compute* $\overline{D}_{3(j+1)+1}$ *and* $\overline{D}_{3(j+1)+2}$ *explicitly.*
3. *we can compute the* $D_{3(j+1)+2}$ *explicitly and it is k-rational.*

Proof. The strategy will be to compute, if possible, $\overline{D}_{3(j+1)}$ (case 1). If it is not possible, then we must consider the other cases. First, we compute $u_{3(j+1)}$ and $w_{3(j+1)}$ as

$$u_{3(j+1)} = u_{3j+2} \cdot \prod_{P_i \in supp(E_{j-1})} (x - x_i), \tag{21}$$

$$w_{3(j+1)} = w_{3j+2} \cdot \prod_{P_i \in supp(E_{j-1})} (y - y_i). \tag{22}$$

Then we will try to find $v_{3(j+1)}$ from the linear system

$$\begin{cases} v_{3(j+1)}(P_i) = 0 & \text{with } P_i \in supp(E_{j-1}) \\ R_y(v_{3j+2}, v_{3(j+1)}) = \lambda u_{3j+2} \ \lambda \text{ a non-zero contant.} \end{cases} \tag{23}$$

We must consider the following cases:

1. Case v_{3j+2} linear (i.e. $v_{3j+2} = b_{10}x + b_{00} + y$). Then $E_{j-1} = P_{01} + P_{02}$ and the system (23) has determinant equal to

$$\begin{matrix} -3y_{01}^2 \cdot v_{3j+2}(P_{01})^2 & \text{if } P_{01} = P_{02} \\ (x_{01} - x_{02}) \cdot v_{3j+2}(P_{01}) \cdot v_{3j+2}(P_{02}) & \text{if } P_{01} \neq P_{02} \end{matrix}$$

then, we have to consider the cases:

1.a. Case $u_{3j+2}(P_{01})$ or $u_{3j+2}(P_{02}) = 0$. Then as u_{3j+2} is of degree 2 we can recover D_{3j+2} without making factorizations, then holds $D_{3(j+1)} = D_{3j+2} + P_{01} + P_{02}$ and we may apply lemma 5.
1.b. Case $v_{3j+2}(P_{01}) = 0$ and $u_{3j+1}(P_{01}) = 0$ (resp. $v_{3j+2}(P_{02}) = 0$ and $u_{3j+1}(P_{02}) = 0$), then the divisor $D_{3(j+1)}$ is a collinear divisor and we can compute the other point M in which v_{3j+2} intercepts C. Then $D_{3j+2} = \sigma M + \sigma^2 M + P_{02}$ (resp. $D_{3j+2} = \sigma M + \sigma^2 M + P_{01}$).
1.c. Case $P_{01} = \sigma P_{02}$, then set $v_{3(j+1)} = (x - x_{01}) \cdot v_{3j+2}$
1.d. Otherwise, the system (23) is solvable.
2. Case v_{3j+2} is a conic (i.e. $v_{3j+2} = b_{20}x^2 + b_{10}x + b_{00} + y$, $b_{20} \neq 0$) and $E_{j-1} = P_{01}$. Then, we begin computing $u_{3(j+1)}$, $w_{3(j+1)}$ as in (21) and (22), respectively. Now, the system (23) has determinant $b_{20} \cdot v_{3j+2}(P_{01})$ and we have the cases:
2.a. if $v_{3j+2}(P_{01}) \neq 0$ we recover $w_{3(j+1)}$ from (23).

2.b. if $v_{3j+2}(P_{01}) = 0$ and $u_{3j+1}(P_{01}) = 0$, clearly $v_{3(j+1)} = v_{3j+2}$.

2.c. if $v_{3j+2}(P_{01}) = 0$ and $u_{3j+2}(P_{01}) = 0$ we look for $w_{3(j+1)}$ in the system

$$
\begin{cases}
v_{3(j+1)}(P_{01}) = 0 \text{ of order two.} \\
R_y(v_{3j+2}, v_{3(j+1)})/(x - x_{01}) = \lambda\left(u_{3j+2}/(x - x_{01})\right), \ \lambda \neq 0
\end{cases}
\tag{24}
$$

2.c.1. In case P_{01} is not a ramification point, the previous system has determinant

$$
b_{20}(6y_{01}^2 x_{01} b_{20} + 3y_{01}^2 b_{10} + p_4^I(x_{01})),
\tag{25}
$$

this expression is equal to zero if and only if v_{3j+2} has a zero of order *two* in P_{01}. If it is the case, then, as u_{3j+2} is of degree *three*, $u_{3j+2}/(x - x_{01})^2$ is linear in x and we can recover (without factorizing) the other point P_2 in D_{3j+2}, and we apply lemma 5 to $D_{3(j+1)} = 3P_{01} + P_2$. Otherwise we solve (24) to find $v_{3(j+1)}$.

2.c.2. In case P_{01} is a ramification point, the determinant of (24) is b_{20}, hence, we can solve for $v_{3(j+1)}$.   □

Combining lemmas (5,6,7) we can construct the algorithm **Algorithm2** (see Table 2 in pag. 24) which is the announced efficient modification to algorithm **Algorithm1** (see Table 1).

Algorithm2 (receives D and returns D_f)
  1-**if** $deg(D) < 4$ **then** set $D_f = D$ and go to **End.**
  2- Set $D_0 = P_1 + P_2 + P_3 + P_4$.
  3- **if** $D_0 \in Div_0^{+,4}$ **then** compute $\overline{D}_0$.
  4- **else** compute $\overline{D}_1, \overline{D}_2$ and go to SubAlg1.
  5- Given $\overline{D}_0$ apply lemma 6 to obtain:
     5.a- D_2 explicitly. if $deg(D) + D_2 < 4$ **then** set $D_f = D + D_2$
        and go to **End.** else set $D_0 = D_2 + E_0$, $D = D - E_0$ and go to 3.
     5.b- $(\overline{D}_1, \overline{D}_2)$ explicitly **then** go to **SubAlg1.**
  SubAlg1 (given $\overline{D}_1$ and $\overline{D}_2$))
    S1 **if** $deg(D) + deg(v_2) < 4$ call **SubFactor**$(D, \overline{D}_2)$
    S2 **else** select E_0, $deg(E_0) + deg(v_2) = 4$, $E_0 \preceq D_0$, set $D = D - E_0$
      and apply lemma 7. We have the following possibilities:
      a) we obtain $\overline{D}_3$. **then** set $\overline{D}_0 = \overline{D}_3$ go to 5.
      b) we obtain $\overline{D}_4$ and $\overline{D}_5$. **then** put $\overline{D}_1 = \overline{D}_4$, $\overline{D}_1 = \overline{D}_4$ and go to S1.
      c) we obtain D_5 explicitly. if $deg(D) + D_5 < 4$ **then** set $D_f = D + D_5$
        and go to **End.** else set $D_0 = D_5 + E_0$, $D = D - E_0$ and go to 3.
  EndSubAlg1
  SubFactor$(D, \overline{D}_2)$
    Using lemma 4 recover D_2, **then** set $D_f = D + D_2$ and go to **End.**
  EndSubAlg1
  End Return(D_f)

Table 2. Algorithm2: the efficient one

Proposition 1. *Given the divisor D, the* **Algorithm 2** *computes the reduced divisor of D making $O(\deg(D))$ operations in k and only one factorization of a polynomial, of degree at most 3 in $k[x]$. Morover, if the ground field k is $\mathbb{F}_q$, the constant c that realizes $O(\deg(D))$ satisfies:*

$$c \leq 2(4\log_2(q))^3. \tag{26}$$

Proof. The fact that the algorithm in Table 2 makes the reduction of D is an inmediate consequence of lemmas (5,6,7). The complexity in every iteration of algorithm is $O(1)$ operations in k, hence, the total cost is $O(\deg(D))$. Moreover, in the worst case, in lemmas (5,6,7) the most expensive computations are solving linear systems of orders (at most) 4×4. Hence, in each iteration of the algorithm we have to solve (at most) 2 linear systems of sizes (at most) 4×4, which give the estimate of (26). $\qquad\square$

Let's illustrate with an example the application of **Algorithm 2**.

Example 1. Let $p = 37$, $k = \mathbb{F}_p$ and $p_4(x) = x^4 + 2x$. The $\mathcal{L}$-polynomial of C_{p_4} is

$$\mathcal{L}(t) = 50653t^6 + 24642t^5 + 6660t^4 + 1225t^3 + 180t^2 + 18t + 1$$

and the cardinal of the group of k-rational points of the Jacobian, $J_p(C_{p_4})$, of C_{p_4} is

$$\#|J_p(C_{p_4})| = \mathcal{L}(1) = 3 \cdot 27793.$$

The group $J_p(C_{p_4})$ is cyclic: the curve C_{p_4} has only one k-rational affine ramification point $R_1 = (0:0:1)$ and the calss $[P_1 - P_\infty]$, where P_1 is any other affine point, generates $J_p(C_{p_4})$. Let $P_1 = (5:29:1)$ then the explicit computation of the reduction of $7 \cdot [P_1 - P_\infty]$ is shown in Table 3. Now, after lemma 4, (applied to $\overline{D}_{11}$) we recover the reduced divisor

D_i	$\overline{D}_i$	E_j
$D_0 = 4 \cdot P_1$	$\overline{D}_0 = (x^4 + 17x^3 + 2x^2 + 18x + 33, x^2 + 14x + 2y + xy + 35, y^4 + 32y^3 + 14y^2 + 13y + 26)$	—
$D_1 = ?$	$\overline{D}_1 = (x^3 + 27x^2 + 2, x^2 + 9x + 21y + 17, y^3 + 15y^2 + 30y + 20)$	—
$D_2 = ?$	$\overline{D}_2 = (x^3 + 9x + 33, x^2 + 9x + 21y + 17, y^3 + 28y^2 + 14y + 26)$	$E_0 = P_1$
$D_3 = ?$	$\overline{D}_3 = (x^4 + 32x^3 + 9x^2 + 25x + 20, 27x^2 + 14x + 3y + xy + 22, y^4 + 36y^3 + 16y^2 + 27y + 23)$	—
$D_4 = ?$	$\overline{D}_4 = (x^3 + 13x^2 + 28x + 7, x^2 + 29x + 24y + 18, y^3 + 19y^2 + 4y + 36)$	—
$D_5 = ?$	$\overline{D}_5 = (x^3 + 19x + 35, x^2 + 29x + 24y + 18, y^3 + 13y^2 + 18y + 1)$	$E_1 = P_1$
$D_6 = ?$	$\overline{D}_6 = (x^4 + 32x^3 + 19x^2 + 14x + 10, 18x^2 + 9x + 33y + xy + 31, y^4 + 21y^3 + 11y^2 + 34y + 8)$	—
$D_7 = ?$	$\overline{D}_7 = (x^3 + 16x^2 + 14x + 8, x^2 + 22x + 33y + 29, y^3 + 31y + 8)$	—
$D_8 = ?$	$\overline{D}_8 = (x^3 + 13x^2 + 32x + 10, x^2 + 22x + 33y + 29, y^3 + 5y^2 + 12y + 31)$	$E_2 = P_1$
$D_9 = ?$	$\overline{D}_9 = (x^4 + 8x^3 + 4x^2 + 35x + 24, 32x^2 + 5x + 11y + xy + 6, y^4 + 13y^3 + 15y^2 + 16y + 26)$	—
$D_{10} = ?$	$\overline{D}_{10} = (x^3 + 11x^2 + 17x + 9, x^2 + 6x + 21y + 19, y^3 + 9y^2 + 26y + 15)$	—
$D_{11} = ?$	$\overline{D}_{11} = (x^3 + 7x^2 + 8x + 18, x^2 + 6x + 21y + 19, y^3 + 34y^2 + 32y + 9)$	$E_3 = 0$

Table 3. Reduction of the divisor $7 \cdot P_1$, $P_1 = (5:29:1)$

$$D_f = P_{1f} + P_{2f} + P_{3f}$$

where

$$P_{1f} = (33\beta^2 + 6\beta + 10 : 5\beta^2 + 17\beta + 1 : 1),$$
$$P_{2f} = (34\beta^2 + 8\beta + 10 : 13\beta^2 + 35\beta + 1 : 1),$$
$$P_{3f} = (7\beta^2 + 23\beta + 10 : 19\beta^2 + 22\beta + 1 : 1),$$

and $k(\beta)$ is an algebraic extention of k defined by the k-irreducible polynomial $z^3 + 2$.

6 Further Remarks

6.1 Improving the Complexity

In fact the complexity estimate given in (26) is an overestimate of the real complexity by the following reasons:

1. Since the coordinates $\overline{D}_i = (u_i, v_i, w_i)$ of a divisor D_i are unique up to constant non-zero factor (*i.e.* we can also recover D_i from (au_i, bv_i, cw_i), $a, b, c \neq 0$) then lemmas (5,6,7) could be reformulated in such a way, that it is not necesary to find inverses of elements in k. Hence, in practice we have only to do *linear algebra* in k consiered as a ring. This ameliorates greatly the bound 26.

2. In several steps of the **Algorithm 2** it goes from $\overline{D}_0$ to $\overline{D}_2$ (resp. to D_5) by solving one linear system (resp. performing only elementary polynomial operations). Clearly, this reduces the complexity of the computations. Computer experiments show that this cases are not infrequent (specially in the case when the divisor to be reduced is a multiple of a point).

3. In the special case when $D = N \cdot P_1$ it is possible to design especial strategies: suppose that in some intermediate step of the algorithm we obtain a divisor D_2 (resp. a D_5) explicitly, i.e. D_2 (resp. D_5) is the reduction of $N_1 \cdot P_1$, for certain $N_1 < N$. Then, it is possible to substitute the original problem (i.e. to find the reduction of $D = NP_1$) by the new one of finding the reduction of $a_1 \cdot D_2 + b_1 \cdot P_1$ (resp. $a_1 \cdot D_5 + b_1 \cdot P_1$), where $N = a_1 \cdot N_1 + b_1$. If N_1 is sufficiently big then, since $\deg(D_2) \leq 3$ (resp. $\deg(D_5) \leq 3$) the original problem is considerably reduced. Proceeding recursively the complexity of computing the reduction of a large multiple of a point could be dramaticaly reduced.

The next example illustrates the discution in 3:

Example 2. (With the same notation of example 1) Let's compute the reduction of $D = 27793 \cdot P_1$, $P_1 = (5 : 29 : 1)$. First we obtain: $35 \cdot (P_1 - P_\infty) \cong (P_{11} + P_{12}) - 2 \cdot P_\infty$, where $P_{11} = (5 : 31 : 1)$ and $P_{12} = (19 : 2 : 1)$, then

$$D = 27793 \cdot (P_1 - P_\infty) \cong 794 \cdot (P_{11} + P_{12}) + 3 \cdot P_1 - 797 \cdot P_\infty.$$

We find that $35 \cdot (P_{11} - P_\infty) \cong (P_{111} + P_{112}) - 2 \cdot P_\infty$ and $35 \cdot (P_{12} - P_\infty) \cong (P_{121} + P_{122}) - 2 \cdot P_\infty$, where $P_{111} = (19 : 20 : 1)$, $P_{112} = (5 : 14 : 1)$, $P_{121} = (19 : 20 : 1)$ and $P_{122} = (13 : 18 : 1)$, then

$$794 \cdot (P_{11} + P_{12} - P_\infty)$$
$$\cong 22 \cdot (P_{111} + P_{112} + P_{121} + P_{122}) + 24 \cdot (P_{11} + P_{12}) - 136 \cdot P_\infty,$$

and the computation of the reduction of $D = 27793 \cdot P_1$ is simplified to the computation of the reduction of the divisor

$$D_1 = 22 \cdot (P_{111} + P_{112} + P_{121} + P_{122}) + 24 \cdot (P_{11} + P_{12}) + 3 \cdot P_1,$$

which is of degree 139. Finally, the reduction of D_1 is $D_f = (0 : 0 : 1)$. Then, as the class $[D_f - P_\infty]$ is a 3-torsion on $J_p(C_{p_4})$, the class $[P_1 - P_\infty]$ is a generator of $J_p(C_{p_4})$ and $J_p(C_{p_4})$ is a cyclic group.

6.2   Comparison with Cantor's Algorithm for Hyperelliptic Curves

In the reduction of small divisors D both algorithms have similar complexities. The more important differences appear in the computation of the reduction of a large divisor D (in particular in the computation of a large multiples of a point).

1. Our algorithm has less memory requirements: Cantor's algorithm (c.f. [1, 12]) associates polynomial coordinates to D, if $D = M \cdot P_1$ with M big, it has to operate (at least in the inicial steps) with large poynomials; in our algorithm it is only necessary to store the point P_1 and the integer M.
2. Computing large multiples of a point our algorithm could be faster than Cantor's algorithm: applying the techniques mentioned in 1 and 3 of 6.1 we can diminish significantly the complexity of computing large multiples of a point. No such technique, is known by the authors for Cantor's algorithm in the hyperelliptic case.

   In summary, we can expect that efficient implementations of our algorithm could be as fast as Cantor's algorithm with the additional advantage of requiring less memory storage.

6.3   Comparison with the General Algorithm of Huang and Ierardi

It is more difficult to compare our algorithm with the fairly general algorithm of Huang and Ierardi (c.f. [11]). In particular, it is hard, at least for us, to estimate from the paper [11] the real complexity of the their algorithms when applied to Picard curves. For intance, it is not easy to estimate how big is the constant involved in their $O(deg(D))$ complexity estimate. Additionally, we don't know any references to effective implementation of their algorithms.

   On the other hand, our algorithm (as Cantor's algorithm) is completelly deterministic: it does not requiere to make probabilistic searches, hence, it has no limitations on the cardinality of ground field k. Moreover, our algorithm is specific for Picard curves. Consequently, it uses (and reflects) special geometric features of this curves that permit us to diminish the complexity: our algorithm handles very efficiently cases in which appear collinearity of divisors, we obtained also efficient techniques to compute multiples of a point, etc. By the above reasons one may expect that our algorithm is faster and better in the Picard curves case.

Acknowledgements

We wish to thank R.-P. Holzapfel, for his valuable comments, discussions and encouragements. Also, to G. Frey and U. Krieger for their valuable comments. This work was partially supported by a DFG grant.

Last but not least, the first and second authors want to thank the third author for his warm hospitality during their stay at U.A.G, Guadeloupe.

References

1. Cantor, D., "*Computing in the Jacobian of a hyperelliptic curve*," Math. of Computation, 48 (1987), 95-101.
2. Estrada Sarlabous. J, "*Higher differentials on Cyclic Curves*,". Math. Nachr. 135 (1988), 311-317.
3. Estrada Sarlabous. J, "*On the Jacobian Varieties of Picard Curves Defined over Fields of Characteristic p*,". Math. Nachr. 152 (1991), 329-340.
4. Estrada Sarlabous. J., "*A finiteness theorem for Picard curves with good reduction*,". Appendix I of *Ball models and some Hilbert Problems* by R.-P. Holzapfel. Lectures in Mathematics. Birkhäuser-Verlag, (1995).
5. Estrada Sarlabous. J, Reinaldo Barreiro. E, Piñeiro Barceló. J.A.,. "*On the Jacobian Varieties of Picard curves: explicit Addition Law and Algebraic Structure*," (to appear in Math. Nachr.), Preprint Nr. 95-5 Humboldt Univ. zu Berlin, 1995.
6. Estrada Sarlabous. J., Piñeiro Barceló. J.A., "*Decoding of codes in Picard curves*,". submitted to Math. Nachrichten 1997. Preprint Nr. 96-30 Humboldt Univ. zu Berlin, 1996.
7. Holzapfel, R.-P., "*Geometry and arithmetic around Euler partial differential equations*,". Dt. Ver. d. wiss., Berlin/Reidel Publ. Comp., Dordrecht (1986).
8. Holzapfel, R.-P., "*On the algebraic value of the Picard modular function*,". Proc. Special. Diff. Equations, Arkata (1991).
9. Holzapfel, R.-P., "*Transcendental Ball Points of Algebraic Picard Integrals*,". Math. Nachr. 162 (1993).
10. Holzapfel, R.-P., *Ball models and some Hilbert problems*. Lectures in Mathematics. Birkhäuser-Verlag (1995).
11. Huang, M.-D and Ierardi, D.J., "*Efficient algorithms for the effective Riemann-Roch problem and for addition in the Jacobian of a curve*,". Proc. of the twenty-first ACM Symp. on the fundations of Computer Science, (May 1991).
12. Koblitz, N., *Hyperelliptic cryptosystems*, Journal of cryptology 1, pp. 139-150.
13. Mumford, D., *Tata Lectures on Theta II. Jacobian theta functions and differential equations*. Progress in Math, Vol.42, Birkhäuser Verlag (1984).

Continued Fractions
in Hyperelliptic Function Fields

T.G. Berry

Departamento de Matematicas Puras y Aplicadas
Universidad Simón Bolívar
Caracas
Venezuela.
berry@usb.ve

Abstract. Known results on hyperelliptic continued fractions, and in particular the Baby-Step Giant-Step algorithm, are obtained using algebro-geometric techniques. The methods used are valid in all characteristics and the proofs are simpler than those based on analogies with real quadratic number fields.

1   Introduction

This article contains virtually no new results. Its purpose is to show how the known properties of hyperelliptic continued fraction expansions, in particular the Baby-Step Giant-Step algorithm, can be derived very easily by algebro-geometric methods. In the published literature hyperelliptic continued fractions are mainly studied by adapting techniques from the theory of real quadratic number fields. The resulting proofs, while not extraordinarily difficult, can be quite intricate, and need slightly different arguments for even and odd characteristic. The geometric arguments work in all characteristics and are (to this author anyway) more transparent than their number-theoretic counterparts.

Notation and Conventions

Let K be an arbitrary field. Everything we define (curves, morphisms between curves, rational functions and divisors on curves etc...) is assumed to be defined over K. Let C be a non-singular hyperelliptic curve of genus p with hyperelliptic double cover $\pi : C \to \mathbf{P}^1$ (so π is a separable degree 2 map). The hyperelliptic involution $C \to C$ is denoted $U \mapsto U^-$, for any object on which it acts. Choose a point ∞ on $\mathbf{P}^1$, and a function $x \in K(\mathbf{P}^1)$ with a simple pole at ∞, and identify $K(\mathbf{P}^1)$ with $K(x)$. Then $K(x) \subset K(C)$ via π. *Norm* refer to norm of this extension. Assume that the pullback under π of the point at infinity on $\mathbf{P}^1$ consists of two distinct points, necessarily unramified, which we call, varying the notation slightly, ∞^+ and ∞^- ; without serious loss of generality, we assume that these points are defined over K. (If not, replace K by an appropriate quadratic extension). The discrete valuations of $K(C)$ corresponding to ∞^+ and ∞^- are denoted v_+, v_- respectively.

2   Continued Fraction Expansions

We recall the theory very briefly. For details and algorithms see [5, 2, 8].

Since ∞^+ is unramified, $1/x$ is a uniformising parameter, and defines an embedding $K(C) \rightarrow K((1/x))$-we call the image of f *the Laurent series* of f (at ∞^+ , understood). Let $f \in K(C)$. The *polynomial part* of f, denoted $\lfloor f \rfloor$ is the principal part of the Laurent series of f, together with the constant term. Thus $\lfloor f \rfloor$ is a polynomial in x, and as such a well-defined rational function on C. The continued fraction expansion (henceforth abbreviated "CFE") of $f \in K(C) \setminus K(x)$ consists of sequences $\{a_i\}, \{p_i\}, \{q_i\} \in K[x]$, $\{g_i\}, \{f_i\} \in K(C)$, defined iteratively by:

$$g_0 = f; a_i = \lfloor g_i \rfloor; g_i = a_i + \frac{1}{g_{i+1}}$$

$$\frac{p_i}{q_i} = [a_0, \dots, a_{i-1}] \quad \text{(in a standard notation for continued fractions. Note the indices!)}$$

$$f_0 = 1; f_i = q_i f - p_i, i \geq 1$$

The a_i are the *partial quotients*, the p_i, q_i the *convergents*, and the g_i, f_i the *reducts* and the *approximants*, respectively, of f.

Let $y \in K(C) \setminus K(x)$ be a function with no finite poles, poles of order $p+1$ at both points at infinity, and whose zero set contains no pair of points twinned in the hyperelliptic involution. Then $K(C) = K(x)(y)$. We assume the affine plane curve defined by the minimal polynomial of y over $K(x)$ is non-singular.

We consider CFE's of functions $f \in K(C)$ of the form $(L + y)/M$, where $L, M \in K[x]$, L is monic, and M divides $\text{Norm}(L + y)$. We call these *standard functions*. (Zuccherato calls them *quadratic irrationalities*, but this seems a bit hard on the other functions in $K(C) \setminus K(x)$). The needed facts on CFE's are summarized in the following proposition (which, but for item 2, holds in fact for an arbitrary function in $K(C) \setminus K[x]$).

Proposition 1 *Let f be a standard function, and consider the CFE of f. For all $i \geq 0$:*

1. $\deg a_i > 0$ *for* $i > 0$.
2. *The reducts, the functions g_i, are standard functions.*
3. $\deg q_1 = 0$, *and for* $i \geq 2$, $\deg q_i = \sum_{j=1}^{i-1} \deg a_j$. *Hence* $\deg q_{i+1} = \deg q_i + \deg a_i$ *and* $\{\deg q_i\}$ *is a strictly increasing sequence.*
4. $f_i = \left(\prod_{j=0}^{i} g_j \right)^{-1}$.
5. *(Diophantine approximation property.) For $i \geq 1$, the f_i satisfy $v_+(f_i) = \deg q_{i+1}$.*

The divisor D of finite poles (i.e poles excluding ∞^+ and ∞^-) of a standard function is what is termed in [3] a *standard finite divisor*, that is, D is an effective divisor, no two points in the support of D are paired in the hyperelliptic involution, and branch points occur with multiplicity at most 1. Conversely, associated to such a divisor D there is a nearly canonical standard function called the *pole function* of D (cf [3]): it is the function $(L+y)/M$ with divisor of finite poles D, with L monic and $\deg L < \deg M$.

A basic theorem is

Theorem 2 *Let f_D be the pole function of a standard finite divisor D, and let $b \in \mathbf{Z}, b \geq 0$. Consider the CFE of f_D. Let k be the index for which $\deg q_k \leq \deg D + b - (g+1) < \deg q_{k+1}$. Then a K-basis of $\mathcal{L}(D + b\infty^-)$ is given by the functions $f_i, 0 \leq i \leq k$, together with functions $x^\alpha f_i$, where $1 \leq \alpha \leq \deg q_{i+1} - \deg q_i - 1$, if $i < k$, and $1 \leq \alpha \leq \deg D + b - (p+1) - \deg q_k$, if $i = k$.*

The proof is not difficult. The dimension $l(D+b\infty^-)$ is given by Riemann-Roch. One checks that the given functions are in $\mathcal{L}(D + b\infty^-)$, that they are linearly independent, and one counts them. For details see [3].

Observations:

1. In general one expects $\deg a_i = 1$, i.e. $\deg q_{i+1} = \deg q_i + 1$, so that the terms $x^\alpha f_i$ can be thought of as correction terms, which occur rarely.
2. The members of the basis have pairwise distinct zero orders at ∞^+ and pairwise distinct pole orders at ∞^-.
3. If $v_-(f_k) < 0$ then (c.f.[3])

$$v_-(f_k) = -(\deg q_k + p + 1 - \deg D) \tag{1}$$

The significance of the index k in the theorem is that it is the index for which $|v_-(f_k)| \leq b < |v_-(f_{k+1})|$.

We shall make considerable use of (1), often without explicit reference.

Consider f_k in the CFE of f_D. For some b, $f_k \in \mathcal{L}(D + b\infty^-)$, so

$$(f_k) = -D' + Z'_k + \text{components at infinity}$$

where $0 \leq D' \leq D$ and Z'_k is the divisor of finite zeros of f_k. It may well happen that $D' < D$, but by adding $D - D'$ to D' and to Z', we can always write

$$(f_k) = -D + Z_k + \text{components at infinity} \tag{2}$$

where $Z_k = Z'_k + D - D'$. In particular, if $v_-(f_k) < 0$, then by Prop.1(4) and (5), and (1) we can write

$$(f_k) = -D + Z_k + \deg q_{i+1}\infty^+ - (\deg q_i + p + 1 - \deg D)\infty^- \tag{3}$$

We shall always write (f_k) in the form (2) or (3). The principal advantage is that, as is easily seen, using Prop.1(4) (c.f. [2])) Z_k is the divisor of finite poles of the function g_k. It is this that makes the CFE useful. Usually in curve theory the objects of basic interest are the spaces $\mathcal{L}(D)$; however, the functions in $\mathcal{L}(D)$ can be difficult to handle. In our case, for example, when k is large p_k, q_k are polynomials of high degree. For many (not all) purposes, it is enough to know the Z_k, and these can be obtained from the g_k, which remain manageable for all $k >> 0$, and which constitute, together with the a_i, the primary data of the CFE.

The following corollary of Theorem 2 is the principal tool used in the study of continued fractions. We continue with the premises and notation of Theorem 2.

Corollary 3 *Suppose there exists a standard finite divisor Z with degree $\leq p$ and and $h \in K(C)$, $h \neq$ constant such that*

$$(h) = -D + Z + a\infty^+ - b\infty^-$$

where $a, b \geq 0$. (Here, we allow D and Z to have points in common.) Then $\exists k$ such that h is a constant multiple of f_k, whence $Z = Z_k$.

Proof. The hypotheses imply that $h \in \mathcal{L}(D + b\infty^-)$. By theorem 2 a basis for this space is given by functions $x^\alpha f_i$ for appropriate indices α and $0 \leq i \leq k$ where k is the index such that $|v_-(f_k)| \leq b < |v_-(f_{k+1})|$. Observe that $a \leq a_k = v_+(f_k)$; indeed the orders of zero of our basis elements at ∞^+ are mutually distinct, so any linear combination, such as h, has order of zero the minimum of the orders of zero of functions occurring in the combination, while a_k is the highest order of zero of basis elements. Also $b_k = |v_-(f_k)| \leq b$ by choice. Thus

$$\left(\frac{f_k}{h}\right) = Z_k - Z + (a_k - a)\infty^+ + (b - b_k)\infty^-$$

has non-negative degree at infinity, so $f_k/h \in \mathcal{L}(Z)$. But Z is a standard finite divisor of degree $\leq p$; by Riemann-Roch $\mathcal{L}(Z) = K$, so $f_k/h =$ constant as was to be proved.

3   Applications

3.1   Reduction Algorithms

The results of this section are the hyperelliptic analogues of the theory of reduction of real quadratic forms.

A further corollary of Theorem 2 is:

Corollary 4 *[Reduction]*

1. *Suppose* $\deg D > p$. *Let k be the index for which $\deg q_k \leq \deg D - (p+1) < \deg q_{k+1}$. Then $\deg Z_k \leq p$ and $f_k \in \mathcal{L}(D)$.*
2. *With k the integer defined above, for all $i > k$ we can write*

$$(f_i) = -D + Z_i + \text{components at infinity}$$

   and $\deg Z_i \leq p$.
3. *If $\deg D \leq p$ then $\deg Z_i \leq p$ for all $i \geq 0$.*

Proof. (1).By Theorem 2, the function $f_k \in \mathcal{L}(D)$, so it only remains to prove $\deg Z_k \leq p$. We have

$$f_k = -D + Z_k + \deg q_{k+1} \infty^+ \ + b \infty^-$$

with $b \geq 0$. Taking degrees, we find $\deg Z_k = \deg D - \deg q_{k+1} - b$. But, by choice, $\deg D - (p+1) < \deg q_{k+1}$, from which $\deg Z_k < p+1-b$ and it follows that $\deg Z_k \leq p$.

(2)Using again Theorem 2, for $i > k$ the function $f_i \notin \mathcal{L}(D)$, hence must have a pole at ∞^- . By the observations following Theorem 2, this implies $|v_-(f_i)| = \deg q_i + p + 1 - \deg D$. Then, taking degrees in (2) gives

$$\deg D = \deg Z_i + \deg q_{i+1} - \deg q_i - (p+1) + \deg D$$

whence $p+1 = \deg Z_i + \deg q_{i+1} - \deg q_i$. Since the $\deg q_i$ form a strictly increasing sequence, it follows that $\deg Z_i \leq p$.

(3) Follows from (2).

Note that the order of zero at ∞^+ of the function f_k used in part (1) is strictly positive, since it is $\deg q_{k+1} > \deg q_k \geq 0$.

Let A be an effective divisor supported only on finite points. We wish now to find a rational function $\sigma \in K(C)$, a standard finite divisor A_{red} of degree $\leq p$, and non-negative integers u, v, such that

$$(\sigma) = -A + A_{red} + u \infty^+ \ + v \infty^-$$

We call this process *total reduction* of the divisor A.

First, by pairing up all hyperelliptic twins (i.e. pairs $P + P^-$) that may occur in A, we can write $A = A_1 + B$ where $B = \pi^* B_1$ for some divisor B_1 on $\mathbf{P}^1$, and A_1 is a a standard finite divisor on C. Then let $G \in K[x]$ be such that $(G) = B_1 - (\deg G)\infty$ on $\mathbf{P}^1$, so that $B_1 = (G) + (\deg G)\infty$; pulling back to C and substituting, we obtain

$$A = (G) + A_1 + (\deg G)(\infty^+ \ + \infty^- \ )$$

If $\deg A_1 \leq p$ we are done. If not, then perform the reduction process described by Cor. 4, starting from the pole function of A_1, (or any other standard function with A_1 as divisor of poles) to find a function $f \in K(C)$ and non-negative integers a, b such that

$$(f) = -A_1 + A_{red} + a \infty^+ \ + b \infty^-$$

where A_{red} is a standard finite divisor of degree $\leq p$, and $a, b, \geq 0$. Substituting for A_1 we obtain

$$A = -(f) + (G) + A_{red} + (\deg G + a)\infty^+ \ + (\deg G + b)\infty^- \qquad (4)$$

whence the sought reduction, with $\sigma = f/G$. Note that we have $2 \deg G + a + b = \deg A - \deg A_{red} \leq \deg A$, whence, recalling that $a > 0$ as remarked after the proof of Cor. 4, we find

$$2 \deg G + b < \deg A \qquad (5)$$

This will be used in the Baby Step-Giant Step algorithm.

The total reduction algorithm can be summarized as follows: Given a standard divisor A, first separate out the divisor B, and find the polynomial $G(x)$. This is usually straightforward. The residual divisor $A_1 = A - B$ will be described by a pole function $f = (L + y)/D$. If $\deg A_1 = \deg D \leq p$, there is no more to do. If not, then compute the CFE of f to level k, for which the inequality of Cor.4(1) holds. Then f_k/G is the reducing function, and $A_{red} = Z_k$.

The reduction algorithm is most often needed to reduce a divisor $A = D_1 + D_2$ where the D_i are standard finite divisors described by standard functions f_{D_i} which have the D_i as finite poles (e.g. they may be the pole functions of the D_i). An auxiliary algorithm, the hyperelliptic analogue of the algorithm for composition of quadratic forms, is needed to obtain, from the f_{D_i}, the polynomial G and a standard function $(L + y)/M$, whose pole divisor is A_1. For this, see, e.g. [5, 8, 8].

3.2   Quasiperiodicity

When the ground field K is finite, hyperelliptic continued fraction expansions of standard functions are periodic (i.e the sequences of partial quotients and reducts are eventually periodic), as in the case of real quadratic irrationals, and the expansions can be studied by the same techniques as in the real case. Over an infinite field, the best one can hope for is quasiperiodicity, i.e. periodicity up to constant multiples. Over a finite field quasiperiodicity coexists with periodicity, and seems the more fundamental, or at least the more useful, phenomenon.

In the following theorem, the ground field K is arbitrary. The notation i s, as always, that of §2.

Theorem 5 *Let D be a standard finite divisor of degree $\leq p$. Then the CFE of f_D is quasiperiodic iff $\infty^+ - \infty^-$ is a torsion divisor. If this is the case then the quasiperiod is bounded above by the order of torsion.*

Proof. (c.f. [2]) Suppose first that the expansion of a given $f = f_D$ is quasiperiodic, with quasiperiod l, so $g_{i+l} = c_i g_i, i \geq 1$. It follows that $Z_{i+l} = Z_i$. Then (c.f. (2)), $(f_{i+l}/f_i) = (f_{i+l}) - (f_i)$ is supported entirely at infinity, which means that $\infty^+ - \infty^-$ is a torsion divisor. Conversely, suppose $\infty^+ - \infty^-$ is torsion of order $R > 0$. Take $h \in K(C)$ with $(h) = R(\infty^+ - \infty^-)$ and let D be any generic finite divisor of degree $\leq p$. Then we can write

$$(h) = -D + D + R(\infty^+ - \infty^-)$$

whence, by Cor.3, there exists for the CFE of f_D an index l such that $h = cf_l$, c constant, and $Z_l = D$. From this it follows that $g_l = cg_1$, c constant which implies the quasiperiodicity of the sequence $\{g_i\}$, with quasiperiod l. Note that the sequence of divisors Z_i is genuinely periodic, with period l. The inequality $l \leq R$ follows from $R = v_+(h) = v_+(f_l) = \deg q_{l+1}$ and $\deg q_i \geq i$ for $i \geq 1$.

3.3 Baby-Step Giant-Step

The Baby-Step Giant Step algorithm in hyperelliptic function fields is a translation of the algorithm of the same name invented by Shanks for real quadratic number fields, for which see [4].

Let D be a standard finite divisor of degree $\leq p$ and let f_i, g_i, p_i, q_i, Z_i be, repectively, the approximants, reducts, and convergents obtained in the CFE of the pole function f_D of D (or any standard function whose finite pole divisor is D). We consider also the CFE of y, with corresponding quantities $y_i, \gamma_i, r_i, s_i, \Phi_i$. For $f \in K(C)$ we write $\delta f = |v_-(f)|$, which avoids confusion over signs, and is consistent with the terminology of [5, 8]. Note that

$$\delta f_{i+1} = \delta f_i + \deg a_i \tag{6}$$

where a_i is the partial quotient, as follows from Prop. 1(3) and (1). Thus there is virtually no overhead in keeping a table of the δf_i.

A baby step of the Baby-Step Giant Step algorithm is just an iterative step in one of the continued fraction expansions mentioned above. A giant step, which we now describe, is an algorithm which takes as input $(Z_i, \delta f_i, \Phi_j, \delta y_j)$ and outputs a pair (Z, δ), such that there exists an index $k \approx i + j$ for which $Z = Z_k, \delta = \delta f_k$ (k remains unknown, in general). In practice, the Z_i, Φ_j are described by the functions g_i, γ_j, so for computational purposes the input can be thought of as $g_i, \delta f_i, \gamma_j, \delta y_j$ and the output is $g = g_k, \delta = \delta_k$. Thus the giant step replaces approximately j baby steps in the CFE of f_D, at the cost of losing sight of the convergents and approximants.

For $i \geq 1$ we can write

$$(f_i) = -D + Z_i + v_+(f_i)\infty^+ - \delta f_i \infty^- \tag{7}$$

$$(y_j) = \Phi_j + v_+(y_j)\infty^+ - \delta y_j \infty^- \tag{8}$$

Adding,

$$(f_i y_j) = -D + Z_i + \Phi_j - (v_+(f_i) + v_+(y_j))\infty^+ - (\delta f_i + \delta y_j)\infty^- \tag{9}$$

Now, using total reduction we have (c.f. (4))

$$Z_i + \Phi_j = (G/f) + Z + (\deg G + a)\infty^+ + (\deg G + b)\infty^-$$

with Z a standard finite divisor of degree $\leq p$, $f \in K(C), G \in K[x]$. Substituting in (9), and collecting divisors of rational functions

$$\left(f\frac{f_i y_j}{G}\right) = -D + Z +$$
$$(v_+(f_i) + v_+(y_j) + \deg G + a)\infty^+ - (\delta f_i + \delta y_j - \deg G - b)\infty^- \tag{10}$$

Suppose that

$$\delta f_i + \delta f_j - \deg G - b \geq 0 \tag{11}$$

Then the hypotheses of Cor.3 are satisfied, and $\exists k$ such that the rational function on the left in (10) is a constant multiple of f_k, and $Z = Z_k$. Moreover, from (10),

$$\delta f_k = \delta f_i + \delta y_j - d \tag{12}$$

where $d = \deg G + b$, so δf_k is known. Thus, finally, output (Z, δ) where δ is given by the right-hand side of (12). We shall say that the pair (Z, δ) is obtained by a giant step, and denote it $(Z_i * \Phi_j, \delta f_i * \delta y_j)$.

Note that $0 \leq d \leq 2p - 1$, as follows from the inequality (5) of §2, using $\deg Z_i, \deg \Phi_j \leq p$, so that we have a bound for the difference between $\delta f_i + \delta y_j$ and $\delta f_i * \delta y_j$.

Practical details of the algorithm are given in [7],[8]. The main application has been to calculation of the regulator of a hyperelliptic function field (c.f. the following section and the papers just cited); there is a cryptographic application in [5].

Using the estimate for d, and the values of $\delta f_i, \delta y_i$, we find a sufficient condition for the constraint (11) to hold is

$$\deg q_i + \deg s_j \geq \deg D - 3$$

a condition vacuously satisfied when $D = 0$, i.e. $f = y$.

Example. The genus 1 case. This is remarkably simple. The following example uses freely the notation established in the present section and also the notation of the reduction and total reduction algorithms (c.f. Cor.4 et. seq). We also use results of continued fraction expansions in the genus 1 case established in §5.

We deal with divisors of degree $\leq p = 1$, and a divisor of degree 1 is a point. Thus let f be the pole function of a point. The expansions of f, y may or may not be periodic (since we make no hypotheses on the groundfield K). If they are periodic, take indices i, j which do not coincide with the periods of f, y, respectively, and $j \geq 1$. Then Z_i and Φ_j both have degree 1 (they have degrees ≤ 1, by Theorem 4(3), and it is a consequence of Prop. 8 that the degree is exactly 1.) Thus $Z_i = P$, $\Phi_j = Q$, for some points $P, Q \in C$. We claim:
If $P = Q^-$ then $(Z_i * \Phi_j, \delta_i * \delta_j) = (0, \delta_i + \delta_j - 1)$
If $P \neq Q^-$ then $(Z_i * \Phi_j, \delta_i * \delta_j) = (S, \delta_i + \delta_j)$
for some point S of the curve.

Indeed, if $P = Q^-$ then (c.f. eqn. (3.3) $Z_i + \Phi_j = P + P^-$; but $P + P^- = (G) + \infty^+ + \infty^-$ for some polynomial $G \in K[X]$ of degree 1. Thus, in the notation of total reduction, $A_{red} = Z_i * \Phi_j = 0$, and, by eqn.(9), $\delta f_i * \delta y_j = \delta f_i + \delta y_j - 1$, which establishes the first assertion. If $P \neq Q^-$ then $P + Q$ is a divisor of degree 2 without hyperelliptic pairs, (thus $G = 0$) and with pole function of the form $h = (L + y)/M$, where $\deg M = 2$ and $\deg L \leq 1$. This function has no poles at infinity. Thus, by Cor. 4(1), a reducing function for $P + Q$ is $h_1 = h - a_0$, where a_0 is a constant. Then $b = v_-(h_1) = 0$, and the second affirmation follows again using eqn (9). The point S is defined by $(h_1) = -P - Q + S + \infty^+$.

4   Calculating the Regulator

The regulator is the order of torsion of the divisor $\infty^+ - \infty^-$, i.e. it is the least integer $R > 0$ such that $R(\infty^+ - \infty^-) \equiv 0$. If K is infinite we assume $R < \infty$. A function $h \in K(C)$ such that $(h) = \pm R(\infty^+ - \infty^-)$ is called a *fundamental unit*.

Suppose then that $(h) = R(\infty^+ - \infty^-)$. Then as shown in §3.2, if D is any standard finite divisor and $\{f_i\}$ the CFE of its pole function then $\exists l | f_l = ch$, for some constant $c \in K$. Thus R can be calculated from any CFE. However, it is best to take $D = 0$, i.e. to look at the CFE of y, because one can take advantage of symmetries in the CFE of y which do not exist in the general case.

We use the results of the previous section with $f = y$. All quantities concerned come from the CFE of y, and are denoted: $a_i, p_i, q_i, g_i, y_i, Z_i$, the g_i being the reducts and the y_i the approximants.

We have, for $i \geq 1$,

$$(y_i) = Z_i + (\deg q_{i+1})\infty^+ - (\deg q_i + p + 1)\infty^- \tag{13}$$

whence

$$R = v_-(y_l) = v_+(y_l) = \deg q_{l+1} = p + 1 + \deg q_l \tag{14}$$

where as in §3.2, $l = \min i | Z_i = 0$ is the quasiperiod.

This is the basic baby step algorithm: develop the CFE of y until hitting l with $Z_l = 0$, keeping track of $\deg q_i$, or, equivalently, of δy_i (notation of §3.3) .

Next we consider some of the structure of this CFE.

Lemma 6. *In the CFE of y, $\deg a_i \leq p+1$ for all i. If $i > 0$ and $\deg a_i = p+1$ then y_i is a fundamental unit and $R = \deg q_{i+1} = \delta y_i$.*

Proof. Take degrees in (13)gives $0 = \deg Z_i + \deg q_{i+1} - \deg q_i - (p + 1)$. The lemma follows since $\deg Z_i \geq 0$ and $\deg a_i = \deg q_{i+1} - \deg q_i$.

The following proposition implies symmetry in the CFE of y up to the quasiperiod. Recall that the action of the hyperelliptic involution is denoted $U \mapsto U^-$.

Proposition 7 *For all $i < l$, $Z_i^- = Z_{l-i}$, and*

$$\deg q_{i+1} + \deg q_{l-i} = \deg q_{l-i+1} + \deg q_i = R - p - 1.$$

Proof. Applying the hyperelliptic involution to (13) we obtain, since the involution interchanges ∞^+ and ∞^-

$$(y_i^-) = Z_i^- + (\deg q_{i+1})\infty^- - (\deg q_i + p + 1)\infty^+ \tag{15}$$

Adding $(y_l) = R(\infty^+ - \infty^-)$ to (15) gives

$$(y_l y_i^-) = Z_i^- + (R - \deg q_i - p - 1)\infty^+ - (R - \deg q_{i+1})\infty^- \tag{16}$$

By lemma 6 $R > \deg q_i + p + 1$. Thus Cor.3 applies to the situation described by (16), and we conclude that, for some j, and some constant c, $y_j = c y_l y_i^-$ and $Z_j = Z_i^-$. Comparing coefficients of ∞^+ and ∞^- in (13) (with j instead of i) and (16) we find

$$\deg q_{i+1} + \deg q_j = \deg q_{j+1} + \deg q_i$$
$$= R - p - 1 \qquad (17)$$

This shows that j, considered as function of i, is strictly decreasing and (using lemma 6) $\leq l - 1$. It follows that $j = l - i$.

Prop. 7 shows that R can be calculated in at most $\lceil l/2 \rceil$ baby steps, by calculating the CFE of y until $Z_i = Z_i^-$ or $Z_i = Z_{i+1}^-$.

Now we consider the Baby-Step Giant-Step algorithm. We write δ_i for δy_i (c.f. §3.3). First calculate t terms of the CFE of y, keeping a table (Z_i, δ_i). We can assume $t < \lceil l/2 \rceil$, otherwise these baby steps would give R. Then, by the periodicity and conjugacy properties of the Z_i discussed above, one knows also the (Z_i, δ_i) in the interval $l - t \leq i \leq l + t$; the situation is illustrated in (18) (19), (20). The symbols a_i-the partial quotients- and $p + 1$ are placed between the Z_i to help with the bookkeeping. The purpose is to have visible the increase in δ_i between i and $i + 1$).

$$Z_0 - p + 1 - Z_1 - a_1 - Z_2 - a_2 - \ldots Z_{t-1} - a_{t-1} - Z_t - a_t - \ldots \qquad (18)$$

The interval $[l - t, l]$ to the left of $Z_l, \delta_l = R$ may be visualized as

$$Z_{l-t} - a_t - Z_{l-t+1} - a_{t-1} - \ldots Z_{l-1} - a_1 - Z_l \qquad (19)$$

where $Z_{l-i} = Z_i^-$, and the interval $[l, l + t]$ to the right of Z_l as

$$Z_l - p + 1 - Z_{l+1} - a_1 - Z_{l+2} - a_{l+2} \ldots Z_{l+t-1} - a_{t-1} - Z_{l+t} \qquad (20)$$

where $Z_{l+i} = Z_i$. We calculate: for $1 \leq i \leq t$,

$$\delta_i = p + 1 + \deg q_i = p + 1 + \sum_{j=1}^{i-1} \deg a_j \qquad (21)$$

$$\delta_{l+i} = \delta_l + p + 1 + \sum_{j=1}^{i-1} \deg a_j = R + \delta_i \qquad (22)$$

$$\delta_{l-i} = \delta_l - \sum_{j=1}^{i} \deg a_j = R - \sum_{j=1}^{i} \deg a_j = R - \delta_{i+1} + (p + 1) \qquad (23)$$

Take $s \leq t$ and define $(\hat{Z}_j, \hat{\delta}_j), j \geq 1$ by

$$(\hat{Z}_1, \hat{\delta}_1) = (Z_s * Z_t, \delta_s * \delta_t)$$
$$(\hat{Z}_{j+1}, \hat{\delta}_{j+1}) = (Z_s * \hat{Z}_j, \delta_s * \hat{\delta}_j)$$

in the notation of (§3.3). Recall also that for each j there is an i such that $(\hat{Z}_j, \hat{\delta}_j) = (Z_i, \delta_i)$. Although we do not in general know i, we do know $\delta_i = \hat{\delta}_j$. If however $\hat{Z}_j$ is recognized as Z_k or Z_k^-, $k \le t$, then we do know i; we have $i = l+k$ or $l-k$ respectively, and R can be calculated from (22) or (23), provided one can be certain that the size of the giant steps is not so large that one has stepped entirely over the segment around Z_l and landed rather in some segment centred on a multiple of l. Clearly one is certain to land in the correct segment for some j, provided the size of the giant steps, $\hat{\delta}_{j+1} - \hat{\delta}_j$ is positive and does not exceed $\delta_{l+t} - \delta_{l-t}$. This gives conditions on s and $t - s$. In fact, by (22) and (23),

$$\delta_{l+t} - \delta_{l-t} = p + 1 + 2\sum_{i=1}^{t-1} \deg a_i + \deg a_t$$

The size of the first giant step is in the interval $[\delta_s + \delta_t - (2p + 1), \delta_s + \delta_t]$ and the size of the subsequent steps is in the interval $[2\delta_s - (2p+1), 2\delta_s]$. Now, $\delta_s = p + 1 + \sum_{i=1}^{s-1} \deg a_i$, so that, for giant steps no larger than $\delta_{l+t} - \delta_{l-t}$ we need

$$2(p + 1 + \sum_{i=1}^{s-1} \deg a_i) \le p + 1 + 2\sum_{i=1}^{t-1} \deg a_i + \deg a_t$$

which is

$$\frac{p+1}{2} \le \sum_{i=s}^{t-1} \deg a_i + \frac{\deg a_t}{2}$$

Since $\deg a_i \ge 1, i \ge 1$ this inequality is certainly satisfied if $t - s \ge (p - 1)/2$. The choice of s remains. The received wisdom is to take s in the order of $\sqrt{B}$, where B is an estimate for R. With these choices R is calculated by $O(\sqrt{B})$ polynomial operations.

5   Genus 1

Throughout this section the genus p of C is 1. CFE's are particularly simple in this case. We recover results of [1, 8] in a characteristic-free fashion.

A distinguished role is played by y, its reducts g_i, and functions $cg_i + h$ where $c \in K$ and $h \in K[x]$. We refer to such functions as *functions in the cycle of y*.

Proposition 8 *Let f be a standard function with finite pole divisor of degree ≤ 1. We consider the CFE of f, with partial quotients a_i.*

1. *If $\infty^+ - \infty^-$ is not torsion, then $\deg a_i = 1$ for all $i \ge 1$.*
2. *If $\infty^+ - \infty^-$ is torsion, so that CFE's are quasiperiodic, with quasiperiod l say, then:*

(a) If $f = y$ then $\deg a_i = 1$ for $1 \leq i \leq l-1$ and $\deg a_l = 2$. More generally if f is a function in the cycle of y then in each quasiperiod of the CFE of f there is just one index j, such that $\deg a_j = 2$ and for all $i \neq j$ in the quasiperiod, $\deg a_i = 1$.

(b) If f is not a function in the cycle of y then $\deg a_i = 1$ for all $i \geq 1$.

The proof is straightforward, using lemma 6.

We find the following, for convergents q_i in the CFE of f, recalling $\deg q_i = \sum_{j=1}^{i-1} \deg a_i, i \geq 1$:

- If $f = y, y_i = q_i f - p_i$ then, for $1 \leq i \leq l$

$$(y_i) = Q_i + i\infty^+ - (i+1)\infty^- \tag{24}$$

- If f_1 is not a reduct of y, then, for $1 \leq i$,

$$(f_i) = -R_i + S_i + i(\infty^+ - \infty^-). \tag{25}$$

Q_i, R_i, S_i being points of C. This gives

Proposition 9 Let C have regulator $R < \infty$. Let f be a standard function with finite pole divisor of degree ≤ 1. If f is in the cycle of y then the CFE of f has quasiperiod $R - 1$, otherwise it has quasiperiod R.

Proof. The results follow from (24) and (25), since l is the least index for which $Q_l = \infty^+$ or $R_l = S_l$.

Now we relate continued fraction expansions with the group law defined by taking ∞^+ as zero for a group law on C. To put this in perspective, let us *start* from an elliptic curve E with O as zero of its group. Let $P \neq O$ be any other K-rational point of E. Then there is a unique involution on E in which P and O are paired; we shall denote E with the new hyperelliptic structure defined by this involution as C, and rename P, O as ∞^-, ∞^+ respectively. In concrete terms, one may suppose E given by a Weierstrass equation, and C described by some other plane model. For Char. $\neq 2$ formulas can be found in [1]; we leave to the reader the pleasure of calculating the analogues in Char. 2.

Recall that $iP = Q$ in the group law on E iff $i(P - O) \equiv Q - O$. The desired connexion between the group law and CFE's in then given by:

Theorem 10 1. P is an R-torsion point on E iff $\infty^+ - \infty^-$ is an R-torsion divisor on C.

2. $iP = Q$ in the group law on E iff Q is a zero of the approximant y_{i-1}, or, equivalently, a pole of the reduct g_{i-1}, of y.

Proof. (1) is immediate from definitions.

(2). Suppose $iP = Q$, so $\exists f \in K(E) = K(C) | (f) = (Q-O) - i(P-O)$. Rewritten in notation appropriate to C, this is

$$(f) = (Q - \infty^+) - i(\infty^- - \infty^+) = Q + (i-1)\infty^+ - i\infty^-$$

This implies, by Cor. 3 (applied with $D = 0$) and (24) that f is a constant multiple of y_{i-1} and $Q = Q_{i-1}$. Moreover $g_{i-1} = y_{i-1}/y_{i-2}$ (c.f. Prop. 1(4)) so, using (24) for i and $i-1$

$$(g_{i-1}) = -Q + Q_{i-2} - \infty^+ + \infty^-$$

The converse is obtained by chasing backwards through this argument.

We note that $g_i \in \mathcal{L}(Q - \infty^+ + \infty^-)$. By Riemann-Roch this space is 1-dimensional, so consists of constant multiples of g_{i-1}. Now it is easy to produce functions in $\mathcal{L}(Q - \infty^+ + \infty^-)$. For example, $f_Q^* = f_Q - < f_Q >$ is such a function, where f_Q is the pole function of Q and $< f_Q >$ denotes the polynomial part of the Laurent expansion of f_Q at ∞^-. (Warning: in [1], a similar function is denoted f_Q). Thus f_Q^* is a constant multiple of g_i.

The discrete log problem on E is: given P and $Q = iP$, determine i. Theorem 10 shows that this is equivalent to the problems:

1. Given Q, determine i so that Q is a zero of y_{i-1}.
2. Given Q, determine i so that Q is a pole of g_{i-1}.
3. Given Q, determine i so that f_Q^* is a constant multiple of g_{i-1}.

This is essentially the result of [8].

References

1. Adams, W. W., Razar, M.J.: Multiples of points on elliptic curves and continued fractions. Proc. Lond. Math. Soc. **41** (1980) 481–498.
2. Berry, T.G.: On periodicity of continued fractions in hyperelliptic function fields. Archiv der Mathematik **55** (1990) 259–266.
3. Berry, T.G.: Construction of linear systems on hyperelliptic curves. Jour. Sym. Comp. **26** (1998)315–327.
4. Cohen, Henri: A course in computational algebraic number theory. Springer-Verlag, Berlin, 1993.
5. Scheidler, R. Stein, A., Williams, H.C.: Key-exchange in Real Quadratic Congruence Function Fields. Des. Codes Crypt. **7** (1996).
6. Stein, A.: Equivalences between elliptic curves and real quadratic congruence function fields. J. Theor. Nombres Bdx, **9** (1997)75–95.
7. Stein, A., Williams, H.C.: Baby Step Giant Step in real quadratic function fields. Preprint (1995). Available from
   http://cacr.math.uwaterloo.ca/ astein/publikationen.html.
8. Zuccherato, R.: The continued fraction algorithm and regulator for quadratic function fields of characteristic 2. J. Algebra **190** (1997) 563–587.

Discrete Logarithms: Recent Progress

Johannes Buchmann[1] and Damian Weber[2]

[1] Technische Universität Darmstadt
Alexanderstraße 10
D–64287 Darmstadt
buchmann@cdc.informatik.tu-darmstadt.de

[2] Institut für Techno– und Wirtschaftsmathematik
Erwin–Schrödinger–Str. 49
D–67663 Kaiserslautern
weber@itwm.uni-kl.de

Abstract. We summarize recent developments on the computation of discrete logarithms in general groups as well as in some specialized settings. More specifically, we consider the following abelian groups: the multiplicative group of finite fields, the group of points of an elliptic curve over a finite field, and the class group of quadratic number fields.

1 Introduction

Within the last few years, due to applications in cryptography, an enormous interest has grown in the question of the actual difficulty of the discrete logarithm (DL) problem in groups. In this paper we summarize recent developments concerning this computational problem.

On the one hand, cryptographers want to find finite groups where this problem is presumably hard – or even better – provably hard. On the other hand, competition arises due to the research in the area of computational number theory where efforts are made to develop efficient algorithms to solve such problems, and exhibiting weaknesses of the corresponding cryptographic protocols.

Since the proposal of the Diffie–Hellman key exchange protocol [13], several other protocols have been developed, whose security depends on the difficulty in solving the DL problem. The DL problem for a group G may be stated as follows.

Given $a, b \in G$, find an $x \in \mathbb{Z}$, such that

$$a^x = b, \tag{1}$$

or prove that such an x does not exist. If x exists, we call the minimal non-negative solution of (1) the discrete logarithm of b with respect to a. We will stick to this notation throughout this article.

Motivated by the existence of subexponential algorithms for $G = GF(p)^*$, the designers of new cryptosystems incorporate other groups into their protocols, in order to avoid subexponential attacks. Though not possible in any case, sometimes the designer may use an arbitrary group with the only precondition that the DL problem be hard (for example the zero–knowledge protocol for DL from [8, 9]). Among these are the group of points of an elliptic curve over a finite field [15] as well as the Jacobian group of a hyperelliptic curve [16]. For none of these a subexponential algorithm is known. For some special cases, which are easily avoided in cryptographic applications, however, faster algorithms are known. This is the case for elliptic curves of trace 1 [30], hyperelliptic curves of large genus [1] and class groups of imaginary quadratic number fields [4].

When trying to solve the DL problem, a first approach is to use generic algorithms which work in any group and use only group operations (multiplication, inversion, equality testing). In section 2 we review both a deterministic and a probabilistic algorithm which produce a solution of (1) after at most $O(\sqrt{|G|})$ group operations. As we will see, both algorithms are optimal in the sense that $O(\sqrt{|G|})$ is shown to be a lower bound for generic algorithms.

In section 3 the group of points of elliptic curves over $GF(p)$ and $GF(2^n)$ is considered. We summarize the results of the probably largest effort to date to attack a DL problem via the implementation of a generic algorithm.

Algorithms of subexponential type can be found if there is an efficient way to produce relations among group elements. That means to form non–trivial power products of elements of a small set which evaluate to the unity element of the group. This is the case, for example, in finite prime fields. The recent practical progress in this setting is surveyed in section 4.

2   Generic Algorithms

Let G be a finite abelian group. In this section we consider algorithms which, given two elements $g, g' \in G$, only make use of three types of operations:

- computing $gg' \in G$,
- computing $g^{-1} \in G$,
- deciding whether $g = g'$.

These are referred to as *group operations*. We denote the identity of G by 1.

2.1   Reducing to Cyclic Groups

It is well known that there are positive integers $m_1, \ldots, m_k$, $k \geq 1$ – the *invariants* of G – where m_i divides m_{i+1} for $1 \leq i < k$ such that

$$G \cong \mathbb{Z}/m_1\mathbb{Z} \times \cdots \times \mathbb{Z}/m_k\mathbb{Z}. \tag{2}$$

The DL problem in the group on the right hand side of (2) can be reduced to the DL problem in each of the $\mathbb{Z}/m_i\mathbb{Z}$. With $m := m_i (1 \leq i \leq k)$, solving the DL problem in $\mathbb{Z}/m\mathbb{Z}$ means solving the congruence

$$ax \equiv b \bmod m$$

which can be done in polynomial time by means of the Euclidean algorithm. At first sight, this suggests that the DL problem is easy in general since (2) describes the structure of any finite abelian group. The problem, however, is that in general neither the invariants of G nor the isomorphism of (2) is known. The difficulty of computing discrete logarithms in G is therefore closely related to the difficulty of finding the invariants of G and the isomorphism of (2).

If computing the invariants is possible, the problem can be reduced further to groups of prime order as we shall see in the next section.

2.2   Reducing to Prime Order Groups

We recall a method by Silver, Pohlig and Hellman, which first has been described for $GF(p)^*$, when $p-1$ is smooth. This method works for arbitrary cyclic groups G of order n and can be used to reduce the DL problem in G to DL problems in prime order groups G_r, where $r|n$.

Let r be any prime dividing n and h an integer with $h \geq 1$. Let r^h be a power of r dividing n; then we are going to compute x modulo r^h.

Suppose (1) is solvable. If $h = 1$, solve $a^{x_0} = b \mod G/G^r$. This group has prime order r. Then $a^{x_0} = a^x c^r$ for some $c \in G$. If g is a generator of G with $g^l = a, g^k = c$, we obtain $lx_0 = lx + kr$ which is equivalent to $x \equiv x_0 \mod r$.

Assume now that we know the value of $x \mod r^{h-1}$, written as

$$x \equiv x_0 + x_1 r + x_2 r^2 + \cdots + x_{h-2} r^{h-2} \mod r^{h-1}.$$

Set

$$a' = a^{n/r},$$

$$b' = \left(\frac{b}{a^{x_0 + x_1 r + x_2 r^2 + \cdots + x_{h-2} r^{h-2}}} \right)^{n/r^h}.$$

Because of

$$a'^r = 1 \quad \text{and}$$
$$b'^r = 1,$$

both a' and b' are members of the unique subgroup G_r of order r in G.

Solving

$$a'^{x_{h-1}} = b' \quad \text{for } x_{h-1},$$

we obtain

$$a'^{x_{h-1}} = \left(\frac{b}{a^{x_0 + x_1 r + x_2 r^2 + \cdots + x_{h-2} r^{h-2}}} \right)^{n/r^h}.$$

With

$$a'^{x_{h-1}} = \left(a^{x_{h-1}} \right)^{\frac{n}{r}} = a^{r^{h-1} x_{h-1} \frac{n}{r^h}},$$

it follows that

$$\left(a^{x_0+x_1r+x_2r^2+\cdots+x_{h-1}r^{h-1}}\right)^{\frac{n}{r^h}} = b^{\frac{n}{r^h}},$$

which is equivalent to

$$x \equiv x_0 + x_1r + x_2r^2 + \cdots + x_{h-2}r^{h-2} + x_{h-1}r^{h-1} \bmod r^h.$$

2.3   Shanks's Baby–Step–Giant–Step Algorithm

The first algorithm, which is deterministic and runs in time $O(\sqrt{n})$ is Shanks's Baby–Step–Giant–Step algorithm [27]. Let $m = \lfloor\sqrt{n}\rfloor+1$, such that $x = x_1m+x_2$ with (unknown) $x_1, x_2 < m$. After computing a set $M := \{a^{mi} \,|\, 0 \leq i \leq m\}$ (the *giant steps*), it remains to decide whether one of the elements $a^{-j}b$, $0 \leq j < m$ lies in M (the *baby steps*). If a match is found, $a^{mi} = a^{-j}b$, therefore $x := mi+j$ satisfies (1). Otherwise, b is not contained in the subgroup generated by a. Obviously, this algorithm requires at most $T(n) = 2\sqrt{n}+T'(\sqrt{n})\sqrt{n}$ group operations, where $T'(m)$ is the time to decide membership for a given element in M. Linear search in M would cause $T(n) > n$, which is slower than the trivial algorithm. With a total ordering of the representation of group elements, however, one may sort M and do binary search on M. There are sorting algorithms which sort m elements within time $O(m \log m)$ (for example heapsort [11]); binary search on m elements consumes time $O(\log m)$. Summing up, we obtain

$$T(n) = C\sqrt{n}\log n$$

for some constant C, which is $O(n^{1/2+\epsilon})$ for all $\epsilon > 0$. Because M has to be stored, this algorithm also consumes $O(\sqrt{n})$ space. We note that there is a time/memory trade–off by reducing the table size $m < \sqrt{n}$ and carrying out $n/m > \sqrt{n}$ membership tests.

By refining this method a considerable theoretical and practical improvement is achieved by Buchmann, Jacobson and Teske (BJT) in [5]. The authors extract the discrete logarithm in time $O(\sqrt{x}+\log\sqrt{x})$ by using a table of $O(\sqrt{x})$ entries. Note that the running time depends on the (unknown) discrete logarithm itself. In the case that the DL does not exist, the running times and space requirements hold for $x := \text{ord}_G a$. In their paper also the practical significance is illustrated on the special case $G = Cl(\Delta)$ of the ideal class group of imaginary quadratic orders. The largest example is given for $\Delta = -4 \cdot (10^{20} + 1)$ where G is the subgroup generated by the ideal over 7 which consists of $1\,856\,197\,104$ elements. Clearly, the smaller x the performance compared to Shanks's original algorithm is the better, as can be seen from table 1.

2.4   Pollard's Probabilistic Algorithms

The advantage of Pollard's algorithms is the use of constant space while preserving an *expected* running time of $O(\sqrt{n})$. In their original version, these algorithms have been proposed for $GF(p)^*$ [22].

x	Shanks	BJT
371 239 423	80 sec	55 sec
742 478 843	91 sec	80 sec
1 113 718 263	106 sec	101 sec
1 484 957 683	113 sec	128 sec
1 856 197 103	106 sec	105 sec

Table 1. Original Shanks and its Refinement (BJT)

The main idea here for computing x is to produce iteratively a sequence of elements $(d_i)_{i \geq 1}, d_i \in G$, where all the d_i's are of the form

$$a^k b^l.$$

So (d_i) will become periodic after at most n iterations. For a sequence producing group elements at random, this can be expected to happen in an expected number of $O(\sqrt{n})$ steps (birthday paradoxon). The computation stops when $d_{i'} \equiv d_i \mod p$ for some pair (i, i') is recognized. Then

$$a^k b^l = d_i = d_{i'} = a^{k'} b^{l'}$$

and

$$a^{k-k'} b^{l-l'} = 1$$

and therefore

$$(k - k' + x(l - l')) \equiv 0 \mod n,$$

which reveals x, provided that $\gcd(l - l', n) = 1$.

There are two methods of finding repetitions in sequences. Pollard's method finds the pair (i, i') by computing the sequence twice as (d_j) and (d_{2j}), waiting for $(d_j) = (d_{2j})$. This will be the case when j is a positive multiple of the period length.

Instead of computing the sequence twice, Brent's algorithm [3] remembers the sequence elements d_{2^i} and compares them to $d_{2^i+j}, 1 \leq j \leq 2^i$. It can be shown that comparison is not needed for $1 \leq j \leq 3 \cdot 2^{i-1}$.

It is worthwhile examining different variants of producing the (d_i), because ideally the (d_i) should behave like a random sequence. In this case the expected length of this sequence can be shown to be $1.25\sqrt{n}$ until an element of G occurs twice. The original sequence of Pollard uses a partition of $G = S_1 \cup S_2 \cup S_3$ with equally sized S_i and is defined by

$$d_{i+1} := \begin{cases} ad_i, & d_i \in S_1 \\ d_i^2, & d_i \in S_2 \\ bd_i, & d_i \in S_3 \end{cases}$$

where an arbitrary product of the form $a^k b^l$ may be chosen as a start value. The sequence of exponents of a produced by this definition are computed by

$e_{i+1} = e_i + 1$ and $e_{i+1} = 2e_i$, starting from some $e_0 = k + xl$. A theoretical result about the "randomness" of that sequence seems not to be known. Teske [31] constructs a sequence where a distribution which is close to uniform can be proven. Her construction partitions G into 20 subsets $S_1, \ldots, S_{20}$ and replaces the definition of d_{i+1} above by

$$d_{i+1} := \begin{cases} m_k d_i, & d_i \in S_k \text{ and } 1 \leq k \leq 16 \\ d_i^2, & d_i \in S_k \text{ and } 17 \leq k \leq 20, \end{cases}$$

where the m_k are initially set to random power products of a and b.

By experiment she determines an average sequence length of $1.596\sqrt{|G|}$ for Pollard's original sequence and $1.292\sqrt{|G|}$ for her newly constructed sequence. The impact on actual computations is illustrated at prime order subgroups of elliptic curves over prime fields. For a subgroup size of 13 decimal digits the average running time over 40 runs of Pollard's original algorithm has been 27.3 minutes in contrast to 22.5 minutes of the improved version. This reflects the stable improvement of about 20% of the total running time observed for all group sizes.

A version which allows for parallel computations has been proposed in [32] and is of practical significance as long as there is no subexponential algorithm available for the group under consideration.

2.5 Shoup's Lower Bound

A recent result of Shoup shows that the two algorithms discussed above are optimal for groups where the group operations themselves are the only computations possible within an algorithm [29]. In fact, there are groups where no algorithms with running time better than $O(\sqrt{n})$ are known (elliptic curves, Jacobians of hyperelliptic curves).

Let G be a cyclic group of order n. Starting from the notion of an oracle which can be asked for the result of the group operations defined at the beginning of section 2, Shoup computes the probability that an algorithm outputs the correct answer to a DL problem in G after m oracle calls. The enumeration of elements of G can be thought of as encodings of the n (distinct) powers of a which are given by a map $\sigma : \mathbb{Z}/n\mathbb{Z} \longrightarrow S$, where S is a set of binary strings representing elements of G uniquely. The problem $a^x = b$ in G is then rewritten as follows. Given $(\sigma(1), \sigma(x))$, find a $y \in \mathbb{Z}/n\mathbb{Z}$ such that $\sigma(y) = \sigma(x)$.

We cite Shoup's main result concerning generic DL algorithms.

Theorem. *Let n be a positive integer whose largest prime divisor is p. Let $S \subset \{0,1\}^*$ be a set of cardinality at least n. Let A be a generic algorithm for $\mathbb{Z}/n\mathbb{Z}$ on S that makes m oracle queries, and suppose that the encoding function σ of $\mathbb{Z}/n\mathbb{Z}$ on S is chosen randomly. The input to A is $(\sigma(1), \sigma(x))$, where $x \in \mathbb{Z}/n\mathbb{Z}$ is random. The output of A is $y \in \mathbb{Z}/n\mathbb{Z}$. Then the probability that $x = y$ is $O(m^2/p)$.*

The conclusion is that for achieving a non-negligible probability for the event of success, one needs $O(\sqrt{p})$ oracle calls; thus a generic DL algorithm performs at least $O(\sqrt{p})$ group operations.

2.6   General Index Calculus

The *index calculus* methods which typically achieve a sub–exponential running time depend on the ability to efficiently generating so called *relations* among group elements. More precisely, if one fixes a (small) subset $S := \{g_1, \ldots, g_k\} \subset G$ and can find elements of the set

$$L := \{(e_1, \ldots, e_k) \mid g_1^{e_1} \cdots g_k^{e_k} = 1\}$$

in an efficient manner, then Shoup's lower bound is not cryptographically relevant for G. We note that L is a lattice in $\mathbb{Z}^k$ and

$$\Phi : \quad \begin{array}{ccc} \mathbb{Z}^k & \longrightarrow & G \\ (e_1, \ldots, e_k) & \mapsto & g_1^{e_1} \cdots g_k^{e_k} \end{array}$$

a homomorphism with kernel L so that

$$\mathbb{Z}^k / L \cong G.$$

From the conditions of section 2.5 we see that the source of producing relations must use some information "outside" G. For example in $GF(p)^*$, we may use the fact that $GF(p)$ is a field, or in class groups of quadratic number fields, we have a reduction theory. Whenever G happens to have an environment with such properties, one can apply the index calculus method, an outline of which is given below.

1. choose factor base:
   fix a set $S := \{g_1, \ldots, g_k\} \subset G$ of group elements, where $g_1 := a$, $g_2 := b$
2. produce relations:
   find relations of the form

$$g_1^{e_i 1} \cdots g_k^{e_i k} = 1, \quad (1 \leq i \leq l)$$

   terminate this step when $\mathrm{rank}(e_{ij}) = k - 1$
3. linear algebra step:
   set $A := (e_{ij})$ and compute a solution to $Ax \equiv 0 \bmod |G|$
4. extract solution:
   Let $x = (x_1, \ldots, x_k)$ be a solution of step 3. Since the rank is $k - 1$, we must have $\lambda \cdot x \equiv (\log g_1, \ldots, \log g_k) \bmod |G|$ for some $\lambda \in \mathbb{Z}$. The logarithms with respect to g_1 – in case they all exist – are then found by setting $\lambda \equiv x_1^{-1} \bmod |G|$.

When analyzing an index calculus variant, several problems have to be addressed. The parameter k is subject to optimization. If k is too small, the time to find relations is probably too large; on the other hand, if k is too big, the linear algebra step consumes too much time. For step 2, the probability to find a relation must be taken into account. In step 3 the linear dependency is found modulo prime divisors of $|G|$, sparse matrix techniques such as Lanczos, Conjugate Gradient can be used. In this case, the running time of this step is $O(k^2 + kw)$, where w is the total number of non–zero entries among the e_{ij}.

3   Elliptic Curves

Today, the best general discrete log algorithms for the group of points of an elliptic curve over a finite field K are the generic ones given in the preceding section. For two special cases, a more efficient way has been found so far. An efficient way to find relations in this group has not been found yet, so the index calculus idea is not applicable. Thus we are left with Shanks's and Pollard's algorithm, but due to the enormous space requirements of the former; only the latter one is actually applicable for larger groups in practice (say $|G| > 10^{15}$).

Let $f(X) = X^3 + a_2 X^2 + a_1 X + a_0 \in K[X]$.

An elliptic curve (EC) over K is defined as the following set of points

$$E = \{(x, y) \mid y^2 = f(x)\} \subset K \times K \} \cup \{\infty\}$$

if $\operatorname{char}(K) \neq 2$. By change of variables this may be transformed to

$$E = \{(x, y) \mid y^2 = x^3 + ax + b\} \subset K \times K \} \cup \{\infty\}$$

for appropriate $a, b \in K$ if $\operatorname{char}(K) \neq 3$.

If $\operatorname{char}(K) = 2$, an elliptic curve is given by

$$E = \{(x, y) \mid y^2 + y = f(x)\} \subset K \times K \} \cup \{\infty\}.$$

The addition law of two points $(x_1, y_1) + (x_2, y_2) = (x_3, y_3) \in E$ is given by the following rule ($\operatorname{char}(K) \notin \{2, 3\}$):

1. if $x_1 \neq x_2$ let $m := (y_2 - y_1)/(x_2 - x_1)$, and $m := (3x_1^2 + a)/2y_1$ otherwise
2. $x_3 := m^2 - x_1 - x_2$
3. $y_3 := -y_1 + m(x_1 - x_3)$.

For $\operatorname{char}(K) \in \{2, 3\}$ similar addition rules hold. With this addition, $(E, +)$ is an abelian group. Let K consist of q elements. By a theorem of Hasse, the number of elements of E is bounded by

$$q + 1 - 2\sqrt{q} \leq \#E \leq q + 1 + 2\sqrt{q}.$$

If $\#E = q + 1 - t$, we call t the *trace* of E.

3.1 General Elliptic Curves

Since 1997, there have been existing several public EC–DL challenges for finite fields [7]. By utilizing the ideas of [32] some of them have already been broken (electronic messages on the number theory net, Harley et al.), see table 2 on the following page.

$\#K$	# group operations
2^{79}	$1.7 \cdot 10^9$
p_{79}	$1.4 \cdot 10^9$
2^{89}	$1.8 \cdot 10^{13}$
p_{89}	$3.0 \cdot 10^{13}$
2^{97}	$2.2 \cdot 10^{13}$
p_{97}	$2.0 \cdot 10^{14}$

Table 2. Solved Elliptic Curve DL Challenges

Here, p_{79}, p_{89}, and p_{97} are 79–, 89–, and 97–bit primes respectively.

We now turn to the two special cases which can be solved efficiently.

3.2 Elliptic Curves of Trace Zero

In 1991, Menezes, Okamoto and Vanstone [21] published a method (MOV–reduction) to reduce the discrete log problem on E over $\mathbf{F}_q$ to the discrete log problem in $(\mathbf{F}_{q^k})^*$ for small k, provided that E is supersingular. A curve of $\mathbf{F}_q$ with $q = p^n$ is called supersingular if its trace is divisible by p. In particular this covers curves over $\mathbf{F}_p$ of order $p+1$. Consequently, this leads to a subexponential discrete log algorithm for supersingular curves.

Koblitz and Balusubramanian, however, showed that it is extremely unlikely for a random curve to be vulnerable by the MOV–reduction [2].

3.3 Elliptic Curves of Trace One

In 1997, Smart [30], and independently Semaev [26], Satoh and Araki [23] found an efficient method for curves E over $\mathbf{F}_p$, if E has p elements. The idea makes use of considering E over $\mathbb{Q}_p$, the p–adic extension of the rationals. In a certain related group to $E(\mathbb{Q}_p)$, there exists a logarithmic map which can be evaluated in polynomial time. It turns out that it suffices to approximate the arithmetic operations in $\mathbb{Q}_p$ by operations in $\mathbb{Z}/p^2\mathbb{Z}$.

4 Prime Fields

We now turn to a special version of the index calculus algorithm of section 2.6.

4.1 Sketch of the Number Field Sieve

Today the Number Field Sieve (NFS) is the asymptotically fastest known method to compute discrete logs in prime fields [14, 24]. Its running time is given by $L_p[1/3, (64/9)^{(1/3)}]$, where

$$L_p[\nu, \delta] = \exp((\delta + o(1))(\log p)^\nu \cdot (\log \log p)^{1-\nu}).$$

This is the same running time as for factoring integers as large as p. This algorithm has been implemented and lead to a record of 85 decimal digits for general p as well as 129 decimal digits for special p. In this algorithm, the computations take place in two number rings $\mathbb{Z}[\alpha_1]$ and $\mathbb{Z}[\alpha_2]$, where α_i are zeros of two polynomials $f_i \in \mathbb{Z}[X]$, $i = 1, 2$ respectively. The number rings are linked to $(\mathbb{Z}/p\mathbb{Z})^*$ by two homomorphisms

$$\varphi_i : \mathbb{Z}[\alpha_i] \longrightarrow \mathbb{Z}/p\mathbb{Z}$$

with

$$\varphi(\alpha_1) = \varphi(\alpha_2).$$

Consequently, the factor bases consist of prime ideals of the ring of integers of $\mathbb{Q}(\alpha_i)$. Note that the factor base members are not element of the group in which the DL has to be computed. An early special case of the NFS, the Gaussian Integer method, was published in [10] which we obtain by setting $\mathbb{Z}[\alpha_1] = \mathbb{Z}$ and $\mathbb{Z}[\alpha_2]$ to be an imaginary quadratic principal ideal ring.

The relations consist of (small) pairs $(c, d) \in \mathbb{Z} \times \mathbb{Z}$, where the ideals $(c + d\alpha_1)$ and $(c + d\alpha_2)$ simultaneously split over the corresponding factor bases. The original number field sieve adaptation uses $\mathbb{Z}[\alpha_1] = \mathbb{Z}$ such that the members of the first factor bases which are (principal) prime ideals of $\mathbb{Z}$ can be interpreted as elements of G (take a generator r of each prime ideal and consider $\varphi_1(r) \in G$). In a generalization of that adaptation where $\mathbb{Z}[\alpha_1] \neq \mathbb{Z} \neq \mathbb{Z}[\alpha_2]$ we can compute the logarithm of $b \in G$ if there exists $\gamma \in \varphi_j^{-1}(b) \in \mathbb{Z}[\alpha_j]$ such that (γ) splits over factor base j for $j = 1$ or $j = 2$ [33]. We observe that the parameters of the DL problem, a and b, have to be small in order to show up as factors of elements of type $c + d\alpha_1$ and $c + d\alpha_2$. For the base a this is not a severe problem because a small generator mod p can be found in polynomial time [28]. The parameter b is usually reduced by finding an expression $b \equiv \prod s_i \bmod p$ with small s_i (see section 4.2).

As above a (sparse) matrix consisting of exponents (e_{ij}) is constructed, (e_{ij}) being the exponent of factor base element number j in relation number i. Reducing to prime order groups leads us to solving the linear algebra problem over $\mathbb{Z}/q\mathbb{Z}$ where $q|p - 1$. Let $(y_3, \ldots, y_k)$ be a solution to $A'y = 0 \bmod q$ where the matrix A' consists of the rows $3, \ldots, k$ of A. Let (c_l, d_l) be the pair which produces relation number l $(1 \leq l \leq k)$.

Then

$$\prod \varphi_1(c_l + d_l\alpha_1)^{y_l} = g_1^{x_1} g_2^{x_2} u^q$$
$$\prod \varphi_2(c_l + d_l\alpha_2)^{y_l} = v^q$$

for some $u, v \in \mathbb{Z}/p\mathbb{Z}$ and $x_1, x_2 \in \mathbb{Z}$, provided that two conditions hold. Firstly, the ideals $\prod \varphi_1(c_l + d_l\alpha_j)^{y_l}$ have to be q-th powers of principal ideals; secondly, the units which are congruent to 1 mod q must be q-th powers. These assumptions, though heuristic, can be reasonably justified, see [24].

Because of $\varphi_1(c_l + d_l\alpha_1) = \varphi_2(c_l + d_l\alpha_2)$, dividing both equations results in a DL solution modulo q:

$$g_1^{x_1} g_2^{x_2} \equiv (v/u)^q \bmod p.$$

In practice, numerous refinements have been discovered during the few years after the first implementation of the NFS [6]. The most effective one surely is the large prime variation, where relations of the form

$$g_1^{e_1} \cdots g_k^{e_k} \cdot h = 1$$

are also accepted, where h (the large prime) is not member of the factor base. Dividing a second relation of this form by the one given above immediately produces an ordinary relation. This has been extended to allowing multiple h's, usually two per factor base. Limiting the number of h's is due to the method they are recognized. After the sieving process, all powers of factor base elements are found. Therefore, allowing one h requires one primality test per relation after the sieving stage. Allowing a rest of the form $h_1 \cdot h_2$ already requires the use of a fast special purpose factoring method to extract h_1 and h_2. This is practical as long as the h's fit in a computer word, so our special purpose factoring method has to split integers of size 2^{64}. In literature, the use of Pollard's $p-1$ method, Shanks's square form factorization method and Lenstra's elliptic curve method are reported. The combination of relations having more than one large prime is non–trivial. The basic case of two large primes in factoring algorithms was settled by [19], extended by [37] to four large primes, and adapted to the discrete log case in [34].

The distribution of the sieving process on many workstations and running the linear algebra step on a massively parallel machine (usually the Lanczos method or the conjugate gradient method) are subject of optimization, too [12, 17, 18].

For general finite fields $GF(p^n)$, the number field sieve ($n < (\log p)^{1/2}$) and the function field sieve ($n > (\log p)^2$) can be employed. For the "gap" between both of them, we are not aware of a subexponential time algorithm. A detailed theoretical survey of both methods and on the current status concerning the gap can be found in [25].

4.2   Experimental Results

Perhaps the most remarkable result of the DL variant of the NFS has probably been the successful solving of McCurley's 129–digit discrete logarithm challenge [36], which McCurley published in his overview paper on the DL problem [20].

In view of the Diffie–Hellman key exchange protocol introduced in [13], Mc-Curley stated a challenge by using the following setup:

$$b_A = 12740218011997394682426924433432284974938204258693$$
$$16216545577352903229146790959986818609788130465951$$
$$66455458144280588076766033781$$

$$b_B = 18016228528745310244478283483679989501596704669534$$
$$66973130251217340599537720584759581769106253806921$$
$$01651848662362137934026803049$$

$$p = (739 \cdot 7^{149} - 736)/3$$

$$q = (p-1)/(2 \cdot 739).$$

The order of the multiplicative group, which is generated by the element 7, splits as follows: $|(\mathbb{Z}/p\mathbb{Z})^*| = 2 \cdot 739 \cdot q$.

- Alice computes (using her secret key x_A) as $7^{x_A} \equiv b_A \pmod{p}$.
- Bob computes (using his secret key x_B) as $7^{x_B} \equiv b_B \pmod{p}$.

Kevin McCurley asked for the common secret key $K \equiv 7^{(x_A \cdot x_B)} \pmod{p}$ which has been computed by Denny and the second author as

$$K = 38127280411190014138078391507929634193998643551018670285056375615 \tag{3}$$
$$04552396692940392210217251405327092887263942637006353279 7740808,$$

by first calculating

$$x_A = 61858690859651883273593331652037904267987643069521713459146222 18 \tag{4}$$
$$49525998156144877820757492182909777408338791850457946749734,$$

the secret key of Alice.

Since the probability of $c + d\alpha$ splitting over a factor base depends on the discriminant of $\mathbb{Z}[\alpha]$, primes of special form – such as McCurley's choice of p – serve as an attractive target for the NFS.

Within a total of ≈ 180 mips years (mips=mega instructions per second), enough large prime relations (with a maximum of two large primes per relation) have been found over two factor bases of 35000 elements each.

Within a total of 1200 CPU hours, distributed on Sparc 4 workstations, the following reduction of b has been found by a combination of trial division and the elliptic curve factoring method:

$$a^{141266132} \cdot b \equiv \frac{t}{v} \pmod{p},$$

where

$$t = 2^3 \cdot 31 \cdot s_1 \cdot s_3 \cdot s_6 \cdot s_8 \cdot s_{10} \cdot s_{11}$$
$$v = 353 \cdot s_2 \cdot s_4 \cdot s_5 \cdot s_7 \cdot s_9 \cdot s_{12}. \tag{5}$$

with

$$s_1 = 603623,$$
$$s_2 = 165073039,$$
$$s_3 = 1571562367,$$
$$s_4 = 1601141623,$$
$$s_5 = 1715568391,$$
$$s_6 = 7575446399,$$
$$s_7 = 13166825869,$$
$$s_8 = 265542836371,$$
$$s_9 = 371303006453,$$
$$s_{10} = 4145488613977,$$
$$s_{11} = 4338202139093,$$
$$s_{12} = 5041332876473.$$

With the aid of 15 solutions of the linear algebra problem, we are able to determine the log of the 15 elements 2, 31, 353 and $s_1, \ldots, s_{12}$. This step costs 911 hours on a Sparc 20 workstation and consumes 30 MB of main memory.

Alternatively, reducing the DL problem to smaller elements can be achieved by means of a recent sieving method [35].

For primes of arbitrary form, the latest official record of a 85–digit p [35] has been superseded by a 90–digit computation in May 1998 (electronic message on the number theory net, Gaussian Integer implementation by Lercier/Joux). In the 85–digit computation, the Gaussian Integer method has been compared to the NFS version with two quadratic polynomials (NFS2Q). With 30 mips years, only 2/3 of the NFS2Q time is needed, and also the time for computing the linear algebra solution has only been 1/3 of the NFS2Q version. The main reason for both observations is that the numbers, which have to be split during the algorithm, are slightly harder to factor over an equally sized factor base. So, more pairs (c, d) have to be tested during the sieving stage, and, during the elimination of large primes, more partials are needed to produce an ordinary relation. The latter obviously increases the number of non–zero entries in the relation matrix.

5   Conclusion

During the past few years there has been considerable progress on the ability to solve discrete logarithm problems. It is a matter of taste whether schemes are considered as secure, when subexponential attacks exist. As sharp lower complexity bounds are typically hard to achieve, it is possible that exponential attacks can be replaced by subexponential attacks, and subexponential attacks by polynomial time attacks. The DL problem, however, will serve as a reliable source for secure protocols as long as the cryptographers are not running out of appropriate groups where no subexponential time algorithm is known. But there is still the possibility that eventually all DL problems will be tackled easily.

We draw the conclusion that apart from DL and factoring, modern cryptography urgently needs further number theoretic problems which can safely be used as a setting for existing and future cryptographic protocols. For achieving provable security, lower bounds in solving the corresponding problems are essential but these are either too difficult to establish or too restricted to a model not sufficiently compliant with reality as in section 2.

References

1. L. M. Adleman, J. DeMarrais, and M.-D. Huang. A subexponential algorithm for discrete logarithms over the rational subgroup of the Jacobians of large genus hyperelliptic curves over finite fields. In *Algorithmic number theory*, number 877 in Lecture Notes in Computer Science, pages 28–40, 1994.
2. R. Balasubramanian and N. Koblitz. The improbability that an elliptic curve has subexponential discrete log problem under the Menezes–Okamoto–Vanstone algorithm. *Journal of Cryptology*, 11:141–145, 1998.
3. R. P. Brent. An improved monte carlo factorization algorithm. *Nordisk Tidskrift for Informationsbehandling (BIT) 20*, pages 176–184, 1980.
4. J. Buchmann and St. Düllmann. On the computation of discrete logarithms in class groups. In *Advances in Cryptology – Crypto '90*, number 537 in Lecture Notes in Computer Science, pages 134–139, 1991.
5. J. Buchmann, M. Jacobson, and E. Teske. On some computational problems in finite abelian groups. *Math. Comp.*, 66(220):1663–1687, 1987.
6. J. Buchmann, J. Loho, and J. Zayer. An implementation of the general number field sieve. In *Advances in Cryptology – Crypto '93*, number 773 in Lecture Notes in Computer Science, 1993.
7. Certicom. ECC challenge. http://www.certicom.com/chal/, 1997.
8. D. Chaum, J.-H. Evertse, and J. van de Graaf. An improved protocol for demonstrating possession of discrete logarithms and some generalizations. In *Advances in Cryptology – Eurocrypt'87*, number 304 in Lecture Notes in Computer Science, pages pp. 127–141, 1988.
9. D. Chaum, J.-H. Evertse, J. van de Graaf, and R. Peralta. Demonstrating possession of a discrete logarithm without revealing it. In *Advances in Cryptology – CRYPTO'86*, number 263 in Lecture Notes in Computer Science, pages pp. 200–212, 1987.
10. D. Coppersmith, A. Odlyzko, and R. Schroeppel. Discrete logarithms in $GF(p)$. *Algorithmica 1*, pages 1–15, 1986.
11. Th. Corman, Ch. Leiserson, and R. Rivest. *Introduction to algorithms*. MIT Press/McGraw–Hill, 1990.
12. Th. F. Denny. *Lösen grosser dünnbesetzter Gleichungssysteme über endlichen Primkörpern*. PhD thesis, Universität des Saarlandes/Germany, 1997.
13. W. Diffie and M. Hellman. New directions in cryptography. *IEEE Trans. Information Theory 22*, pages pp. 472–492, 1976.
14. D. Gordon. Discrete logarithms in $GF(p)$ using the number field sieve. *SIAM J. Discrete Math.*, 6:124–138, 1993.
15. N. Koblitz. Elliptic curve cryptosystems. *Math. Comp.*, 48:203–209, 1987.
16. N. Koblitz. Hyperelliptic cryptosystems. *Journal of Cryptology*, 1:139–150, 1989.

17. M. LaMacchia and A. Odlyzko. Solving large sparse linear systems over finite fields. In *Advances in Cryptology – Crypto '90*, number 537 in Lecture Notes in Computer Science, pages 109–133, 1990.
18. R. Lambert. *Computational aspects of discrete logarithms*. PhD thesis, University of Waterloo/Canada, 1996.
19. A. K. Lenstra and M.S. Manasse. Factoring with two large primes. *Math. Comp.*, 63:77–82, 1994.
20. K. S. McCurley. The discrete logarithm problem. In *Cryptology and Computational Number Theory*, number 42 in Proc. Symp. in Applied Mathematics, pages 49–74. American Mathematical Society, 1990.
21. A. Menezes, T. Okamoto, and S. A. Vanstone. Reducing elliptic curve logarithms to logarithms in a finite field. In *Proceedings of the 23rd Annual ACM Symposium on the Theory of Computing,* pages 80–89, 1991.
22. J. M. Pollard. Monte carlo methods for index computation (mod p). *Math. Comp.*, 32:918–924, 1978.
23. T. Satoh and K. Araki. Fermat quotients and the polynomial time discrete log algorithm for anomalous elliptic curves. preprint.
24. O. Schirokauer. Discrete logarithms and local units. *Phil. Trans. R. Soc. Lond. A 345*, pages 409–423, 1993.
25. O. Schirokauer, D. Weber, and Th. F. Denny. Discrete logarithms: the effectiveness of the index calculus method. In H. Cohen, editor, *Algorithmic Number Theory – ANTS II*, number 1122 in Lecture Notes in Computer Science, 1996.
26. I. A. Semaev. Evaluation of discrete logarithms on some elliptic curves. *Math. Comp.*, 67:353–356, 1998.
27. D. Shanks. Class number, a theory of factorization and genera. In *Proc. Symposium Pure Mathematics*, volume 20, pages 415–440. American Mathematical Society, 1970.
28. V. Shoup. Searching for primitive roots in finite fields. In *Proc. 22nd Annual ACM Symp. on Theory of Computing (STOC)*, pages 546–554, 1990.
29. V. Shoup. Lower bounds for discrete logarithms and related problems. In *Advances in cryptology – Eurocrypt'97*, number 1233 in Lecture Notes in Computer Science, pages 256–266, 1997.
30. N. P. Smart. The discrete logarithm problem on elliptic curves of trace one. *Journal of Cryptology*. to appear.
31. E. Teske. Speeding up pollard's rho method for computing discrete logarithms. In *Algorithmic Number Theory – ANTS III*, number 1423 in Lecture Notes in Computer Science, 1998.
32. P. van Oorschot and M. Wiener. Parallel collision search with cryptanalytic applications. *Journal of Cryptology*. to appear.
33. D. Weber. Computing discrete logarithms with the number field sieve. In H. Cohen, editor, *Algorithmic Number Theory – ANTS II*, number 1122 in Lecture Notes in Computer Science, 1996.
34. D. Weber. *On the computation of discrete logarithms in finite prime fields*. PhD thesis, Universität des Saarlandes/Germany, 1997.
35. D. Weber. Computing discrete logarithms with quadratic number rings. In *Eurocrypt'98*, number 1403 in Lecture Notes in Computer Science, 1998.
36. D. Weber and Th. Denny. The solution of McCurley's discrete log challenge. In *Advances in Cryptology – CRYPTO'98*, number 1462 in Lecture Notes in Computer Science, 1998.
37. J. Zayer. *Faktorisieren mit dem Number Field Sieve*. PhD thesis, Universität des Saarlandes/Germany, 1995.

One-weight $\mathbf{Z}_4$-linear Codes

Claude Carlet

INRIA Projet CODES, BP 105, 78153 Le Chesnay Cedex, France; and GREYC, Université de Caen, France.

Abstract. We show that, for every ordered pair of nonnegative integers (k_1, k_2), there exists a unique (up to equivalence) one-weight $\mathbf{Z}_4$-linear code of type $4^{k_1} 2^{k_2}$. We derive an upper bound and a lower bound on the highest minimum distance between some extended one-weight $\mathbf{Z}_4$-linear codes and the Reed-Muller codes of order 1 and same lengths.

1 Introduction

A binary code is $\mathbf{Z}_4$-linear if it is equivalent (i.e. permutation-equivalent) to the image by the Gray map of a linear code over $\mathbf{Z}_4$ (cf. [8]). The Gray map is the mapping from $\mathbf{Z}_4$ to $GF(2)^2$ defined by:

$$0 \to (0,0),\ 1 \to (0,1),\ 2 \to (1,1),\ 3 \to (1,0),$$

coordinatewisely extended to a mapping from $\mathbf{Z}_4{}^m$ to $GF(2)^{2m}$.

The Gray map is distance preserving:
define the Lee weights of the elements $0, 1, 2$ and 3 of $\mathbf{Z}_4$ to be respectively $0, 1, 2$ and 1; define the Lee weight of a quaternary word to be the sum of the Lee weights of its coordinates, then the Hamming distance between the images by the Gray map of two quaternary words is equal to the Lee weight of their difference (their Lee distance). Consequently, any $\mathbf{Z}_4$-linear code is distance invariant.

In [8] (see page 303) is recalled that any linear code over $\mathbf{Z}_4$ is permutation-equivalent to a code with generator matrix:

$$G = \begin{bmatrix} I_{k_1} & A & B \\ 0 & 2I_{k_2} & 2C \end{bmatrix}.$$

The quaternary code has then cardinality $4^{k_1} 2^{k_2}$. Its associated $\mathbf{Z}_4$-linear code is said to have type $4^{k_1} 2^{k_2}$.

A code is called *one-weight* if all its nonzero codewords have the same Hamming weight. We know that, for every k, there exists a unique (up to equivalence) one-weight binary linear code of dimension k such that any two columns in its generator matrix are linearly independent. We generalize in section 3 this result to $\mathbf{Z}_4$-linear codes.

We call distance between two codes C_1 and C_2 of same length the minimum Hamming distance between any element a of C_1 and any element b of C_2, such that $(a, b) \neq (0, 0)$.

There exists an upper bound on the distance between any extended one-weight linear code and the Reed-Muller code of order 1 with same length: cf. [5, 18] (but the results are not presented in these papers in terms of one-weight codes). We study in section 4 an upper bound and a lower bound on the highest distance between some extended one-weight $\mathbf{Z}_4$-linear codes of type 4^k and the Reed-Muller code of order 1 and length 2^{2k+1}.

2   The extended one-weight linear codes and their distances to the Reed-Muller codes of order 1

We reduce ourselves to the binary case, since we focus later on $\mathbf{Z}_4$-linear codes, but Proposition 1 is valid for codes over general Galois fields. The characterization of one-weight linear codes is well-known (cf. [16]).

Lemma 1. *Let C be a linear binary code of length n such that:*

$$\forall j = 1, \dots, n, \ \exists\, x = (x_1, \dots, x_n) \in C \mid x_j \neq 0$$

(i.e. the generator matrix of C has no column equal to the zero vector). Then the sum $\sum_{x \in C} w(x)$ of the Hamming weights of all the elements of C is equal to:

$$\frac{n\, |C|}{2}.$$

Proposition 1. *Let k be any positive integer. There exists a unique (up to equivalence) binary linear code of dimension k such that:*
- any two columns in its generator matrix are linearly independent,
- all the nonzero words in this code have same weight.

The rows of the generator matrix of the code C characterized by Proposition 1 are all the coordinate functions on $(GF(2)^k)^*$ (or more precisely the lists of values taken on $(GF(2)^k)^*$ by the coordinate functions; we will not distinguish in the sequel between a Boolean function and the word equal to its list of values). Thus, C is the simplex code of length $2^k - 1$. If we extend it with a zero coordinate (i.e. a parity-check), all its nonzero words are balanced (i.e. take the values 0 and 1 equally often) and if we identify $GF(2)^k$ to the Galois field $GF(2^k)$, the words of this extended code are all the functions $tr(ax)$, $a \in GF(2^k)$, where tr is the trace function from $GF(2^k)$ to $GF(2)$.

For every choice of a permutation F on $GF(2^k)$ such that $F(0) = 0$, we have a corresponding extended one-weight linear code, whose elements are the Boolean functions:

$$x \in GF(2^k) \rightarrow tr(aF(x)); \ a \in GF(2^k),$$

and any extended one-weight linear code is of this form.
The distance between such code and the Reed-Muller code of order 1 depends

on the choice of permutation F. The determination of this distance presents an interest from both viewpoints of coding theory and cryptography. Recall that the distance between the code $\{tr(aF(x)); \; a \in GF(2^k)\}$ and the Reed-Muller code of order 1 is an essential criterion for the design of cryptographic schemes using F, to prevent them from linear attacks (cf. [13]).

An upper bound on the distance between any extended one-weight linear code of dimension k and the Reed-Muller code of order 1 and length 2^k has been derived by Sidelnikov [18] and re-discovered by Chabaud and Vaudenay [5]. This distance is at most

$$2^{k-1} - 2^{\frac{k-1}{2}}. \tag{1}$$

This bound improves upon the bound derived from Parseval's relation, that is similar to (1), but with $2^{k-1} - 2^{\frac{k}{2}-1}$ in the place of $2^{k-1} - 2^{\frac{k-1}{2}}$; recall that Parseval's relation, for a Boolean function f, is:

$$\sum_{b \in GF(2^k)} \left(\sum_{x \in GF(2^k)} (-1)^{f(x)+tr(bx)} \right)^2 = 2^{2k}. \tag{2}$$

The idea of the proof of bound (1) is simple:

since the Reed-Muller code of order 1 contains the all-one vector, since all its other nonzero words have weight 2^{k-1}, and since the nonzero words of the extended one-weight linear code of dimension k associated to permutation F have weight 2^{k-1} as well, the distance between these codes is equal to the minimum distance between any function $tr(aF(x))$, $a \neq 0$ and any function $\ell(x) = tr(bx) + \epsilon$; $b \in GF(2^k)$, $\epsilon \in GF(2)$. For every $a \neq 0$ and every b, the minimum distance between the function $tr(aF(x))$ and the functions $\ell(x) = tr(bx) + \epsilon$; $\epsilon \in GF(2)$ is equal to:

$$2^{k-1} - \frac{1}{2} |\mu_F(a, b)|,$$

where:

$$\mu_F(a, b) = \sum_{x \in GF(2^k)} (-1)^{tr(aF(x)+bx)}.$$

The inequality:

$$\max_{a \in GF(2^k)^*, b \in GF(2^k)} \mu_F^2(a, b) \geq \frac{\sum_{a \in GF(2^k)^*, b \in GF(2^k)} \mu_F^4(a, b)}{\sum_{a \in GF(2^k)^*, b \in GF(2^k)} \mu_F^2(a, b)} \tag{3}$$

gives the desired conclusion, since we have:

$$\sum_{a \in GF(2^k), b \in GF(2^k)} \mu_F^2(a, b) = 2^{3k},$$

thanks to Parseval's relation and to the fact that $\mu_F(0, b) = \begin{cases} 2^k & \text{if } b = 0 \\ 0 & \text{otherwise.} \end{cases}$,

and since:

$$\sum_{a \in GF(2^k), b \in GF(2^k)} \mu_F{}^4(a, b) =$$

$$\sum_{a \in GF(2^k), b \in GF(2^k)} \left(\sum_{x_1, \dots, x_4 \in GF(2^k)} (-1)^{tr(a(F(x_1) + \dots + F(x_4)) + b(x_1 + \dots + x_4))} \right) =$$

$$2^{2k} \left| \left\{ (x_1, \dots, x_4) \in GF(2^k)^4 \mid \begin{cases} F(x_1) + \dots + F(x_4) = 0 \\ x_1 + \dots + x_4 = 0 \end{cases} \right\} \right| \geq$$

$$2^{2k}(3 \cdot 2^{2k} - 2^{k+1}).$$

Bound (1) is tight for every odd k. The mappings which reach it are called *almost bent* (AB). Unfortunately (cf. [4, 5]) there exist very few known AB mappings. All of them are straightforwardly related to the power functions x^{2^i+1} and $x^{2^{2i}-2^i+1}$, $(i, k) = 1$. There exists also a conjecture by Welch (cf. [14]) that $x^{2^{\frac{k-1}{2}}+3}$ reaches the bound.

3   The one-weight $\mathbf{Z}_4$-linear codes

The following Lemma can be found in [21]:

Lemma 2. *Let $\hat{C}$ be a linear quaternary code of length n such that:*

$$\forall j = 1, \dots, n, \ \exists x = (x_1, \dots, x_n) \in \hat{C} \mid x_j \neq 0.$$

Then the sum $\sum_{x \in \hat{C}} w_L(x)$ of the Lee weights of all the elements of $\hat{C}$ is equal to $n \, |\hat{C}|$.

Thus, the result of Lemma 1 is still valid for $\mathbf{Z}_4$-linear codes, since the length of the $\mathbf{Z}_4$-linear code C associated to $\hat{C}$ is $2n$.

Remark 1. Lemma 2 can be generalized to the $\mathbf{Z}_{2^m}$-linear codes, defined in [3]. The sum of the weights of such code of length $n \, 2^{m-1}$ is equal to $n \, |\hat{C}| \, 2^{m-2}$.

Proposition 2. *Let k_1 and k_2 be two nonnegative integers, $k_1 + k_2 > 0$. There exists a unique (up to equivalence) $\mathbf{Z}_4$-linear code C_{k_1, k_2} of type $4^{k_1} 2^{k_2}$ such that:*
- no column in the generator matrix of the associated linear code over $\mathbf{Z}_4$ is the zero vector and any two columns in this matrix are different,
- all the nonzero words in this code have same weight.
The code C_{k_1, k_2} has length $2^{2k_1+k_2+1} - 2$.

Proof. Let C be a $\mathbf{Z}_4$-linear code of length $2n$ and type $4^{k_1} 2^{k_2}$ with one unique nonzero weight γ. Let $\hat{C}$ be the linear quaternary code associated to C, in the generator matrix of which the last k_2 lines have even coordinates. According to Lemma 2, we have, denoting by $w(x)$ the Hamming weight of a word x of C:

$$\sum_{x \in C} w(x) = n |C|.$$

On the other hand, this sum is also equal to $(|C| - 1)\gamma$. We deduce:

$$(4^{k_1} 2^{k_2} - 1)\gamma = n \, 4^{k_1} 2^{k_2}.$$

Since $(4^{k_1} 2^{k_2} - 1)$ and $4^{k_1} 2^{k_2}$ are co-prime, there must exist an integer r such that $n = r(4^{k_1} 2^{k_2} - 1)$ and $\gamma = r \, 4^{k_1} 2^{k_2}$. But, because of the condition on $\hat{C}$, n can not be greater than $(4^{k_1} 2^{k_2} - 1)$. Thus, $r = 1$, $\gamma = 4^{k_1} 2^{k_2}$ and the columns of the generator matrix of $\hat{C}$ are all the nonzero words equal to the concatenation of a quaternary word of length k_1 and of a quaternary word of length k_2 with even coordinates.

Conversely, the image by the Gray map of such code is one-weight.

Remark 2. The hypothesis of Proposition 2 is not equivalent to that of Proposition 1. There exist indices $i_1 \neq i_2$ such that the coordinates of indices i_1 and i_2 in any codeword of C_{k_1,k_2} are equal to each other, since there exist indices i such that the coordinate of index i in any codeword of $\hat{C}_{k_1,k_2}$ is even.

Remark 3. Proposition 2 can be also generalized to $\mathbf{Z}_{2^m}$-linear codes. The unique (up to equivalence) $\mathbf{Z}_{2^m}$-linear code of type $2^{k_m} 4^{k_{m-1}} \ldots (2^m)^{k_1}$ has length

$$2^{k_m + 2k_{m-1} + \ldots + m \, k_1 + m - 1} - 2^{m-1}.$$

Accordingly to the proof of Proposition 2, a generator matrix of the quaternary code $\hat{C}_{k_1,k_2}$ can be defined in the following way: its columns are all the nonzero words of length $k_1 + k_2$ over $\mathbf{Z}_4$ whose k_2 last coordinates are even. These columns can be labelled by the nonzero ordered pairs (i, j) of $\{0, \cdots, 2^{k_1} - 1\} \times \{0, \cdots, 4^{k_2} - 1\}$: the column of index (i, j) is the concatenation of the i-th word of length k_1 over $\mathbf{Z}_4$ and of the j-th even word of length k_2. Thus, the one-weight $\mathbf{Z}_4$-linear code of type $4^{k_1} 2^{k_2}$ can be expressed by means of that of type 4^{k_1} and of the simplex code of dimension k_2:

Proposition 3. *The codewords of the one-weight $\mathbf{Z}_4$-linear code C_{k_1,k_2} of type $4^{k_1} 2^{k_2}$ are the words $(a_i + b_j, a'_i + b_j)_{i=0\ldots4^{k_1}-1; j=0\ldots2^{k_2}-1; (i,j) \neq (0,0)}$, where:*

- *$a_0 = a'_0 = 0$;*
- *$(a_i, a'_i)_{i=0\ldots4^{k_1}-1}$ ranges over the one-weight $\mathbf{Z}_4$-linear code C_{k_1} of type 4^{k_1};*
- *$b_0 = 0$;*
- *$(b_j)_{j=0\ldots2^{k_2}-1}$ ranges over the simplex code of length 2^{k_2}.*

There is a characterization of the type 4^k case which is similar to the binary case. We first recall some background on Galois rings.

. The Gray map can be extended to (or, more exactly viewed as) a mapping from the set of all the $\mathbf{Z}_4$-valued functions on a given set $\mathcal{E}$, to the set of all the Boolean functions on $\mathcal{E} \times GF(2)$: any $\mathbf{Z}_4$-valued function can be written $2f(x) + g(x)$, where f and g take the values 0 and 1 only; the image of the function $2f(x)+g(x)$ by the Gray map is the function $(x, \epsilon) \in \mathcal{E} \times GF(2) \rightarrow f(x) \oplus \epsilon g(x)$, where f and g are considered as valued in $GF(2)$.

. Let $\mathcal{R}$ be the Galois ring of order 4^k. The Teichmuller set $\mathcal{T}$ is the set of all squares in $\mathcal{R}$ and has cardinality 2^k (cf. [8, 12]). The Galois ring $\mathcal{R}$ is equal to $\mathcal{T} + 2\mathcal{T}$. There exists in $\mathcal{T}$ an element ξ such that $\xi^{2^k} = \xi$ and $\mathcal{T} = \{0, 1, \xi, \xi^2, \dots, \xi^{2^k-2}\} = \{\xi^i; i = \infty, 0, 1, 2, \dots, 2^k - 2\}$, which leads to the natural bijection $\theta^i \rightarrow \xi^i$ between the Galois field $GF(2^k)$ and $\mathcal{T}$ (where θ is a primitive element of the field $GF(2^k)$), which respects multiplication. The addition in $GF(2^k)$ corresponds in $\mathcal{T}$ via this bijection to the addition mod $2\mathcal{T}$ (we shall speak more simply of the addition mod 2). We shall call *projection modulo 2* on $GF(2^k)$ the mapping $\xi^i + 2\xi^j \rightarrow \theta^i$.

The set of the units of $\mathcal{R}$ is $\mathcal{R}^* = \mathcal{T}^* + 2\mathcal{T}$, where $\mathcal{T}^* = \mathcal{T} - \{0\}$. We denote by φ the Frobenius automorphism on $\mathcal{R}$:

$$\varphi(u + 2v) = u^2 + 2v^2, \; u, \, v \in \mathcal{T}$$

and by Tr the trace function from $\mathcal{R}$ to $\mathbf{Z}_4$:

$$Tr(z) = z + \varphi(z) + \dots + \varphi^{k-1}(z), \, z \in \mathcal{R}.$$

Since the columns of the generator matrix of $\hat{C}_k$ are all the nonzero quaternary words of length k, the family of all the coordinate functions on $(\mathbf{Z}_4^k)^*$ generates $\hat{C}_k$. Thus:

Proposition 4. *The unique (up to equivalence) one-weight $\mathbf{Z}_4$-linear code of type 4^k is the set of the images by the Gray map of the following functions on $\mathcal{R} \setminus \{0\}$:*

$$x + 2y \in \mathcal{R} \setminus \{0\} \mapsto Tr(a(x + 2y)); \; a \in \mathcal{R}.$$

Indeed, all of these functions belong to C_k and their number is 4^k.

4   Bounds on the minimum nonzero distances between the doubly extended one-weight $\mathbf{Z}_4$-linear codes and the Reed-Muller codes of order 1

"The" one-weight $\mathbf{Z}_4$-linear code of type $4^{k_1} 2^{k_2}$ has length $2^{2k_1+k_2+1} - 2$. For every choice of a permutation on $\{1, \dots, 2^{2k_1+k_2+1}-2\}$, we have a corresponding one-weight $\mathbf{Z}_4$-linear code.

We shall reduce ourselves to the type 4^k only; for every choice of a permutation on $(\mathcal{R} \setminus \{0\}) \times GF(2)$, we have a corresponding one-weight $\mathbf{Z}_4$-linear code;

we shall view the permutations on $(\mathcal{R} \setminus \{0\}) \times GF(2)$ as permutations on the union of two copies $\mathcal{R}_1 \setminus \{0\}$ and $\mathcal{R}_2 \setminus \{0\}$ of $\mathcal{R} \setminus \{0\}$.

We shall also consider only the permutations which act separately and in the same manner on $\mathcal{R}_1 \setminus \{0\}$ and $\mathcal{R}_2 \setminus \{0\}$, i.e. the one-weight $\mathbf{Z}_4$-linear codes whose elements are the images by the Gray map of the functions on $\mathcal{R} \setminus \{0\}$:

$$Tr(aF(x + 2y)); \; a \in \mathcal{R} \tag{4}$$

where F is a permutation on $\mathcal{R} \setminus \{0\}$. This case is clearly the most natural one to investigate for cryptographic applications. It seems also natural to extend F to the whole ring $\mathcal{R}$ by $F(0) = 0$. We call *doubly extended* one weight $\mathbf{Z}_4$-linear code the image by the Gray map of the functions of relation (4).

Remark 4. Clearly, the permutations on $\mathcal{R}_1 \bigcup \mathcal{R}_2$ are not all as in (4), but our framework includes those permutations which also exchange $\mathcal{R}_1$ and $\mathcal{R}_2$, since $-Tr(aF(x+2y))$ is equal to $Tr(a(-F(x+2y)))$ and since $-F$ is a permutation if F is one.

The more general case of the one-weight $\mathbf{Z}_4$-linear codes corresponding to the permutations on $\mathcal{R}_1 \bigcup \mathcal{R}_2$ which act the same way on each copy $\mathcal{R}_i$ and then exchange some of the elements of $\mathcal{R}_1$ with the elements of same index of $\mathcal{R}_2$ can be studied the same way and leads to the same bounds.

<u>Notation</u>: We denote by Λ_F the minimum distance between the images by the Gray map of the functions $Tr(aF(x+2y)); \; a \in \mathcal{R} - \{0\}$ and the Reed-Muller code of order 1 and length 2^{2k+1}.

We have $\Lambda_F = \min(\Lambda'_F, \Lambda''_F)$, where Λ'_F (resp. Λ''_F) is the minimum distance between the images by the Gray map of the functions $Tr(aF(x + 2y)); \; a \in 2T^\star$ (resp. $a \in \mathcal{R}^\star$) and this code.

Any element of the Reed-Muller code of order 1 and length 2^{2k+1} has the form $\ell(x, y, \epsilon) = tr(bx + cy) + \eta\epsilon + \eta'; \; b, c \in GF(2^k), \; \eta, \eta' \in GF(2)$, and is the image by the Gray map of the function $Tr(2bx + 2cy) + 2\eta' + \eta$. Thus, Λ_F is the minimum Lee distance between the functions $Tr(aF(x + 2y)); \; a \in \mathcal{R} - \{0\}$ and the functions $Tr(2bx + 2cy) + \eta + 2\eta', \; b, c \in T, \; \eta, \eta' \in \{0, 1\}$.

4.1   An upper bound on Λ_F

In Proposition 5 below, we use the notion of *bent* function over a finite field. This notion has been introduced by Nyberg (cf. [15]) under the name of perfect non-linear mapping (in a wider framework), see also Ambrosimov [1] and Logachev et al. [11]. Its definition in the particular framework which is ours is the following: a mapping G from $GF(2^k)^2$ to $GF(2^k)$ is bent if for every $a \in GF(2^k), \; a \neq 0$ and every $(b, c) \in GF(2^k)^2$ the number:

$$\mu_G(a, (b, c)) = \sum_{x,y \in GF(2^k)} (-1)^{tr(aG(x,y)+bx+cy)}$$

is equal to $\pm 2^k$ (i.e. the Boolean function $tr(aG)$ is bent in the sense of [6]). Two general examples of such bent mappings, which are directly deduced from the two general classes of bent functions studied by Dillon in [6], are $G(x, y) = \sigma\left(x\, y^{2^k-2}\right)$ and $G(x, y) = x\, \sigma(y)$, where σ is, in each case, any permutation on $GF(2^k)$.

For every $a \in GF(2^k)^*$, $b, c \in GF(2^k)$, we have (cf. [6]) :

$$\sum_{x \in GF(2^k)} (-1)^{tr(a\sigma(x\, y^{2^k-2})+bx+cy)} = 2^k\, (-1)^{tr(a\sigma(c\, b^{2^k-2}))} \tag{5}$$

and

$$\sum_{x \in GF(2^k)} (-1)^{tr(ax\sigma(y)+bx+cy)} = 2^k\, (-1)^{tr\left(c\sigma^{-1}\left(\frac{b}{a}\right)\right)}. \tag{6}$$

Proposition 5. *Let F be a mapping from $\mathcal{R}$ to itself. We have:*

$$\Lambda'_F \le 2^{2k} - 2^k. \tag{7}$$

Λ'_F *is equal to $2^{2k} - 2^k$ if and only if the function $F' = F \bmod 2$, considered as a mapping from $GF(2^k)^2$ to $GF(2^k)$ is bent.*

Proof. Since the Lee weight of any element u of $\mathbf{Z}_4$ is equal to $1 - \frac{1}{2}(i^u + i^{-u})$, the Lee distance between the functions $Tr(aF(x + 2y))$ and $Tr(2bx + 2cy) + 2\eta' + \eta$ is equal to:

$$2^{2k} - \frac{(-1)^{\eta'}}{2} \nu_F(a, b, c, \eta), \tag{8}$$

where:

$$\nu_F(a, b, c, \eta) = \sum_{x, y \in \mathcal{T}} i^{\,Tr(aF(x+2y)+2bx+2cy)-\eta} + \sum_{x, y \in \mathcal{T}} i^{\,Tr(-aF(x+2y)+2bx+2cy)+\eta}.$$

If a is in $2\mathcal{T}$, say $a = 2a'$ then we have:

$$\nu_F(a, b, c, \eta) = \begin{cases} 2\mu_{F'}(a', (b, c)) & \text{if } \eta = 0 \\ 0 & \text{otherwise.} \end{cases}$$

According to Parseval's relation (2), the number $\max_{a', b, c \in GF(2^k)} (\mu_{F'}{}^2(a', (b, c)))$ is greater than or equal to 2^{2k}. This bound is tight and the mappings F' which reach it are the bent mappings.

Corollary 1. *Let F be a mapping from $\mathcal{R}$ to itself. We have:*

$$\Lambda_F \le 2^{2k} - 2^k.$$

We find here again the bound (1) for mappings from $GF(2^{2k+1})$ to itself. But this upper bound can not be reached by the mappings F considered here: these mappings F being permutations, the functions $Tr(aF(x + 2y))$, $a \in 2\mathcal{T}^*$ are balanced Boolean functions, which implies that the associate mappings F' can not be bent.

However, we know (cf. [7]) that the largest distance between the balanced functions and the Reed-Muller code of order 1 and length 2^{2k} is reasonably near to the covering radius of this code (but unknown).

4.2 A lower bound on $\max_F \Lambda_F$

In the sequel, we shall use the notation $tr(v)$, $v \in \mathcal{R}$. This notation means $tr(v')$ where v' is the element of $GF(2^k)$ image of v by the projection modulo 2.

We first recall the value of the sum $\sum_{x \in \mathcal{T}} i^{Tr(ax)}$ for any $a \in \mathcal{R}$, which will be useful in the sequel. The following Lemma can be deduced from the results of Yang et al. [21].

Lemma 3. *The number:*

$$\lambda = \sum_{x \in \mathcal{T}} i^{Tr(x)}$$

is equal to $\pm 2^{\frac{k}{2}} e^{i\pi Tr(1)/4}$.
For any a in $\mathcal{R}$, the sum:

$$\sum_{x \in \mathcal{T}} i^{Tr(ax)}$$

is equal to 2^k if $a = 0$, to 0 if $a \in 2\mathcal{T}^*$ and to $\dfrac{2^k}{\lambda} i^{Tr((1+u)^2)} = \overline{\lambda} i^{Tr((1+u)^2)}$ if $a \in \mathcal{T}^*(1 + 2u)$, $u \in \mathcal{T}$.

Remark 5. Lemma 3 is valid even for $u \in \mathcal{R}$ since the values of $\sum_{x \in \mathcal{T}} i^{Tr((1+2u)x)}$ and $Tr((1 + u)^2)$ do not change if we replace $u \in \mathcal{T}$ by any other element of $\mathcal{R}$ congruent to u modulo 2.
If u belongs to $\mathcal{T}$, then $Tr((1 + u)^2) = Tr(1 - u)$, since $Tr(u^2) = Tr(u)$.

Proposition 6. *Assume that k is odd and that σ is an almost bent permutation on $GF(2^k)$ such that $\sigma(0) = 0$. Set $r = \dfrac{2^{k+1} - 1}{3}$. The mappings $x \to x^r$ and $x \to x^3$ are inverses one of each other.*
Set $F(x + 2y) = x\,y^{2^k - 2} + \delta_0(x)y^r + 2[xy + \delta_0(y)\sigma(x)]$, where δ_0 is the Dirac symbol ($\delta_0(x) = 1$ if $x = 0$; $\delta_0(x) = 0$ otherwise). Then, F is a permutation on $\mathcal{R}$ and $\Lambda_F \geq 2^{2k} - 2^{k+1} - 2^{\frac{k}{2}+2}$.

Proof. It is a simple matter to verify that F is a permutation, by checking that:
– the image of $\mathcal{T}^* + 2\mathcal{T}^*$ is equal to $\mathcal{T}^* + 2\mathcal{T}^*$,
– the image of $2\mathcal{T}^*$ is equal to $\mathcal{T}^*$,
– the image of $\mathcal{T}^*$ is equal to $2\mathcal{T}^*$,
– and the image of 0 is equal to 0.
According to relation (8), the Lee distance between the functions $Tr(aF(x+2y))$ and $Tr(2bx+2cy)+\eta+2\eta'$ is equal to $2^{2k}-(-1)^{\eta'}\mathcal{R}e\left(\sum_{x,y \in \mathcal{T}} i^{Tr(aF(x+2y)+2bx+2cy)-\eta}\right)$,
where $\mathcal{R}e$ denotes the function 'real part'.

Let us first assume that a is even $(a = 2a' \neq 0)$.

$\mathcal{R}e\left(\displaystyle\sum_{x,y\in T} i^{\,Tr(aF(x+2y)+2bx+2cy)-\eta}\right)$ is then equal to:

$$\begin{cases} 0 \text{ if } \eta = 1 \\ \sum_{x,y\in GF(2^k)}(-1)^{\,tr(a'F'(x+2y)+bx+cy)} \text{ if } \eta = 0 \end{cases}$$

where $F' = F \bmod 2$. We have:

$$\sum_{x,y\in GF(2^k)}(-1)^{\,tr(a'F'(x+2y)+bx+cy)} =$$

$$\sum_{x,y\in GF(2^k)}(-1)^{\,tr(a'x\,y^{2^k-2}+bx+cy)} - \sum_{y\in GF(2^k)}(-1)^{\,tr(cy)} + \sum_{y\in GF(2^k)}(-1)^{\,tr(a'y^r+cy)}.$$

Thus, according to (5):

$$\sum_{x,y\in GF(2^k)}(-1)^{\,tr(a'F'(x+2y)+bx+cy)} =$$

$$2^k(-1)^{\,tr(a'c\,b^{2^k-2})} - 2^k\delta_0(c) + \sum_{y\in GF(2^k)}(-1)^{\,tr(a'y^r+cy)}.$$

$2^k(-1)^{\,tr(a'c\,b^{2^k-2})} - 2^k\delta_0(c)$ is equal to 0 or to $\pm 2^k$ and $\displaystyle\sum_{y\in GF(2^k)}(-1)^{\,tr(a'y^r+cy)} =$

$\displaystyle\sum_{y\in GF(2^k)}(-1)^{\,tr(a'y+cy^3)}$ is equal to 0 or to $\pm 2^{\frac{k+1}{2}}$, since the power function x^3 is almost bent.

We deduce:

$$\left| \sum_{x,y\in GF(2^k)}(-1)^{\,tr(a'F'(x+2y)+bx+cy)} \right| \leq 2^k + 2^{\frac{k+1}{2}}$$

and we have: $\Lambda'_F \geq 2^{2k} - 2^k - 2^{\frac{k+1}{2}}$.

Let us consider now the case where a is a unit, say $a = a'(1+2u)$, $a' \in T^*$ $u \in T$.

We have:

$$\sum_{x,y\in T} i^{\,Tr(aF(x+2y)+2bx+2cy)-\eta} = \sum_{x\in T, y\in T^*} i^{\,Tr(a(x\,y^{2^k-2}+2xy)+2bx+2cy)-\eta} -$$

$$\sum_{y\in T} i^{\,Tr(2cy)-\eta} + \sum_{y\in T} i^{\,Tr(ay^r+2cy)-\eta} + \sum_{x\in T} i^{\,Tr(2a\sigma(x)+2bx)-\eta}.$$

Moreover, changing x into xy in the first of these three last sums, we have:

$$\sum_{x\in T, y\in T^*} i^{\,Tr(a(x\,y^{2^k-2}+2xy)+2bx+2cy)} - \sum_{y\in T} i^{\,Tr(2cy)} =$$

$$\sum_{x\in T, y\in T^*} i^{\,Tr(a(x+2xy^2)+2bxy+2cy)} - 2^k\,\delta_0(c) =$$

$$\sum_{x\in T}\left(i^{\,Tr(ax)}\sum_{y\in T}(-1)^{tr\left((ax+b^2x^2+c^2)y^2\right)}\right) - 2^k\,\delta_0(c) - \sum_{x\in T} i^{\,Tr(ax)}.$$

The sum $\sum_{y\in T}(-1)^{tr\left((ax+b^2x^2+c^2)y^2\right)}$ equals 0 if $ax+b^2x^2+c^2 \neq 0 \bmod 2$. Using Lemma 3, we deduce:

$$\sum_{x\in T, y\in T^*} i^{\,Tr(a(x\,y^{2^k-2}+2xy)+2bx+2cy)} - \sum_{y\in T} i^{\,Tr(2cy)} =$$

$$2^k\sum_{x\in T\ |\ ax+b^2x^2+c^2=0\bmod 2} i^{\,Tr(ax)} - 2^k\,\delta_0(c) - \overline{\lambda}\,i^{\,Tr(1-u)}.$$

If $c = 0$, then the equation $ax + b^2x^2 + c^2 = 0 \bmod 2$ admits 0 as solution, plus a second solution if $b \neq 0$; if $c \neq 0$, then the equation $ax + b^2x^2 + c^2 = 0 \bmod 2$ admits zero or two solutions if $b \neq 0$, and 1 if $b = 0$. We deduce, in both cases:

$$\left| \sum_{x\in T, y\in T^*} i^{\,Tr(a(x\,y^{2^k-2}+2xy)+2bx+2cy)-\eta} - \sum_{y\in T} i^{\,Tr(2cy)-\eta} \right| \leq 2^{k+1} + 2^{\frac{k}{2}}.$$

We have also:

$$\left| \mathcal{R}e\left(\sum_{y\in T} i^{\,Tr(ay^r+2cy)-\eta}\right)\right| = \left| \mathcal{R}e\left(\sum_{y\in T} i^{\,Tr(ay+2cy^3)-\eta}\right)\right| \leq 2^{\frac{k+1}{2}},$$

according to [9], and:

$$\left| \sum_{x\in T} i^{\,Tr(2a\sigma(x)+2bx)-\eta}\right| \leq 2^{\frac{k+1}{2}},$$

since σ is an almost bent mapping.
Thus:

$$\left| \mathcal{R}e\left(\sum_{x,y\in T} i^{\,Tr(aF(x+2y)+2bx+2cy)-\eta}\right)\right| \leq 2^{k+1} + 2^{\frac{k}{2}} + 2^{\frac{k+1}{2}} + 2^{\frac{k+1}{2}} \leq 2^{k+1} + 2^{\frac{k}{2}+2}.$$

4.3   More on the case "a not even"

We have seen in subsection 4.1 that the suitable permutations F for crypto-graphic use must be such that Λ'_F is near to the upper bound $2^{2k} - 2^k$. We study in this subsection the behaviour of these permutations (among which the permutation of Proposition 6) with respect to Λ''_F.

Proposition 7. *Let F be a permutation on $\mathcal{R}$. Assume that Λ'_F is equal to $2^{2k} - 2^k - K$, $0 \leq K \leq 2^k$. Then:*

$$\Lambda''_F \leq 2^{2k} - \sqrt{2^{2k-1} + 2^k - 15K}.$$

Proof. We keep the notations of the proof of proposition 5. We have:

$$\max_{a\in\mathcal{R}^*;\, b,c\in\mathcal{T};\, \eta\in\{0,1\}} \left(\nu_F{}^2(a,b,c,\eta)\right) \geq$$

$$\frac{\sum_{a\in\mathcal{R}^*;\, b,c\in\mathcal{T};\, \eta\in\{0,1\}} \left(\nu_F{}^4(a,b,c,\eta)\right)}{\sum_{a\in\mathcal{R}^*;\, b,c\in\mathcal{T};\, \eta\in\{0,1\}} \left(\nu_F{}^2(a,b,c,\eta)\right)}. \tag{9}$$

We shall compute the value of the lower sum and a lower bound on the value of the upper sum. We have:

$$\sum_{a\in\mathcal{R};\, b,c\in\mathcal{T};\, \eta\in\{0,1\}} \nu_F{}^2(a,b,c,\eta) =$$

$$\sum_{a\in\mathcal{R};\, b,c\in\mathcal{T};\, \eta\in\{0,1\}} \quad \sum_{x_1,x_2,y_1,y_2\in\mathcal{T};\, e_1,e_2\in\{-1,1\}} I_{a,b,c,\eta,x_1,x_2,y_1,y_2,e_1,e_2}$$

where

$$I_{a,b,c,\eta,x_1,x_2,y_1,y_2,e_1,e_2} = i^{Tr(a(e_1 F(x_1+2y_1)+e_2 F(x_2+2y_2))+2b(x_1+x_2)+2c(y_1+y_2)-(e_1+e_2)\eta}.$$

We know that, for any v in $\mathcal{R}$ and any even τ in Z_4:

$$\sum_{a\in\mathcal{R}} i^{Tr(av)} = \begin{cases} 2^{2k} & \text{if } v = 0 \\ 0 & \text{if } v \neq 0 \end{cases} ; \quad \sum_{b\in\mathcal{T}} i^{Tr(2bv)} = \begin{cases} 2^k & \text{if } v = 0 \bmod 2 \\ 0 & \text{if } v \neq 0 \bmod 2 \end{cases} ;$$

$$\sum_{\eta\in\{0,1\}} i^{\tau\eta} = \begin{cases} 2 & \text{if } \tau = 0 \\ 0 & \text{if } \tau \neq 0 \end{cases}. \tag{10}$$

Since $e_1 + e_2$ is always even, we deduce that $\displaystyle\sum_{a\in\mathcal{R};\, b,c\in\mathcal{T};\, \eta\in\{0,1\}} \nu_F{}^2(a,b,c,\eta)$ is

equal to:

$$2^{4k+1} \left| \left| \left\{ (x_1,x_2,y_1,y_2,e_1,e_2) \;\middle|\; \begin{cases} e_1 F(x_1+2y_1)+e_2 F(x_2+2y_2) = 0 \\ x_1+x_2 &= 0 \bmod 2 \\ y_1+y_2 &= 0 \bmod 2 \\ e_1+e_2 &= 0 \end{cases} \right\} \right| \right|,$$

that is to: 2^{6k+2}.

We have also:

$$\sum_{a\in\mathcal{R};\,b,c\in\mathcal{T};\,\eta\in\{0,1\}} \nu_F{}^4(a,b,c,\eta) =$$

$$\sum_{a\in\mathcal{R};\,b,c\in\mathcal{T};\,\eta\in\{0,1\};e_1,\dots,e_4\in\{-1,1\}} \;\; \sum_{x_1,\dots,x_4,y_1,\dots,y_4\in\mathcal{T}} J_{a,b,c,\eta,e_1,\dots,e_4,x_1,\dots,x_4,y_1,\dots,y_4}$$

where

$$J_{a,b,c,\eta,e_1,\dots,e_4,x_1,\dots,x_4,y_1,\dots,y_4} =$$

$$i^{\,Tr(a(e_1 F(x_1+2y_1)+\dots+e_4 F(x_4+2y_4))+2b(x_1+\dots+x_4)+2c(y_1+\dots+y_4))-(e_1+\dots+e_4)\eta}.$$

Thus, $\sum_{a\in\mathcal{R};\,b,c\in\mathcal{T};\,\eta\in\{0,1\}} \nu_F{}^4(a,b,c,\eta)$ is equal to 2^{4k+1} times:

$$\left| \left\{ (x_1,\dots,x_4,y_1,\dots,y_4,e_1,\dots,e_4) \left|\; \begin{cases} e_1 F(x_1+2y_1)+\dots+e_4 F(x_4+2y_4)=0 \\ x_1+\dots+x_4 = 0 \bmod 2 \\ y_1+\dots+y_4 = 0 \bmod 2 \\ e_1+\dots+e_4 = 0 \bmod 4 \end{cases} \right. \right\} \right|.$$

Two cases occur for $e_1,\dots,e_4$: either they are equal to each others (case 1) or two of them are equal to 1 and the two others are equal to -1 (case 2).

In the case 1, if two elements among $(x_1,y_1),\dots,(x_4,y_4)$ are equal to each other, say $(x_{i_1},y_{i_1}) = (x_{i_2},y_{i_2})$ as well as the two remaining elements, $(x_{i_3},y_{i_3}) = (x_{i_4},y_{i_4})$, then the condition reduces to $F(x_{i_1}+2y_{i_1}) = F(x_{i_3}+2y_{i_3}) \bmod 2$, which leads for each of these 2 possible values of $(e_1,\dots,e_4)$ to $3\cdot 4^k(2^k-1)+4^k$ solutions (since F is a permutation).

In the case 2, say $e_{i_1} = e_{i_2} = 1$ and $e_{i_3} = e_{i_4} = -1$ (which corresponds to 6 possible values of $(e_1,\dots,e_4)$), if $(x_{i_1},y_{i_1}) = (x_{i_3},y_{i_3})$ and $(x_{i_2},y_{i_2}) = (x_{i_4},y_{i_4})$ or if $(x_{i_1},y_{i_1}) = (x_{i_4},y_{i_4})$ and $(x_{i_2},y_{i_2}) = (x_{i_3},y_{i_3})$, then the condition is satisfied; the number of such $(x_1,\dots,x_4,y_1,\dots,y_4)$ is:

$$2\cdot 4^k(4^k-1)+4^k;$$

and if $(x_{i_1},y_{i_1}) = (x_{i_2},y_{i_2})$ and $(x_{i_3},y_{i_3}) = (x_{i_4},y_{i_4})$, then the condition is satisfied if and only if $F(x_{i_1}+2y_{i_1}) = F(x_{i_3}+2y_{i_3}) \bmod 2$. The number of such additional $(x_1,\dots,x_4,y_1,\dots,y_4)$ is:

$$4^k(2^k-1),$$

since F is a permutation. We deduce that $\sum_{a\in\mathcal{R};\,b,c\in\mathcal{T};\,\eta\in\{0,1\}} \nu_F{}^4(a,b,c,\eta)$ is greater than or equal to:

$$2^{4k+1}\left[2\cdot\left(3\cdot 4^k(2^k-1)+4^k\right)+6\cdot\left(2\cdot 4^k(4^k-1)+4^k+4^k(2^k-1)\right)\right] =$$

$$2^{4k+2}\left(3\cdot 2^{4k+1}+3\cdot 2^{3k+1}-2^{2k+3}\right).$$

According to Parseval's relation (2), we have:

$$\sum_{a\in T;\, b,c\in T;\, \eta\in\{0,1\}} \nu_F{}^2(2a,b,c,\eta) = 4 \sum_{a,b,c\in GF(2^k)} \mu_{F'}{}^2(a,(b,c)) = 2^{5k+2}.$$

According to the hypothesis on F, we have:

$$\max_{a\in 2T^*;\, b,c\in T} |\nu_F(2a,b,c,\eta)| = 2^{k+1} + 2K.$$

Thus, $|\nu_F(2a,b,c,\eta)|$ is equal to 0 if $\eta = 1$ and is smaller than or equal to $2^{k+1} + 2K$ if $\eta = 0$. We deduce:

$$\sum_{a\in T^*;\, b,c\in T;\, \eta\in\{0,1\}} \nu_F{}^4(2a,b,c,\eta) \le (2^{k+1}+2K)^4(2^k-1)2^{2k},$$

and, since:

$$(2^{k+1}+2K)^4 \le 2^{4k+4} + 15\,K\cdot 2^{3k+4}$$

and

$$\nu_F(0,b,c,\eta) = \begin{cases} 2^{2k+1} & \text{if } b=c=\eta=0 \\ 0 & \text{otherwise} \end{cases}$$

we have:

$$\sum_{a\in T;\, b,c\in T;\, \eta\in\{0,1\}} \nu_F{}^4(2a,b,c,\eta) \le 2^{8k+4} + (2^{6k+4} + 15K\,2^{5k+4})(2^k-1).$$

Thus, we finally deduce from relation (9):

$$\max_{a\in\mathcal{R}^*;\, b,c\in T;\, \eta\in\{0,1\}} \left(\nu_F{}^2(a,b,c,\eta)\right) \ge$$

$$\frac{2^{4k+2}\left(3\cdot 2^{4k+1} + 3\cdot 2^{3k+1} - 2^{2k+3}\right) - 2^{8k+4} - (2^{6k+4} + 15K\,2^{5k+4})(2^k-1)}{2^{6k+2} - 2^{5k+2}} =$$

$$\frac{2^{3k+1} + 2^{2k+1} - 2^{k+2} - 60\,K(2^k-1)}{2^k-1} = 2^{2k+1} + 2^{k+2} - 60\,K$$

and the proof is complete.

Remark: If we focus only on Λ_F'', it is possible to find a permutation F such that Λ_F'' is close to $2^{2k} - 2^k$: it is proved in [2] that for any k, $x - y$ ranges once over $\mathcal{R}^*$, when x and y range over T and $x \ne y$.

Consider any permutation σ on T. The following mapping is a permutation on $\mathcal{R}$:

$$F:\ x + 2y \to \begin{cases} x - y & \text{if } x \ne y \\ 2\sigma(x) & \text{otherwise} \end{cases}.$$

Assume now that σ, viewed as a permutation on $GF(2^k)$, is an almost bent mapping (k odd). Then, we have:

$$\Lambda_F'' \geq 2^{2k} - 2^k - 2^{\frac{k+1}{2}}.$$

Indeed, we have:

$$\sum_{x,y \in \mathcal{T}} i^{Tr(aF(x+2y)+2bx+2cy)-\eta} =$$

$$\sum_{x,y \in \mathcal{T} | x \neq y} i^{Tr(a(x-y)+2bx+2cy)-\eta} + \sum_{x \in \mathcal{T}} i^{Tr(2a\sigma(x)+2bx+2cx)-\eta} =$$

$$i^{-\eta} \left(\sum_{x \in \mathcal{T}} i^{Tr(ax+2bx)} \right) \overline{\left(\sum_{y \in \mathcal{T}} i^{Tr(ay+2cy)} \right)} +$$

$$\sum_{x \in \mathcal{T}} i^{Tr(2a\sigma(x)+2bx+2cx)-\eta} - \sum_{x \in \mathcal{T}} i^{Tr(2bx+2cx)-\eta}.$$

Thus, according to Lemma 3, if $a \in \mathcal{R}^*$, say $a = a'(1+2u)$, $a' \in \mathcal{T}^*$, $u \in \mathcal{T}$, then $\sum_{x,y \in \mathcal{T}} i^{Tr(aF(x+2y)+2bx+2cy)-\eta}$ is equal to:

$$2^k i^{-\eta} \left(i^{Tr((1+u+b')^2-(1+u+c')^2)} \right) + i^{-\eta} \sum_{x \in \mathcal{T}} (-1)^{tr(a\sigma(x)+(b+c)x)} - 2^k i^{-\eta} \delta_0(b-c),$$

where $b' = \frac{b}{a'}$ and $c' = \frac{c}{a'}$.
Therefore:

$$\left| \sum_{x,y \in \mathcal{T}} i^{Tr(aF(x+2y)+2bx+2cy)-\eta} \right| \leq 2^k + 2^{\frac{k+1}{2}},$$

(tight bound) since if $b = c$, then $i^{Tr((1+u+b')^2-(1+u+c')^2)} = 1$.

Concluding Remark

There is a gap between the upper bound: $\Lambda_F \leq 2^{2k} - 2^k$ and the lower bound: $\max_F (\Lambda_F) \geq 2^{2k} - 2^{k+1} - 2^{\frac{k}{2}+2}$.
If we can find in the future a permutation F such that Λ_F is near $2^{2k} - 2^k$, then it will be intersting to develop such constructions on $\mathbf{Z}_4$ for cryptographic purpose. In any case, it would be useful to get more precise bounds on Λ_F.

Acknowledgement

The author thanks Pascale Charpin for helpful discussions.

References

1. A.S. Ambrosimov *Properties of bent functions of q-valued logic over finite fields* Discrete Math. Appl. vol 4, n° 4, p. 341-350 (1994)
2. A. Bonnecaze and I. Duursma, *Translates of $\mathbf{Z}_4$-linear codes*, IEEE Trans. on Inf. Theory., vol. 43, p. 1218-1230 (1997)
3. C. Carlet. $\mathbf{Z}_{2^k}$-linear codes, *IEEE Trans. Inform. Theory* vol. 44, n° 4, p. 1543-1547 (1998).
4. C. Carlet, P. Charpin and V. Zinoviev. *Codes, bent functions and permutations suitable for DES-like cryptosystems*, Designs Codes and Cryptography, to appear (1998)
5. F. Chabaud and S. Vaudenay, *Links between differential and linear cryptanalysis* In Advances in Cryptology EUROCRYPT'94, Perugia, Italy, Lecture Notes in Computer Science No. 950, p. 356-365, Springer-Verlag, 1995.
6. J. F. Dillon, *Elementary Hadamard Difference sets*, Ph. D. Thesis, Univ. of Maryland (1974).
7. H. Dobbertin, *Construction of bent functions and balanced Boolean functions with high nonlinearity*, Fast Software Encryption (Proceedings of the 1994 Leuven Workshop on Cryptographic Algorithms), Lecture Notes in Computer Science 1008, p. 61-74 (1995).
8. A. R. Hammons Jr., P. V. Kumar, A. R. Calderbank, N. J. A. Sloane and P. Solé, *The Z_4-linearity of Kerdock, Preparata, Goethals and related codes* IEEE Transactions on Information Theory, vol 40, p. 301-320, (1994)
9. Tor Helleseth, P. V. Kumar, Oscar Moreno and A.G. Shanbhag *Improved estimates via exponential sums for the minimum distance of $\mathbf{Z}_4$-linear trace codes* IEEE Transactions on Information Theory, vol 42, p. 1212-1216, (1996)
10. P. V. Kumar, T. Helleseth and A. R. Calderbank *An upper bound for Weil exponential sums over Galois rings and applications* IEEE Transactions on Information Theory, vol 41, p. 456-468, (1995)
11. O. A. Logachev, A. A. Salnikov and V. V. Yashchenko *Bent functions on a finite abelian group*, Discrete Math. Appl., Vol. 7, p. 547-564 (1997)
12. B.R. MacDonald,, *Finite rings with identity*, Marcel Dekker, NY, 1974
13. M. Matsui, *Linear cryptanalysis method for DES cipher*, EUROCRYPT'93 Advances in Cryptography, Lecture Notes in Computer Science 765, p. 386-397 (1994).
14. Y. Niho, *Multi-valued cross-correlation functions between two maximal linear recursive sequences*, Ph. D. Thesis, USCEE Rep. 409 (1972).
15. K. Nyberg, *Perfect non-linear S-boxes*, Advances in Cryptology, EUROCRYPT' 91, Lecture Notes in Computer Science 547, p. 378-386, Springer Verlag (1992)
16. W. W. Peterson, *Error-correcting codes*, Amsterdam, North Holland 1977.
17. R. A. Rueppel *Analysis and design of stream ciphers* Com. and Contr. Eng. Series, Berlin, Heidelberg, NY, London, Paris, Tokyo 1986
18. V. M. Sidelnikov, *On the mutual correlation of sequences*, Soviet Math. Dokl. 12, p. 197-201 (1971).
19. F. J. Mac Williams and N. J. Sloane, *The theory of error-correcting codes*, MIT Press, Cambridge, MA 1961.
20. P. Shankar, *On BCH codes over arbitrary integer rings*, IEEE Trans. on Inf. Theory, vol. 25, p. 480-483 (1979).
21. K. Yang, T. Helleseth, P. V. Kumar and A. G. Shanbhag, *On the weight hierarchy of Kerdock codes over $\mathbf{Z}_4$*, IEEE Trans. on Inf. Theory, vol. 42, p. 1587-1593 (1996).

Efficient Algorithms for the Jacobian Variety of Hyperelliptic Curves $y^2 = x^p - x + 1$ Over a Finite Field of Odd Characteristic $p^\star$

Iwan Duursma[1],[**] and Kouichi Sakurai[2],[***]

[1] University of Limoges
Mathematics Department
123 avenue Albert Thomas
87060 LIMOGES Cedex,France
`duursma@unilim.fr`
[2] Kyushu University, Computer Science Dept.
6-10-1 Hakozaki, Higashi-ku, Fukuoka 812-8581, JAPAN
`sakurai@csce.kyushu-u.ac.jp`

Abstract. We develop efficient algorithms for the Jacobian of the hyperelliptic curve defined by the equation $y^2 = x^p - x + 1$ over a finite field F_{p^n} of odd characteristic p. We first determine the zeta function of the curve which yields the order of the Jacobian. We also investigate the Frobenius operator and use it to show that, for field extensions F_{p^n} of degree n prime to p, the Jacobian has a cyclic group structure. We furthermore propose a method for faster scalar multiplication in the Jacobian by using efficient operators other than the Frobenius that have smaller eigenvalues.

1 Introduction

ALGORITHMIC ASPECTS OF JACOBIAN VARIETIES.

The development of algorithms for abelian varieties [PIL90], [AH96] is an important topic in algorithmic number theory with many applications. Elliptic curves produce intractable discrete logarithm problems suitable for public public-key cryptosystems [KO88,KO89]. Koblitz [KO88,KO89] furthermore investigated the Jacobians of hyperelliptic curves defined over finite fields as a suitable source of finite abelian groups suitable for cryptographic discrete logarithm problems.

The basic algorithmic problems on Jacobian varieties over finite fields are

1. Counting the order of the Jacobian [PIL90,AH96]
2. Developing fast algorithms for addition and scalar multiplication on the Jacobian [CA87,KO89]

---

[*] Presented at the International Conference on Coding Theory and Cryptography, Guanajuato, Mexico, 1998 (Extended Abstract: Version 8th of March 1999)
[**] Work done while with AT&T Labs-Research, Florham Park NJ, U.S.A.
[***] Work done while visiting Columbia University, Computer Science Department.

3. Investigating the Frobenius action on the Jacobian [Ko89]

These problems have been deeply investigated for the case of genus one, i.e., elliptic curves. In this paper, we consider the three problems for the special class of hyperelliptic curves defined by the equation $y^2 = x^p - x + d$ over a field F_{p^n} of odd characteristic p. They fall into two different classes, for d a quadratic residue or non-residue modulo p, respectively. Note that the genus of the curve is $(p-1)/2$, which is related to the characteristic. The curves were investigated in [Duu93] for their relation with Aurifeuillian identities [Luc1878].

OUR RESULTS

We investigate the structure of the Jacobian of the hyperelliptic curve, and we develop efficient algorithms associated with the Jacobian.

1. We determine the zeta function of the curve $y^2 = x^p - x + d$ over the base field F_p and give an explicit formula (from which the complex zeros of the zeta function have been eliminated) for the order of the Jacobian over extension fields F_{p^n}.
2. For each of the eigenspaces under the Frobenius action on the Jacobian, we compute the eigenvalue as an integer modulo the exponent of the Jacobian. This gives us an alternative way for computing the order of the Jacobian. We use the eigenvalues and the fact that they are all different to prove that the group structure of the Jacobian is cyclic for constant field extensions F_{p^n} of odd degree n prime to p.
3. We give explicit formulas for faster exponentiation in the Jacobian. In particular, we present as a new technique the use of other automorphisms, different from the Frobenius map. In this way, we obtain maps with smaller eigenvalues that lead to faster scalar multiplication.

TOWARDS APPLICATIONS IN CRYPTOGRAPHY

Koblitz [Ko98] independently investigated the curve $y^2 = x^3 - x + 1$ (the case $p = 3$) and presents an efficient Digital Signature Algorithm (DSS) [NIST93] based on the curve $y^2 = x^3 - x + 1$. Our results show that the other curves in the family have a similar potential to be used in DSS or other applications. It remains to choose particular curves and to develop implementations for them.

2   Preliminaries

2.1   Jacobian varieties

Let $\mathbf{K}$ be an arbitrary field and let $\bar{\mathbf{K}}$ denotes its algebraic closure. We define a hyperelliptic curve C of genus g over $\mathbf{K}$ to be an equation of the form $v^2 + h(u)v = f(u)$, where $h(u)$ is a polynomial of degree at most g and $f(u)$ is a monic polynomial of degree $2g + 1$.

The points $P(x, y)$, for $x, y \in \bar{K}$ with $y^2 + h(x)y = f(x)$, generate the free group of divisors. A divisor D is a finite formal sum of $\bar{\mathbf{K}}$-points $D = \sum_P m_P P$, $m_P \in \mathbf{Z}$. We define the degree of D to be $\deg(D) = \sum m_P$. The divisors form an additive group $\mathbf{D}$, in which the divisors of degree 0 form a subgroup $\mathbf{D}^0$. A divisor D is called principal if there exists a rational function f such that m_P gives the order of vanishing of f in P. The principal divisors are of degree zero and form a subgroup $\mathbf{P}$ of $\mathbf{D}^0$. The Jacobian is defined as the quotient $\mathbf{J}(\mathbf{K}) = \mathbf{D}^0/\mathbf{P}$.

For hyperelliptic curves, addition in the Jacobian can be done explicitly with Cantor's algorithm [CA87]. For elliptic curves, addition follows the famous chord and tangent method. For general hyperelliptic curves, addition is isomorphic to Gauss' addition law for quadratic forms. For a curve of the form $v^2 = f(u)$, Cantor's algorithm represents divisors by pairs of polynomials $(\alpha(u), \beta(u))$ such that $\alpha | \beta^2 - f$ and $\deg \beta < \deg \alpha \leq g$. In particular there exists a polynomial $\gamma(u)$ such that $\alpha X^2 + 2\beta XY + \gamma Y^2$ is a quadratic form of discriminant $4\beta^2 - 4\alpha\gamma = 4f$. It is well-known that the group law of the Jacobian corresponds with Gauss' group law for quadratic forms of discriminant $4f$.

2.2   Discrete logarithm problem for Jacobian varieties

Let $\mathbf{F}_q$ be a finite field with q elements. The discrete logarithm problem on $\mathbf{J}(\mathbf{F}_{q^n})$ is the problem, given two divisors D_1 and D_2 defined over $\mathbf{F}_{q^n}$, of determining an integer $m \in \mathbf{Z}$ such that $D_2 = mD_1$ if such m exists [KO89].

The discrete logarithm problem is believed to be hard and many cryptographic applications are based on its intractability. On the other hand such applications ask that exponentiation, the computation of D_2 given m and D_1, is efficient. The Jacobians of hyperelliptic curves have an explicit addition law and they are an important source for discrete logarithm problems.

2.3   Computing the order of Jacobian varieties

The computation of the order of the Jacobian makes use of Weil's theorem. We recall Weil's theorem and give a brief description of its application to counting points on Jacobians. Beth and Schaefer use the theorem in the construction of elliptic cryptosystems [BESC91] and Koblitz uses the theorem in his construction of cryptosystems from hyperelliptic curves defined over finite fields [KO88,KO89].

Let $C/\mathbf{F}_q$ be a hyperelliptic curve defined over a finite field of q elements by an equation of the form $v^2 = f(u)$, for f of odd degree. The genus g of the curve follows from $\deg(f) = 2g + 1$. For an extension $\mathbf{F}_{q^n}$ of $\mathbf{F}_q$ let $N_n = \sharp C(\mathbf{F}_{q^n})$ be the number of rational points of the curve. These consist of the affine solutions (u, v) to the equation $v^2 = f(u)$ and one additional point at infinity. Weil's theorem proves the existence of numbers $\alpha_1, \alpha_2, \ldots, \alpha_{2g}$ such that, for $n > 0$,

$$N_n = q^n + 1 - \sum \alpha_i^n.$$

Equivalently, the power series

$$Z(t) \;=\; \exp\left(\sum_{n=1}^{\infty} N_n \frac{t^n}{n}\right)$$

represents a rational function, with numerator

$$P(t) \;=\; \prod_{i=1}^{2g} (1 - \alpha_i t),$$

and denominator $(1 - t)(1 - qt)$. It follows that the zeta-function $Z_n(t)$ for the curve $C/\mathbf{F}_{q^n}$ is related to $Z(t) = Z_1(t)$ via

$$Z_n(t^n) \;=\; \prod_{\zeta^n = 1} Z_1(\zeta t). \tag{1}$$

We will make use of the Riemann hypothesis for curves which says that the numbers α_i have absolute value

$$|\alpha_i| \;=\; q^{1/2}. \tag{2}$$

Once $Z(t)$ has been found, the order of the Jacobian $J(\mathbf{F}_{q^n})$ can be determined with the formula

$$\sharp J(\mathbf{F}_{q^n}) \;=\; \prod_{i=1}^{2g} (1 - \alpha_i^n) \tag{3}$$

3   Our discussed hyperelliptic curves

For use in hyperelliptic cryptosystems, we consider the class of hyperelliptic curves defined by the equation

$$C^d/\mathbf{F}_p : y^2 = x^p - x + d,$$

where p is an odd prime number and d an integer modulo p. For given p, we distinguish up to isomorphy the three cases $d = 0$, and d a nonzero quadratic residue or non-residue.

Definition 1. *Let d^+ and d^- denote a fixed quadratic residue and non-residue modulo p. We define curves*

$$\begin{aligned}
C^0/\mathbf{F}_p &: y^2 = x^p - x, \\
C^+/\mathbf{F}_p &: y^2 = x^p - x + d^+, \\
C^-/\mathbf{F}_p &: y^2 = x^p - x + d^-.
\end{aligned}$$

The curves have genus $(p-1)/2$. We will consider the curves over finite extensions $\boldsymbol{F}_q$ of $\boldsymbol{F}_p$. The curves C^+ and C^- are isomorphic over the extension $\boldsymbol{F}_{p^2}$. All three curves are isomorphic over the extension $\boldsymbol{F}_{p^p}$.

Over the extension field $\boldsymbol{F}_{p^n}$ of $\boldsymbol{F}_p$, for n an odd prime number different from p, the Jacobian of the curves C^+ and C^- has a cyclic group structure. Moreover, elements in the Jacobian that are defined over $\boldsymbol{F}_{p^n}$ but not over $\boldsymbol{F}_p$ have order 1 modulo $2pn$, unless n divides $p^p - (-1/p)$, when elements may have order n. Thus, in general the order of the quotient group $J(\boldsymbol{F}_{p^n})/J(\boldsymbol{F}_p)$ contains no elements of small order. In many cases the order is prime or almost prime. So that the discrete logarithm problem defined on the group is a priori hard to solve. The group is of further interest for cryptosystems because scalar multiplication is very efficient. For this, we will express the multiplier in base p and use explicit formulas for multiplication by $m = 0, 1, \dots, p-1$ and $m = p, p^2, \dots$.

3.1 Zeta functions for the curves C^d over $\boldsymbol{F}_{p^2}$

Many properties, such as order of the Jacobian and eigenvalues of the Frobenius operator, follow from the zeta function. The expression of the eigenvalues as exponential sums can be found in Hasse [Has34,HD34]. For the curves C^d, we use the explicit form given in [Duu93]. Over a quadratic constant field $\boldsymbol{F}_{p^2}$, the zeta functions are given by the following lemmas.

Lemma 2. For the curve $C^0/\boldsymbol{F}_{p^2}$, all reciprocal zeros are equal to $(-1/p)p$.

PROOF. See [Duu93]. ■

Let $\bar{p} = (-1/p)p$.

Lemma 3. For the curve $C^\pm/\boldsymbol{F}_{p^2}$, the zeta function has numerator $P_2(t) = \Phi_p(\bar{p}t)$.

PROOF. See [Duu93]. ■

3.2 Aurifeuillian identities

The numerator $P_1(t)$ of the zeta function over the constant field $\boldsymbol{F}_p$ relates to the numerator $P_2(t)$ via $P_2(t^2) = P_1(t)P_1(-t)$. It follows from the lemma that the polynomial $\Phi_p(\bar{p}t^2)$ has a non-trivial factorization over the integers. For we have $\Phi_p(\bar{p}t^2) = P_2(t^2) = P_1(t)P_1(-t)$. The factorization of $\Phi_p(\bar{p}t^2)$ is a special case of an identity of type

$$\Phi_n(X) = F(X)^2 - kXG(X)^2. \tag{4}$$

Identities of this type were first studied by Aurifeuille [Luc1878]. They show that the irreducible polynomial $\Phi_n(X)$ factors with the argument $X = kt^2$. In our case $n = 2p$ and $k = -\bar{p}$. More general choices for n and k and properties of the polynomials F and G can be found in [Ste87], [Bre93].

For $p = 3, 5$,

$$\begin{aligned}
\Phi_6(X) &= X^2 - X + 1 &&= (X+1)^2 - 3X. \\
\Phi_{10}(X) &= X^4 - X^3 + X^2 - X + 1 &&= (X^2 - 3X + 1)^2 + 5X(X-1)^2.
\end{aligned}$$

After substitution of $X = 3t^2, X = -5t^2$, respectively, we find the zeta functions

$$\begin{aligned}
C^\pm/\boldsymbol{F}_3 &: & P^\pm(t) &= 3t^2 + 1 \pm 3t. \\
C^\pm/\boldsymbol{F}_5 &: & P^\pm(t) &= 25t^4 + 15t^2 + 1 \pm 5t(5t^2 + 1).
\end{aligned}$$

3.3   Zeta functions for the curves C^d over F_p

In general, $P^\pm(t)$ is obtained as a factor of $\Phi_p(\bar{p}t^2) = P_2(t^2) = P_1(t)P_1(-t)$. The factorisation is the special case $n = 2p$ and $k = -\bar{p}$ in the Aurifeuillian identity (4), for $X = kt^2$. Using the results of [STE87] with these particular values yields the following.

Theorem 4. *Let ζ denote a fixed primitive p-th root of unity and let*

$$\sqrt{\bar{p}} = -(-1/p) \sum_{a=1}^{p-1} (a/p) \zeta^a. \tag{5}$$

For the curves $C^+/\boldsymbol{F}_p$ and $C^-/\boldsymbol{F}_p$, the numerator of the zeta-function is given by

$$P^+(t) = \prod_{a=1}^{p-1} (1 - (a/p)\zeta^a \sqrt{\bar{p}}t), \quad P^-(t) = \prod_{a=1}^{p-1} (1 + (a/p)\zeta^a \sqrt{\bar{p}}t),$$

respectively.

PROOF.   The factorization is given in [STE87]. The convention for $\sqrt{\bar{p}}$ in (5) guarantees that the factors $P^+(t)$ and $P^-(t)$ yield $2p + 1$ and 1 rational points respectively, for the curves $C^+/\boldsymbol{F}_p$ and $C^-/\boldsymbol{F}_p$ respectively.   ∎

For a fast computation over the integers of the two factors we implement ideas from [BRE93] in Algorithm 1.

3.4   Class numbers

Now that we have the zeta function of the curve $C^d/\boldsymbol{F}_p$ we can compute the order of the Jacobian of the curve for any extension $\boldsymbol{F}_q$ of $\boldsymbol{F}_p$ by following the method outlined in Subsection 2.3. As in [KO89], we are primarily interested in extensions of degree a prime number. In this case, the method leads to an explicit formula.

Proposition 5. *Let J^d be the Jacobian of the curve $C^d/\boldsymbol{F}_p$ and let $n > 0$ be an odd integer prime to p. Over the extension $\boldsymbol{F}_{p^n}$ of $\boldsymbol{F}_p$, the Jacobian J^d has number of rational points*

$$\sharp J^d(\boldsymbol{F}_{p^n}) = P^\pm(\bar{p}^{(n-1)/2}), \qquad \text{for } \pm = (nd/p).$$

Algorithm 1 Computing the zeta function for $C/\mathbf{F}_p : y^2 = x^p - x + 1$.

Input:	Odd prime number p.
Step1	Set $\bar{p} := (-1/p)p$.
Step2	Compute sums of powers of reciprocal zeros.
	$r_0 := p - 1$.
	For $i = 1, 3, \dots, p - 2$ do
	$\qquad r_i := -(-i/p)\bar{p}^{(i+1)/2}, \quad r_{i+1} := -\bar{p}^{(i+1)/2}$.
Step3	Transform power sums into symmetric sums
	$c_0 := 1, \quad P := 1$.
	For $k = 1, 2, \dots, p - 1$ do
	$\qquad c_k := r_k$
	$\qquad$ For $j = 1, 2, \dots, k - 1$ do
	$\qquad\qquad c_k := c_k - c_j r_{k-j}, \quad c_k := c_k / k$.
	$\qquad P := P - c_k t^k$.
Output:	Reciprocal zeta polynomial $P(t) = 1 + pt + \cdots + p^{(p-1)/2} t^{p-1}$.

PROOF. Formula (1) and Theorem 4 gives, for n odd and prime to p,

$$\sharp J^d(\mathbf{F}_{p^n}) = \prod_{a=1}^{p-1}(1 - ((d/p)(a/p)\zeta^a \sqrt{\bar{p}})^n)$$

$$= \prod_{a=1}^{p-1}(1 - (d/p)(a/p)\zeta^{an}\sqrt{\bar{p}}\bar{p}^{(n-1)/2})$$

$$= \prod_{a=1}^{p-1}(1 - (nd/p)(a/p)\zeta^a\sqrt{\bar{p}}\bar{p}^{(n-1)/2})$$

$$= P^{\pm}(\bar{p}^{(n-1)/2}), \qquad \text{for } \pm = (nd/p).$$

■

3.5 Almost prime for cryptographic applications

For the application to cryptosystems we need a large cyclic subgroup of prime order in $J^d(\mathbf{F}_{p^n})$. Shanks' baby-step giant-step method [OD85] and Pohlig-Hellman method [PH78] have a running time that is proportional to the square root of the largest prime factor of $\sharp J$. Therefore, we can avoid these attacks by choosing curves such that $\sharp J$ is divisible by a large prime factor. Numerical experiments show that the order $\sharp J$ is prime or almost prime in many cases of (p, n), so that the induced cryptosystems are secure against the first attack [OD85,PH78].

In general the group contains a subgroup $J^d(\mathbf{F}_p)$. As in [Ko89], we take n prime to avoid other trivial factors in the group order of $J^d(\mathbf{F}_{p^n})$. And for such n we check whether the sequence $\sharp J^d(\mathbf{F}_{p^n})/J^d(\mathbf{F}_p)$ contains 'analogues of Mersenne primes'.

Example 1. Using the formula in the proposition with the zeta functions

$$C^\pm/\mathbf{F}_3 \quad : \quad P^\pm(t) = 3t^2 + 1 \pm 3t.$$
$$C^\pm/\mathbf{F}_5 \quad : \quad P^\pm(t) = 25t^4 + 15t^2 + 1 \pm 5t(5t^2 + 1).$$

we get

$$C^\pm/\mathbf{F}_3 \quad : \quad h^\pm = 3^n + 1 \pm 3^{(n+1)/2}.$$
$$C^\pm/\mathbf{F}_5 \quad : \quad h^\pm = 5^{2n} + 3 \cdot 5^n + 1 \pm (5^{(3n+1)/2} + 5^{(n+1)/2)}).$$

The group order h_n/h_1 is a prime in each of the following cases

$$p = 3, (d/p) = +1 : n = 5, 11, 31, 37, 47, 53, 97, 163, 167, \ldots$$
$$p = 3, (d/p) = -1 : n = 5, 7, 11, 17, 19, 79, 163, \ldots$$
$$p = 5, (d/p) = +1 : n = 3, 7, 13, 17, 29, 41, \ldots$$
$$p = 5, (d/p) = -1 : n = 3, 107 \ldots$$

Table 1 gives parameters (p, n) for which the size of the Jacobian does not exceed 400 bits. In each case, either p is small or n is small and the table is divided into two parts accordingly. Tables 2 and 3 give those (p, n) for which the order of the Jacobian is almost prime. For each pair (p, n) we give the bitsize of the largest prime factor and the cofactor of this prime factor in h_n/h_1. Table 2 includes the cases where p is small, and Table 3 the remaining cases with n small. The case $(d/p) = +1$ (resp. $(d/p) = -1$) appears on the left (resp. right) in each table. Only Jacobians with a prime factor of at least 40 bits are included. Most of the examples with $p = 3$ and $p = 5$ given above appear in the tables, all with cofactor 1. For $p = 3$ and $(d/p) = +1$,

$$h_{73}/h_1 = 11009154363058726777293427189333 \cdot 877.$$

This example is represented in Table 2 under $p = 3, n = 73$ together with the bitsize 103 of the large prime factor and the cofactor 877.

4   On the Frobenius action

We will determine the eigenvalues of the Frobenius action on the points of the Jacobian J^d. Their complex value is given by the reciprocal roots of the zeta function of the curve C^d. On the other hand, we can determine the eigenvalues as modular integers. To see this, let P generate a cyclic subgroup of $J^d(\mathbf{F}_{p^n})$ on which the Frobenius action is well-defined. That is to say, the group is defined over $\mathbf{F}_p$ and so with the point P also contains ϕP. For example, this is the case if we take P of large prime order l; if l^2 exceeds the order of the Jacobian then

		50	60	70	80	90	100	120	140	160	180	200	240	280	320	360	400
p=	3	n= 31	37	43	47	53	61	73	83	97	113	-	151	173	199	227	251
	5	7	11	13	17	19	-	23	29	31	37	43	47	59	67	73	83
	7	5	-	-	-	-	11	13	-	17	19	23	-	31	37	41	47
	11	1	3	-	-	5	-	-	7	-	-	-	13	-	17	19	23
	13	1	-	3	-	-	-	5	-	7	-	-	-	11	-	-	17
n=	1	p= 19	23	29	31	-	37	43	47	53	61	-	73	83	97	107	113
	3	7	11	13	-	-	17	19	-	23	-	-	31	-	37	43	47
	5	7	-	-	-	11	-	13	-	-	17	19	-	23	-	29	31
	7	5	-	-	-	-	-	-	11	13	-	-	17	19	-	23	-
	11	3	5	-	-	-	7	-	-	-	-	-	-	13	-	17	-

Table 1. Prime pairs (p, n) for which $\sharp J(F_{p^n})$ does not exceed a given bitsize.

there is a unique subgroup of order l that with P also contains ϕP. Then there exist an integer ρ such that

$$\phi P = \rho P.$$

Knowledge of ρ is important for fast multiplication in the Jacobian. This can be attained by expressing the multiplier m in mP in base ρ. We will determine ρ for the Jacobian J^d of the curve C^d.

Lemma 6. *For a prime power factor l dividing the order of the Jacobian $J^d(F_{p^n})$, the roots of unity modulo l of order dividing pn are generated by $\bar{p}$ (mod ℓ).*

PROOF. The order of the Jacobian $J^d(F_{p^n})$ is h^+ or h^-, and by Theorem 4 we have

$$h^+ h^- = P^+(\bar{p}^{(n-1)/2}) P^-(\bar{p}^{(n-1)/2}) = \Phi_p(\bar{p}\bar{p}^{n-1}) = \frac{\bar{p}^{pn} - 1}{\bar{p}^n - 1}.$$

Modulo a prime power factor ℓ the pn-th roots of unity appear among the pn power of $\bar{p}$. Note that modulo an arbitrary factor, the roots of unity can be computed with the Chinese remainder theorem. In general these are not of the form a power of $\bar{p}$. ∎

Proposition 7. *For $d \in F_p$ and for n odd prime to p, let $C^d/F_{p^n} : y^2 = x^p - x + d$ have Jacobian J^d. The Frobenius $\phi : [(a, b) - \infty] \mapsto [(a^p, b^p) - \infty]$ acts on elements of $J^d(F_{p^n})$ as scalar multiplication by a common scalar,*

$$\phi D = ((-1/p)p)^{pk} D,$$

where k is such that $2kp \equiv 1 \pmod{n}$.

PROOF. We compute the eigenvalue under the assumption that the point ϕP is an element of the group generated by D. As remarked above, this is certainly the case for element D of large prime order (the situation that is of interest in cryptography). We will see in Theorem 10 that the assumption is justified in general.

The Frobenius is of order n and therefore the scalar ρ is an n-th root of unity modulo the order of the Jacobian $J^d(\mathbf{F}_{p^n})$. With the Chinese remainder theorem it suffices to prove the claim modulo a prime power factor of the order of the Jacobian. We see from the lemma that, modulo a given prime power factor of the order, the n-th roots are generated by $\bar{p}^p$. And we may assume that ρ is of the form $\bar{p}^{kp}$. On the other hand, the Frobenius eigenvalue is a reciprocal zero

p	n	bitsize	cofactor(+)	p	n	bitsize	cofactor(−)
3	31	46	1	3	41	43	2337001
3	37	55	1	3	43	42	1549 · 27091
3	47	71	1	3	47	66	283
3	53	81	1	3	53	68	48973
3	73	103	877	3	73	96	741973
3	79	87	14221 · 2275201	3	79	125	1
3	89	117	1069 · 2137	3	163	258	1
3	97	150	1	3	167	236	8017 · 44089
3	103	141	524683	3	193	305	1
3	149	219	15199	3	239	378	1
3	163	255	1	5	13	49	131
3	167	261	1	5	17	64	1531
3	193	271	251287 · 12739	5	19	73	2851
5	13	54	1	5	61	258	1884901
5	17	72	1	5	83	361	1209311
5	19	60	4113691	7	11	58	463 · 333103
5	29	128	1	7	17	97	239 · 973897 · 2381
5	31	107	311 · 5624951	7	23	176	1289
5	37	154	2591	7	29	237	1
5	41	184	1	7	43	337	171571
5	43	183	1291	7	47	372	92779
5	71	280	71 · 17041 · 7484111	11	7	80	23577401
7	11	54	258576319	11	17	267	1871
7	17	80	377231 · 953 · 20123377	13	3	45	1
7	19	137	8513	13	5	65	131 · 156131
11	17	265	1123				
13	5	77	1171				
13	7	131	1				
13	17	324	710581301				

Table 2. Parameters (p, n), for $p = 3, 5, 7, 11, 13$, with a large prime factor in the order of the Jacobian of $C^{\pm}$. Given are the bitsize of the large prime factor and the cofactor in h_n/h_1 (Example 1)

p	n	bitsize	cofactor($+$)		p	n	bitsize	cofactor($-$)
31	1	51	$373 \cdot 62869$		31	1	45	$145577 \cdot 1613$
41	1	102	83		37	1	85	149
43	1	88	$86689 \cdot 1291$		43	1	89	$6709 \cdot 947$
53	1	133	141829		67	1	198	1
61	1	179	1		79	1	244	1
101	1	295	$157561 \cdot 607 \cdot 5657$		83	1	237	$499 \cdot 9463$
17	3	57	103		97	1	315	1
19	3	43	$166783 \cdot 17443$		17	3	58	409
23	3	75	$691 \cdot 7591$		19	3	69	457
47	3	219	$283 \cdot 1129 \cdot 76423$		19	7	230	1
13	5	77	1171		23	7	272	$22541 \cdot 9661$
17	7	166	$16661 \cdot 17851$		19	11	343	$419 \cdot 3442436911$

Table 3. Parameters (p, n), for $n = 1, 3, 5, 7, 11$, with a large prime factor in the order of the Jacobian of $C^{\pm}$. Given are the bitsize of the large prime factor and the cofactor in h_n/h_1 (Example 1)

of the polynomial $P^d(t)$, whose reciprocal is the characteristic polynomial of the Frobenius. Thus it appears among

$$(d/p)(a/p)\zeta^a \sqrt{\bar{p}}, \quad a = 1, 2, \dots, p - 1,$$

where ζ is a p-th root of unity and $\sqrt{\bar{p}}$ is defined by (3). To decide which one of these values is a n-th root of unity, we solve

$$\zeta^{2a}\bar{p} \equiv \bar{p}^{2kp} \pmod{\bar{p}^{pn} - 1}.$$

Both sides are pn-th roots of unity and for $\zeta = \bar{p}^n$ we find

$$2an + 1 \equiv 2kp \pmod{pn},$$

which is equivalent to the two conditions $2an \equiv -1 \pmod{p}$ and $2kp \equiv 1 \pmod{n}$. ∎

4.1 An alternative way of computing class numbers

Observe that the common eigenvalue ρ depends on the constant field $\boldsymbol{F}_{p^n}$ but not on the curve C^d. We use this observation to obtain a different way to compute the class numbers that avoids the computation of the zeta function.

Lemma 8. *Let the square root of $\bar{p}$ be defined as in (5),*

$$\sqrt{\bar{p}} = -(-1/p) \sum_{a=1}^{p-1} (a/p)\zeta^a.$$

Modulo the order h of $J^d(\boldsymbol{F}_{p^n})$, we have, for $\zeta = \bar{p}^r$,

$$\sqrt{\bar{p}} \equiv (-2rd/p)((-1/p)p)^{(pn+1)/2} \pmod{h}.$$

PROOF. From the proof of the proposition we see

$$(d/p)(a/p)\zeta^a\sqrt{\bar{p}} \equiv \bar{p}^{kp} \pmod{h},$$

for $2an \equiv -1 \pmod{p}$ and $2kp \equiv 1 \pmod{r}$. Therefore

$$\sqrt{\bar{p}} \equiv (d/p)(-2n/p)\bar{p}^{kp-an} \equiv (-2nd/p)\bar{p}^{(pn+1)/2} \pmod{h}.$$

∎

Proposition 9. *The class numbers h^+ and h^- have product $h^+h^- = \Phi_p(\bar{p})$. Moreover,*

$$h^\pm \Big| \sum_{a=1}^{p-1}(a/p)((-1/p)p)^{an} \pm (2n/p)((-1/p)p)^{(pn+1)/2}.$$

Therefore h^+ and h^- are coprime and they can be computed as the greatest common divisor of $\Phi_p(\bar{p})$ and the appropriate multiple.

4.2 Group structure of the Jacobian

By choosing a subgroup of prime order in $J^d(\boldsymbol{F}_{p^n})$ we are of course automatically assured of working in a cyclic subgroup. But we will see that the group $J^d(\boldsymbol{F}_{p^n})$ is in general cyclic, for any odd n prime to p.

Theorem 10. *For $d \neq 0$, the set of Frobenius eigenvalues for $C^d/\boldsymbol{F}_p$ is given modulo $((-1/p)p)^{pn} - 1$ by*

$$(n/p)(s/p)((-1/p)p)^{(p-s)n+1)/2}, \qquad s = \pm 2, \pm 4, \ldots, \pm(p-1).$$

The Frobenius value in Proposition 7 is attained for $ns \equiv 1 \pmod{p}$.

PROOF. By the lemma,

$$\sqrt{\bar{p}} \equiv (-2nd/p)((-1/p)p)^{(pn+1)/2} \pmod{h}.$$

The full set of eigenvalues is therefore

$$(-2an/p)((-1/p)p)^{(pn+1)/2+an}, \quad \text{for } a = 1, 2, \ldots, p-1.$$

After substitution of $a = -s/2$, they take the given form. The variable s runs over the nonzero residue classes modulo p. The given choice of representatives minimizes the absolute values. ∎

Corollary 11. *The group $J^d(\boldsymbol{F}_{p^n})$, for n odd prime to p, is cyclic.*

PROOF. The theorem shows that the eigenvalues are all different and therefore that two elements of $J^d(\boldsymbol{F}_{p^n})$ have a common eigenvalue if and only if they belong to the same cyclic component. By Proposition 7 elements of $J^d(\boldsymbol{F}_{p^n})$ have one common eigenvalue. ∎

Example 2. For $p = 3, 5$, we find for the Frobenius eigenvalues modulo $\bar{p}^{pn} - 1$,

$$p = 3 : (n/p) \times \ \{-(-3)^{(n+1)/2}, \ (-3)^{(5n+1)/2}\}.$$
$$p = 5 : (n/p) \times \ \{(5)^{(n+1)/2}, \ -(5)^{(3n+1)/2}, \ -(5)^{(7n+1)/2}, \ +(5)^{(9n+1)/2}\}.$$

5   Efficient computation in the Jacobian

In general, fast multiplication is obtained by using that the cyclic group used for the discrete logarithm has a well-defined Frobenius action σ. The cyclic group then has a common eigenvalue ρ, i.e. $\sigma P = \rho P$, for all elements P of the subgroup. The benefits for fast multiplication depend on the size of ρ.

The powers of the Frobenius map have eigenvalues $\bar{p}^p, \bar{p}^{2p}, \dots, \bar{p}^{pn} = 1$. By using other automorphisms of the curve, in particular the translation $\tau : (x, y) \mapsto (x + 1, y)$ of order p, we obtain a map with the much smaller eigenvalue $\bar{p}$.

Theorem 12. *Let (a, b) be a point on the curve $C^d/\overline{\mathbf{F}_p} : y^2 = x^p - x + d$, and let ∞ denote the point at infinity. We assume $b \neq 0$. Then*

$$p\left[(a, b) - \infty\right] = [(a', b') - \infty], \quad \text{for } a' = a^{p^2} + d^p + d, \ b' = (-1/p)b^{p^2}.$$

PROOF.   Note that the function $x - a'$ has divisor $(a', b') + (a', -b') - 2\infty$. It therefore suffices to show that the function $f = b^p y - (a^p - x + d)^{(p+1)/2}$ has divisor $p(a, b) + (a', -b') - (p + 1)\infty$. In solving $f = 0$, we may eliminate y and solve for x first.

$$
\begin{aligned}
0 &= (b^2)^p y^2 - (a^p - x + d)^{p+1} \\
&= (a^{p^2} - a^p + d^p)(x^p - x + d) - (a^{p^2} - x^p + d^p)(a^p - x + d) \\
&= (a^{p^2} - x^p + d^p + x^p - a^p)(a^p - x + d + x^p - a^p) \\
&\quad - (a^{p^2} - x^p + d^p)(a^p - x + d) \\
&= (x^p - a^p)(a^{p^2} + d^p + d - x).
\end{aligned}
$$

Substitution of $x = a$ in $f = 0$ yields $y = b$. Substitution of $x = a'$ in $f = 0$ yields

$$b^p y = (a^p - a^{p^2} - d^p)^{(p+1)/2} = (-b^2)^{p(p+1)/2},$$

and $y = -b'$.   ∎

This means that we can describe the multiple mP of a point P in base p,

$$mP = e_{pn}p^{pn-1}P + \cdots + e_1 pP + e_0 P,$$

and compute the multiple via fast evaluation of the right hand side. Multiplication of a point $P = [x - a, b]$ by $m = 0, 1, \dots, p - 1$ is trivial,

$$m[x - a, b] = [(x - a)^m, b], \quad \text{for } m = 1, 2, \dots, (p - 1)/2.$$

$$m[x - a, b] = [x - a', b'] + [(x - a)^{p-m}, -b], \quad \text{for } m = (p + 1)/2, \dots, p - 1.$$

and the evaluation reduces to $\log(pn)$ additions.

Thus, our proposed algorithm is described in Algorithm 2. Note that points of the Jacobian are represented by a pair of polynomials $(\alpha(x), \beta(x))$ as in [CA87], see Section 2.1.

Algorithm 2	Multiplication with base p for the cases
	(E) The group $E(\mathbf{F}_q)$ of an elliptic curve $E/\mathbf{F}_p$.
	(J) The Jacobian $J(\mathbf{F}_q)$ of the curve $C/\mathbf{F}_p : y^2 = x^p - x + d$.
Input:	(E) Point $P(a,b)$, integer m.
	(J) Reduced divisor $P(a,b) = \gcd(a(x), y - b(x))$, integer m.
Step1	Express m in base p.
	$m = e_0 + e_1 p + e_2 p^2 + \cdots + e_{n-1} p^{n-1}$.
Step2	Compute $P, 2P, \ldots, (p-1)P$.
	$P_0 := O, \quad P_1 := P$.
	For $i = 2, 3 \ldots, p-1$ do
	$\qquad P_i := P_{i-1} + P$.
Step3	Compute mP.
	$Q(\alpha, \beta) := P_{e_{n-1}}$.
	For $i = n-2, n-3, \ldots, 0$ do
	$\qquad$ (E) $Q := t\phi Q - \phi^2 Q$.
	$\qquad$ (J) $Q := \phi^2(\alpha(x - 2d), (-1/p)\beta(x))$.
	$\qquad Q := Q + P_{e_i}$.
Output:	Multiple $Q = mP$ of P by the integer m.

6   On embedding the Jacobian over F_q into $F_{q^k}^*$

The curve $C^d : y^2 = x^p - x + d$ is supersingular and there exists an index k, such that the subgroup $J(\mathbf{F}_q)$ of $\mathbf{F}_q$-rational points of the Jacobian embeds into the multiplicative group of the extension field $\mathbf{F}_{q^k}$. The embedding is realized with the Tate-pairing [FR94]. Precisely, we show the following.

Proposition 13. *The embedding index for the curve* $C^d : y^2 = x^p - x + d$ *with* $d \neq 0$ *is* p *for* $p \equiv 1 \pmod 4$ *and* $2p$ *for* $p \equiv 3 \pmod 4$.

PROOF. To embed the group $J(\mathbf{F}_q)$ via the Tate-pairing into $\mathbf{F}_{q^k}^*$ it is necessary that the group $J(\mathbf{F}_q)$ is cyclic and that its order divides the order $q^k - 1$ of $\mathbf{F}_{q^k}^*$. By Theorem 4 we have, for $q = p^n$ with n odd and not divisible by p,

$$h^+ h^- = P^+(\bar{p}^{(n-1)/2}) P^-(\bar{p}^{(n-1)/2}) = \Phi_p(\bar{p}\bar{p}^{n-1}) = \frac{\bar{p}^{pn} - 1}{\bar{p}^n - 1}.$$

Therefore $h^{\pm}$ divides the order of $\mathbf{F}_{p^{pn}}^*$, $\mathbf{F}_{p^{2pn}}^*$, respectively, for $p \equiv 1, 3 \pmod 4$, respectively. On the other hand, the subextensions $\mathbf{F}_{p^n}^*$ (resp. $\mathbf{F}_{p^{2n}}^*$) are not divisible by $h^{\pm}$, whose order of magnitude is $p^{p(n-1)/2}$. $\blacksquare$

7   Concluding remarks

We have obtained a class of hyperelliptic curves suitable for applications in public key cryptosystems. The class includes the supersingular curve $y^2 = x^3 - x + 1$

that was presented independently by Koblitz [Ko98]. For the particular curve, Koblitz gives an efficient elliptic curve DSA implementation that is twelve times faster than for a general elliptic curve. We are analysing the possible gain for DSA implementations with other members in the class.

Acknowledgments

The second author would like to thank Zvi Galil and Moti Yung for their hospitality while he was visiting the Columbia University Computer Science Department.

References

[AH96] L.M. ADLEMAN and M. HUANG, "Counting rational points on curves and abelian varieties over finite fileds," *Proc. of 2nd ANTS , LNCS*, vol. 1122, Springer-Verlag, (1996), 1-25.

[ADH94] L.M. ADLEMAN, J. DEMARRAIS and M. HUANG, "A Subexponential Algorithm for Discrete Logarithm over the Rational Subgroup of the Jacobians of Large Genus Hyperelliptic Curves over Finite Fields", *Proc. of ANTS1, LNCS*, vol. 877, Springer-Verlag, (1994), 28-40

[BeSc91] T. BETH and F. SCAEFER, "Non supersingular elliptic curves for public key cryptosystems", *Advances in Cryptology - EUROCRYPT '91, Lecture Notes in Computer Science*, **547**, pp.316-327 (1991).

[BRE93] Brent, R.P.: On computing factors of cyclotomic polynomials, Mathematics of Computation 61, No.203, (1993), 131-149.

[CA87] D.G. CANTOR, "Computing in the Jacobian of a Hyperelliptic Curve", *Math. Comp*, **48**, No.177 (1987), 95-101

[CMT97] J. CHAO, N. MATSUDA, and S. TSUJII, "Efficient construction of secure hyperelliptic discrete logarithm problems", *Proc. of ICICS'97*, LNCS 1334, Springer-Verlag, (1997), 292-301.

[CTT94] J. CHAO, K. TANAKA, and S. TSUJII, "Design of elliptic curves with controllable lower boundary of extension degree for reduction attacks", *Advances in Cryptology - Crypto'94*, Springer-Verlag, (1994), 50-55.

[DH76] W. DIFFIE and M. HELLMAN, "New directions in cryptography," *IEEE Trans. on IT* **22**, (1976), 644-654.

[DUU93] Iwan Duursma, "Class numbers for some hyperelliptic curves". Arithmetic, geometry and coding theory (Luminy, 1993), 545–52, de Gruyter, Berlin, 1996.

[ELG85] T.. ELGAMAL "A public key cryptosystem and a signature scheme based on discrete logarithms", *IEEE Trans. on IT* **31**, (1985), 469-472.

[FRE97] G. FREY, "Aspects of DL-systems based on hyperelliptic curves", *Keynote Lecture in Waterloo-Workshop on Elliptic Curve Discrete Logarithm Problem*, 4th of Nov. (1997).

[FR94] G. FREY and H.G. RÜCK, "A Remark Concerning m-Divisibility and the Discrete Logarithm in the Divisor Class Group of Curves", *Math. Comp*, **62**, No.206 (1994), 865-874

[GLV98] R. GALLANT, R. LAMBERT and S. VANSTONE, "Improving the parallelized Pollard lambda search on binary anomalous curves", http://grouper.ieee.org/groups/1363/contrib.html, (1998)

[Has34] Hasse, H.: Theorie der relativ-zyklischen algebraischen Funktionenkörper, insbesondere bei endlichem Konstantenkörper. Crelle 172 (1934), 37-54.

[HD34] Hasse, H., and Davenport, H.: Die Nullstellen der Kongruenzzetafunktionen in gewissen zyklischen Fällen. Crelle 172 (1934), 151-182.

[Ko87] N. KOBLITZ, "Elliptic curve cryptosystems", *Mathematics of Computation*, **48** (1987), 203–209.

[Ko88] N. KOBLITZ, "A Family of Jacobians Suitable for Discrete Log Cryptosystems", *Advances in Cryptology - Crypto'88*, Springer-Verlag, (1990), 94-99

[Ko89] N. KOBLITZ, "Hyperelliptic Cryptosystems", *J.Cryptology*, **1** (1989), 139-150

[Ko91] N. KOBLITZ, "CM-curves with good cryptographic properties", *Advances in Cryptology - Crypto'91*, Springer-Verlag, (1992), 279-287.

[Ko97] N. KOBLITZ, "A Very Easy Way to Generate Curves over Prime Fields for Hyperelliptic Cryptosystems", *Crypto'97* Rump Talk (1997): The journal version is in J. Buhler and N. Koblitz, "Lattice basis reduction, Jacobi sums and hyperelliptic cryptosystems,' Bull. Austral. Math. Soc. (1998).

[Ko98] N. KOBLITZ, "An elliptic curve implementation of the finite fiel digital siganture algorithm," To appear in Proc. Crypto'98; a draft available from the author (1998).

[Luc1878] Lucas, E.: Sur la série récurrente de Fermat, Bul. Bibl. Storia Sc. Mat. e Fis. 11 (1878), 783-789.

[Lov92] R. LOVORN, "Rigorous, subexponential algorithms for discrete logarithms over finite fields", *Ph.D Thesis*, Univ. of Georgia, (1992).

[Men93] A.J. MENEZES, "Elliptic curve public key cryptosystems", *Kluwer Academic Publishers*, (1993).

[Mil85] V. MILLER, "Uses of elliptic curves in cryptography", *Lecture Notes in Computer Science*, **218** (1986), 417–426. (Advances in Cryptology – CRYPTO '85.)

[Miy92] A. MIYAJI, "Elliptic curve over F_p suitable for cryptosystems", *Advances in Cryptology - Asiacrypt'92*, Springer-Verlag, (1993), 479-491.

[Miy93] A. MIYAJI, "Elliptic curve cryptosystems immune to any reduction into the discrete logarithm problem", *IEICE Trans.*, Fundamentals, **E76-A** (1993), pp. 50–54.

[MOV93] A.J. MENEZES, T. OKAMOTO and S.A. VANSTONE, "Reducing elliptic curve logarithm to logarithm in a finite field", *IEEE Trans. on IT*, **39**, (1993), 1639-1646

[MS92] W. Meier and O. Staffelbach, "Efficient multiplication on certain nonsupersingular elliptic curves," *Advances in Cryptology - Crypto'92*, Springer-Verlag, (1993), 333-344.

[NIST93] NIST, "Digital signature standard", *FIPS Publication*, 186, 1993.

[Od85] A. ODLYZKO, "Discrete logarithm and their cryptographic significance", *Advances in Cryptology - Eurocrypto'84*, Springer-Verlag, (1985), 224-314

[Pil90] J. PILA, "Frobenius maps of abelian varieties and finding roots of unity in finite fields", *Math. Comp*, **55**, No.206 (1990), 745-763.

[PH78] S.C. POHLIG and M.E. HELLMAN, "An improved algorithm for computing logarithms over $GF(p)$ and its cryptographic significance", *IEEE Trans. on IT*, **24**, (1978), 106-110

[RSA78] R. RIVEST, A. SHAMIR, and L. ADLEMAN, "A method for obtaining digital signatures and public-key cryptosystems", *Comm. ACM.* 21, (1978), pp.120-126.

[Ru97] H.G. RÜCK, "On the discrete logarithms in the divosor class group of curves", To appear in *Math. Comp.* (1997)

[SA97] T. SATOH and K. ARAKI, "Fermat Quotients and the Polynomial Time Discrete Log Algorithm for Anomalous Elliptic Curves", *preprint*, (1997)

[SCH85] R. SCHOOF, "Elliptic curves over finite fields and the computation of square root mod p", *Math. Comp*, **44**, (1985), 483-494.

[SEM95] I.A. SEMAEV, "Evaluation of discrete logarithms on some elliptic curves", to appear in *Math. Comp.*, (1995).

[SMA97] N.P. SMART, "The Discrete Logarithm Problem on Elliptic Curves of Trace One", *preprint*, (1997)

[SCH62] A. Schinzel, "On primitive factors of $a^n - b^n$," Proc. Cambridge Philos. Soc. 58 (1962), 555-562.

[SOL97] J.A.Solians, "An improved algorithm for arithmetic on a family of elliptic curves," Proc. CRYPTO'97.

[STE87] P. Stevenhagen,: "On Aurifeuillian factorizations," Nederl. Akad. Wetensch. Indag. Math. 49 (1987), 451-468.

[VOL97] J.F. VOLOCH, "The discrete logarithm problem on elliptic curves and descents", *preprint*, (1997)

[WEI48] A. Weil, "Sur les courbes algébriques et les variétés qui s'en déduisent," Hermann, Paris, 1948.

[WZ98] M. WIENER AND R. ZUCCHERATO, "Faster Attacks on Elliptic Curve Cryptosystems," `http://grouper.ieee.org/groups/1363/contrib.html`, (1998)

On Weierstrass Semigroups and One-point Algebraic Geometry Codes

J.I. Farrán[*]

Departamento de Matemática Aplicada a la Ingeniería
E.T.S. Ingenieros Industriales – Universidad de Valladolid
Paseo del Cauce s/n - 47011 Valladolid – SPAIN
e-mail: ignfar@eis.uva.es

Abstract. We present two different algorithms to compute the Weierstrass semigroup at a point P together with functions for each value in this semigroup from a plane model of the curve. The first one works in a quite general situation and it is founded on the Brill-Noether algorithm. The second method works in the case of P being the only point at infinity of the plane model, what is very usual in practice, and it is based on the Abhyankar-Moh theorem, the theory of approximate roots and an integral basis for the affine algebra of the curve. This last way is simpler and has an additional advantage: one can easily compute the Feng-Rao distances for the corresponding array of one-point algebraic geometry codes, this thing be done by means of the Apéry set of the Weierstrass semigroup. Everything can be applied to the problem of decoding such codes by using the majority scheme of Feng and Rao.

1 Introduction

Feng and Rao introduced in [8] a majority scheme for the so called *one-point algebraic geometry codes*, what gives nowadays the most efficient decoding algorithm for algebraic geometry codes. Moreover, this procedure corrects up to half the so called *Feng-Rao designed minimum distance*, that is a lower bound for the minimum distance of these codes which is better that the *Goppa designed minimum distance*.

However, this idea is not new, since Goppa himself suggested in [10] that one can use the Weierstrass semigroup at the point P in order to obtain a lower bound of the Goppa distance for the one-point code $C_\Omega(D, mP)$. This was explicitly stated for example in [9]. Since, by definition, the Feng-Rao distance is closely related to the Weierstrass semigroup, one can look for lower bounds for such distance by using properties of this semigroup. In addition, and from "decoding reasons", the knowledge of the corresponding rational functions associated to the semigroup is also desirable.

More precisely, let $\tilde{\chi}$ be a non-singular projective algebraic curve defined over a finite field $\mathbb{F}$ such that $\tilde{\chi}$ is irreducible over $\overline{\mathbb{F}}$. In order to define the so called

[*] Partially supported by DIGICYT PB94-1111-C02-01

algebraic geometry codes (AG codes in short), take $\mathbb{F}$-rational points $P_1, \dots, P_n$ on the curve and a $\mathbb{F}$-rational divisor G having disjoint support with the divisor $D \doteq P_1 + \dots + P_n$, and then consider the linear well-defined maps

$$ev_D \;:\; \mathcal{L}(G) \longrightarrow \mathbb{F}^n$$
$$f \mapsto (f(P_1), \dots, f(P_n))$$

$$res_D \;:\; \Omega(G - D) \longrightarrow \mathbb{F}^n$$
$$\omega \mapsto (res_{P_1}(\omega), \dots, res_{P_n}(\omega))$$

where the $\mathbb{F}$-vector spaces of finite dimension

$$\mathcal{L}(H) \doteq \{ f \in \mathbb{F}(\tilde{\chi}) \mid (f) + H \geq 0 \} \cup \{0\}$$

$$\Omega(H) \doteq \{ \omega \in \Omega(\tilde{\chi}) \mid (\omega) \geq H \} \cup \{0\}$$

are defined for any $\mathbb{F}$-rational divisor H on $\tilde{\chi}$. Thus one defines the linear codes

$$C_L \equiv C_L(D, G) \doteq Im(ev_D)$$

$$C_\Omega \equiv C_\Omega(D, G) \doteq Im(res_D)$$

The above constructed codes have asymptotically excellent parameters, namely they are the only known family of codes whose parameters are asymptotically better than the Varshamov-Gilbert bound, provided that q is a square and $q \geq 49$ (see [19]), and this is the main reason to study such codes. The length of both codes is obviously n, and one has $(C_\Omega) = C_L^\perp$ by the *residues theorem*. On the other hand, given D and G as above there exists a differential form ω such that $C_L(D, G) = C_\Omega(D, D - G + (\omega))$ and thus it suffices to deal with the codes of type C_Ω.

Thus, denote by $k = k(C)$ and $d = d(C)$ respectively the dimension over $\mathbb{F}$ and the minimum distance of the linear code $C = C_\Omega(D, G)$, $d(C)$ being the minimum value of non-zero entries of a non-zero vector of C. If $2g - 2 < deg\,G < n$, the Riemann-Roch formula gives the estimates

$$\begin{cases} k = n - deg\,G + g - 1 \\ d \geq deg\,G + 2 - 2g \doteq d^* \end{cases}$$

where d^* is called Goppa distance of C (see [20] for further details).

The above estimates only depend on the degree of G, assumed that $\tilde{\chi}$ and n are fixed. This leads us to consider the special case $G = mP$, P being a $\mathbb{F}$-rational point of the curve which is not in the support of the divisor D. In this case, the so called one-point AG codes $C_m \doteq C_\Omega(D, mP)$ can be decoded by the majority scheme of Feng and Rao, which yields so far the most efficient decoding algorithm for this kind of codes (see [8]).

In order to implement this decoding method, one has to fix for every non-negative integer i a function f_i in $\mathbb{F}(\tilde{\chi})$ with an only pole at P of order i for those values of i for which it is possible, i.e. for i in the Weierstrass semigroup $\Gamma = \Gamma_P$

of $\tilde{\chi}$ at P. For a received word $\mathbf{y} = \mathbf{c} + \mathbf{e}$, where $\mathbf{c} \in C_m$, one can consider the unidimensional and bidimensional syndromes respectively given by

$$s_i(\mathbf{y}) \doteq \sum_{k=1}^{n} e_k \, f_i(P_k)$$

$$s_{i,j}(\mathbf{y}) \doteq \sum_{k=1}^{n} e_k \, f_i(P_k) \, f_j(P_k)$$

Notice that the set $\{f_i \mid i \le m, \ i \in \Gamma\}$ is actually a basis for $\mathcal{L}(mP)$ and hence

$$C_m = \{\mathbf{y} \in \mathbb{F}^n \mid s_i(\mathbf{y}) = 0 \ \text{for} \ i \le m\}$$

Thus we can calculate $s_i(\mathbf{y})$ for $i \le m$ from the received word $\mathbf{y}$ as $s_i(\mathbf{y}) = \sum_{k=1}^{n} y_k \, f_i(P_k)$, and such syndromes are called *known*.

In fact, it is well-known that if one had a high enough number of syndromes $s_{i,j}(\mathbf{y})$ for $i + j > m$ we could know the emitted word $\mathbf{c}$, and all the above syndromes can be computed by a majority voting (see [8] and [17]). The complexity of this algorithm is lower than the usual algorithms for general AG codes, and moreover the Weierstrass semigroup at P gives an estimate for $d(C)$ which is better than the Goppa bound. Such bound is the so called Feng-Rao distance, defined by

$$\delta_{FR}(m) \doteq \min\{n_s \mid s \in \Gamma \ \text{and} \ s \ge m\}$$

where $n_s \doteq \sharp\{(i,j) \in \Gamma \times \Gamma \mid i + j = s\}$ for every $s \in \Gamma$.

Apart from finding all the closed singular points and all the $\mathbb{F}$-rational points of χ, what can be done by means of Gröbner bases computation, and also apart from computing the order of a rational function at a rational point and evaluating such function at this point if possible, what can be done for instance from the resolution tree of a plane model of the curve at such point by successive blowing-ups (see [12]), the main problem in practice is the computation of Γ and the functions f_i achieving the values of the semigroup Γ in order to carry out the Feng-Rao procedure.

Our aim is just to solve this last problem from the knowledge of a (possibly singular) plane model for the smooth curve by using geometric techniques, that is, from a geometric point of view. More precisely, in section **2** we study a method based on the Brill-Noether algorithm, which works in a quite general situation, and in section **3** we study an alternative method for the case of P being the only point at infinity of the plane model, which is founded on the Abhyankar-Moh theorem. This second method is not so general but it is more simple, and has the advantage of computing the Feng-Rao distance in an effective way. We conclude the paper giving in section **4** two examples where we compare both methods.

2   Weierstrass semigroups and adjoints

For a given plane curve χ, the computation of a basis for $\mathcal{L}(G)$, G being a rational divisor over $\tilde{\chi}$, is reduced, by the Brill-Noether theorem, to compute bases for spaces of adjoints of a suitable degree n. In fact, Goppa himself already mentioned in [10] the Brill-Noether theory as a way to construct AG codes in general. This theory can be done effective from the desingularization of χ, lazy parametrizations of the rational branches at all the singular points of the plane curve and testing virtual passing conditions (see [6], or [11] for an alternative method).

Now, in order to compute the Weierstrass semigroup Γ_P at P and a rational function f_l with a unique pole at P of order l for any fixed $l \in \Gamma_P$, we need not actually carry out the whole mentioned algorithm until we get a basis of $\mathcal{L}(mP)$ for a suitable m, but we can determine such semigroup and those functions by using only a part of the steps of the algorithm, as you can see in [7]. However, since the explicit description of all the steps of this algorithm and their effective solution would take long, we will give a procedure which assumes such a basis has been previously computed.

First of all, we need a bound $\tilde{l}$ for the values in Γ_P which will be used in the Feng and Rao procedure (see [8]). Then, assume that a basis $\{h_1, \ldots, h_s\}$ of $\mathcal{L}(\tilde{l}P)$ over $\mathbb{F}$ has been already computed, and that $\tilde{l}$ is not a gap. We give a triangulation method which works by induction on the dimension s as follows:

(1) Compute the pole orders $\{-v_P(h_i)\}$ at P, and assume that the functions $\{h_i\}$ are ordered so that these pole orders are increasing in i.

(2) At least the function h_s satisfies $-v_P(h_s) = \tilde{l}$ and we set $f_{\tilde{l}} \doteq h_s$. If any other h_j satisfies the same condition, there exists a non-zero constant λ_j in $\mathbb{F}$ such that $-v_P(h_j - \lambda_j h_s) < \tilde{l}$; then we change such functions h_j by $g_j \doteq h_j - \lambda_j h_s$ and set $g_k \doteq h_k$ for all the others. The result now is obviously another basis $\{g_1, \ldots, g_s\}$ of $\mathcal{L}(\tilde{l}P)$ over $\mathbb{F}$ but with only one function $g_s = f_{\tilde{l}}$ whose pole at P has maximum order $\tilde{l}$.

(3) Since the functions g_i are linear independent over $\mathbb{F}$ and $-v_P(g_i) < \tilde{l}$ for $i < s$, one has obtained a basis $\{g_1, \ldots, g_{s-1}\}$ of $\mathcal{L}(l'P)$ over $\mathbb{F}$, where l' denotes the largest non-gap such that $l' < \tilde{l}$. But now the dimension is $s - 1$ and we can continue by induction.

The result of the above procedure is a function f_l for each non-gap $l \leq \tilde{l}$, and in fact it can be used to compute the Weierstrass semigroup up to an integer $\tilde{l}$, since the maximum gap l' such that $l' \leq \tilde{l}$ is just $\max\{-v_P(h_1), \ldots, -v_P(h_s)\}$, in the above notations, and so on by induction.

The limitation of this general method is just the computation of the Feng-Rao distance, what is in general a complex problem of arithmetic semigroups. If we compute an arbitrary generator system (for example, the set of all the primitive elements of Γ_P, which is always contained in the set of the fist $g + 2$ non-gaps, as you can see in [16]), the problem is the effective description of the elements of Γ_P in terms of such generators, which is not even unique in general. As we will see later, this problem becomes easier by considering special generators, but then we do not have in general a reasonable bound for the largest element in such system, unless the semigroup is a special one. This leads us to consider an alternative way to compute Weierstrass semigroups, when P is the only branch at infinity of a plane model, and where the Abhyankar-Moh theorem together with the theory of Apéry systems allows us to compute easily the Feng-Rao distance.

3   Weierstrass semigroups and approximate roots

Let $\tilde{\chi}$ be again a non-singular projective algebraic curve defined over a finite field $\mathbb{F}$ and which is absolutely irreducible. Let χ be now a plane model for $\tilde{\chi}$, and assume that the hypothesis

(H1) χ has a unique branch at infinity

is satisfied, i.e. there exist a birational morphism

$$n \; : \; \tilde{\chi} \to \chi \subseteq \mathbb{P}^2$$

and a line $L \subset \mathbb{P}^2$ defined over $\mathbb{F}$ such that $L \cap \chi$ consists of only one point P and $\tilde{\chi}$ has only one branch at P. Notice that both P and the branch at P are defined over the underlying finite field $\mathbb{F}$, since χ does. Thus there is only one point of $\tilde{\chi}$ over P, which will be denoted by $\overline{P}$.

Set $\tilde{\varUpsilon} = \tilde{\chi} \setminus \{\overline{P}\}$ and $\varUpsilon = \chi \setminus \{P\}$. One has the two following additive subsemigroups of $\mathbb{N}$:

$$\varGamma_P \doteq \{-v_{\overline{P}}(f) \mid f \in \mathcal{O}_{\tilde{\chi}}(\tilde{\varUpsilon})\}$$
$$S_P \doteq \{-v_{\overline{P}}(f) \mid f \in \mathcal{O}_{\chi}(\varUpsilon)\}$$

Notice that $\varGamma_P$ is just the Weierstrass semigroup of $\tilde{\chi}$ at $\overline{P}$ and it contains S_P, but they are different unless the curve χ is non-singular in the affine part. Moreover, $\mathbb{N} \setminus \varGamma_P$ has g elements, g being the genus of $\tilde{\chi}$, and $\varGamma_P \setminus S_P$, which is also finite, will be computed below.

The first question to solve is the description of the semigroup S_P. In order to do that, we state the Abhyankar-Moh theorem, where the hypothesis

(H2) *char* $\mathbb{F}$ does not divide either *deg* χ or $e_P(\chi)$

is assumed. This result provides us with a set of generators for S_P with nice arithmetic properties (see [1] or [15]).

Theorem 3.1 (Abhyankar-Moh). *Assumed that (H1) and (H2) are satisfied by χ, then there exist an integer h and a sequence of integers $\delta_0, \ldots, \delta_h \in S_P$ which generate S_P such that:*

(I) $d_{h+1} = 1$ *and* $n_i > 1$ *for* $2 \leq i \leq h$, *where* $d_i \doteq gcd\,(\delta_0, \ldots, \delta_{i-1})$ *for* $1 \leq i \leq h+1$ *and* $n_i \doteq d_i/d_{i+1}$ *for* $1 \leq i \leq h$.
(II) $n_i \delta_i$ *is in the semigroup generated by* $\delta_0, \ldots, \delta_{i-1}$ *for* $1 \leq i \leq h$.
(III) $n_i \delta_i > \delta_{i+1}$ *for* $1 \leq i \leq h-1$.

Such semigroups are a particular case of telescopic semigroups, and their main arithmetic property is that every $n \in S_P$ can be easily written in an unique way in the form

$$n = \sum_{i=0}^{h} \lambda_i \delta_i \qquad [\star]$$

with $\lambda_0 \geq 0$ and $0 \leq \lambda_i < n_i$ for $1 \leq i \leq h$ (see [14] or [15]). Apéry in [4] and Angermüller in [3] worked with a slightly different semigroup, that is the semigroup of values of a branch. The type of semigroup in this case is very similar to the given by the Abhyankar-Moh theorem, but with the property

(III)* $n_i \delta_i < \delta_{i+1}$ for $1 \leq i \leq h - 1$

instead of **(III)**.

Now we will say how to obtain these generators of S_P in a constructive way together with functions in $B \doteq \mathcal{O}_\chi(\Upsilon)$ having poles of order equal to those generators (and hence one will have functions in B with poles of order any element in S_P by using the arithmetic properties of such generators). For it, we need first the concept of approximate root.

Definition 3.2.   *Let S be a ring, $G \in S[Y]$ a monic polynomial of degree e and $F \in S[Y]$ a monic polynomial of degree n with $e | n$. Then G will be called an approximate b-th root of F if $\deg (F - G^b) < n - e = e(b - 1)$.*

Now the main remark is that for every monic polynomial $F \in S[Y]$ of degree n and for every b divisor of n which is a unit in S, there exists a unique approximate b-th root of F, and it can be computed very efficiently (see [7]).

Thus, let the affine plane model of the curve given by the equation

$$F = F(X, Y) = Y^m + a_1(X) Y^{m-1} + \ldots + a_m(X)$$

and suppose that *char* $\mathbb{F}$ satisfies the assumption of the Abhyankar-Moh theorem. Up to a change of variables in the form $X' = X + Y^n$, $Y' = Y$, we can actually assume that *char* $\mathbb{F}$ does not divide the total degree m of χ. On the other hand, denote the approximate d-th root of F with respect to the coefficient ring $S = \mathbb{F}[X]$ by $app(d, F)$. Thus, the so called *algorithm of approximate roots* computes the generators given by the Abhyankar-Moh theorem as follows:

$$F_0 = X, \, \delta_0 = d_1 = m, \, F_1 = Y, \, \delta_1 = \deg_X Res_Y(F, F_1)$$

$$n > 1 \Rightarrow \begin{bmatrix} d_n = g\,c\,d\,(\delta_0, \delta_1, \ldots, \delta_{n-1}) \\ F_n = app(d_n, F) \\ \delta_n = \deg_X Res_Y(F, F_n) \end{bmatrix}$$

The procedure stops at the first $h \geq 1$ with $d_{h+1} = d_{h+2}$, what happens just when $d_{h+1} = 1$, since the point at infinity is unibranch (see [2] and [7]).

As a consequence, the generators of S_P given by the Abhyankar-Moh theorem and the corresponding functions can be easily computed in terms of approximate roots of F and resultants of polynomials. In particular, we can compute a rational function with an only pole at P of order n for every $n \in S_P$. In fact, if $n = \sum_{i=0}^{h} \lambda_i \delta_i$ with $\lambda_0 \geq 0$ and $0 \leq \lambda_i < n_i$ for $1 \leq i \leq h$, then $f_n = \prod_{i=0}^{h} F_i^{\lambda_i}$ is the searched function, where F_i are the polynomials which are obtained in the algorithm of approximate roots. In particular, this also allows us to compute a basis of the space $\mathcal{L}(lP)$ for every $l \in \Gamma_P$.

Now the remaining part of the method is the computation of $\Gamma_P \backslash S_P$ with the corresponding functions, what can be done effective by means of the following

Lemma 3.3. *Let A and B be the respective affine coordinate $\mathbb{F}$-algebras of $\tilde{\Upsilon}$ and Υ, i.e. $A = \mathcal{O}_{\tilde{\chi}}(\tilde{\Upsilon})$ and $B = \mathcal{O}_{\chi}(\Upsilon)$; then one has:*

$$\sharp(\Gamma_P \setminus S_P) = dim_{\mathbb{F}}(A/B)$$

Proof:

Take a basis $\{h_1, \ldots, h_l\}$ of A/B over $\mathbb{F}$, which can be calculated either in algebraic terms with the aid of the *integral basis algorithm* (see [13] or [18]) or in geometric terms from the desingularization of the affine part of the curve χ. Now we will show a *triangulation procedure* to find the values in $\Gamma_P \setminus S_P$ as well as functions which provide these values.

Set $B^i \doteq B + \mathbb{F}h_1 + \ldots + \mathbb{F}h_i$, for $0 \leq i \leq l$; we will proceed by induction, so let $0 \leq i < l$ and suppose we have found functions $g_1, \ldots, g_i$ which are linearly independent over $\mathbb{F}$ with

$$\Gamma_P^i \doteq S_P \cup \{-\upsilon_{\overline{P}}(g_1), \ldots, -\upsilon_{\overline{P}}(g_i)\} \subseteq \Gamma_P$$
$$-\upsilon_{\overline{P}}(g_j) \notin \Gamma_P^{i-1}$$
$$B + \mathbb{F}g_1 + \ldots + \mathbb{F}g_i = B^i$$

Now look at h_{i+1}; if $-\upsilon_{\overline{P}}(h_{i+1}) \notin \Gamma_P^i$, then set $g_{i+1} = h_{i+1}$ and go on. Otherwise, there exists $f \in B^i$ with

$$\upsilon_{\overline{P}}(h_{i+1}) = \upsilon_{\overline{P}}(f)$$
$$-\upsilon_{\overline{P}}(h_{i+1} - f) < -\upsilon_{\overline{P}}(h_{i+1})$$

Thus we can repeat the process with $h_{i+1} - f$ replacing to h_{i+1}; since $h_{i+1} \notin B^i$, one obtains in a finite number of steps a function g_{i+1} such that

$$g_{i+1} \equiv h_{i+1} \pmod{B^i} \quad \text{and} \quad -\upsilon_{\overline{P}}(g_{i+1}) \notin \Gamma_P^i$$

At the end of the procedure l different elements in $\Gamma_P \setminus S_P$ will be added, and then $\sharp(\Gamma_P \setminus S_P) \geq dim_{\mathbb{F}}(A/B)$. The equality follows immediately from the formula $A = B^l = B + \mathbb{F}g_1 + \ldots + \mathbb{F}g_l$.

$\square$

In order to complete this section, we show how to calculate the Feng-Rao distance from the above computations. First we have to present some basic tools for arbitrary semigroups.

Definition 3.4. *Let* $S \subseteq \mathbb{N}$ *a semigroup with* $\sharp(\mathbb{N} \setminus S) < \infty$ *and* $0 \in S$; *for* $m \in S$ *define*

$$\delta_{FR}(m) \doteq min\{N_s \mid s \geq m, \ s \in S\}$$

where $N_s \doteq \sharp\{(a, b) \in S^2 \mid a + b = s\}$ *for every* $s \in S$.

Definition 3.5. *Let* $S \subseteq \mathbb{N}$ *a semigroup with the same hypothesis as in the previous definition; for* $n \in S \setminus \{0\}$ *define the Apéry set of* S *related to* n *as the set whose elements are the numbers*

$$a_i \doteq min\{m \in S \mid m \equiv i \ (mod \ n)\}$$

for $0 \leq i \leq n - 1$ [1].

In the sequel, the index i will be considered as an element in $\mathbb{Z}/(n)$. Thus, one has a disjoint union

$$S = \bigcup_{i=0}^{n-1} (a_i + n\mathbb{N})$$

and therefore the set $\{a_1, \dots, a_{n-1}, n\}$ is a generator system for the semigroup S, which is called the Apéry (generator) system of S related to n.

Moreover, if $i, j \in \mathbb{Z}/(n) \equiv \mathbb{Z}_n$ then $a_i + a_j = a_{i+j} + \alpha_{i,j} n$ with $\alpha_{i,j} \geq 0$, by definition of the Apéry set. With this notation, every $m \in S$ can be written in a unique way as $m = a_i + ln$, with $i \in \mathbb{Z}_n$ and $l \geq 0$; so we can associate to m two *coordinates* $(i, l) \in \mathbb{Z}_n \times \mathbb{N}$.

Apéry *relations* are very useful to compute N_m . In fact, for $0 \leq i \leq n - 1$ and $h \geq 0$ one can define $B_i^{(h)} \doteq \sharp\{\alpha_{k,i-k} \leq h \mid k \in \mathbb{Z}_n\}$ and then one has the following result.

Proposition 3.6. $N_m = B_i^{(0)} + B_i^{(1)} + \dots + B_i^{(l)}$

Proof:

If $\alpha_{k,i-k} = h \leq l$, it has been considered $l - h + 1$ times in the sets defining $B_i^{(h)}, B_i^{(h+1)}, \dots, B_i^{(l)}$, but also the equality $l_1 + l_2 = l - \alpha_{k,i-k}$ holds for $l - h + 1$ possible pairs l_1, l_2 . $\square$

Thus, N_m is *increasing in* l, and it suffices to calculate the *minimum in* i in order to obtain the corresponding Feng-Rao distance, according to the following result.

Theorem 3.7. *With the above notations, for each* $j \in \mathbb{Z}_n$ *take* $m_j = a_j + t_j n$, *where* t_j *is the minimum integer such that* $t_j \geq max \left(\dfrac{a_i - a_j}{n} + l, 0 \right)$. *Then one has*

$$\delta_{FR}(m) = min\{N_{m_j} \mid j \in \mathbb{Z}_n\}$$

---

[1] We could actually remove $a_0 = 0$ since it does not add any information about S.

Proof:

The formula is quit clear if one realizes that m_j is the minimum element of S with first Apéry coordinate equal to j such that $m_j \geq m$, using the above remark on the number N_m.

$\square$

Thus, computing the Feng-Rao distance is easy if we have the Apéry set related to an element of an arbitrary semigroup. This gives an effective algorithm to compute the Feng-Rao distance of a Weierstrass semigroup when computed by the method given in this section, because of the two following remarks:

(i) The Apéry set of the semigroup S_P related to $m = \delta_0 = deg\,\chi$ is just the set of all the elements of the form $\displaystyle\sum_{k=1}^{h} \lambda_k \delta_k$ with $0 \leq \lambda_k < n_k = d_k/d_{k+1}$ for $1 \leq k \leq h$, since using the property $[\star]$ one has that all these elements are different modulo m, minimum in S with this condition and the number of such elements is exactly m.

(ii) Now the Apéry set related to $m = deg\,\chi$ for the Weierstrass semigroup $\Gamma_P = S_P + b_1\mathbb{N} + \ldots + b_l\mathbb{N}$ where $l = dim_{\mathbb{F}}(A/B)$ and b_i being computed as in lemma 3.3, and the corresponding Apéry relations, can be easily obtained from those of S_P in at most l steps, each of them involving only the elements $a_j + \lambda\,b_k$ with $0 \leq j \leq n-1$ and $0 \leq \lambda \leq n-1$.

4   Examples and conclusions

The choice of the method to use in order to compute a Weierstrass semigroup depends on the situation. More precisely, the Brill-Noether method works in a general situation, but the implementation is complicate and it does not give a nice description of the semigroup (namely, an Apéry system in order to calculate the Feng-Rao distance of certain one-point AG code). On the other hand, the Abhyankar-Moh method gives such a description of Γ_P and the algorithm works in a very simple way, but it requires some additional hypothesis on the plane model: it must have an only rational branch P at infinity which is defined over the base field $\mathbb{F}$ and the characteristic of $\mathbb{F}$ must not divide at the same time to the degree of the plane model and the multiplicity of P, what is not always fulfilled. If moreover the plane model has no other singular points at the affine part the curve, the algorithm of approximate roots directly yields the Weierstrass semigroup, and then the algorithm can be very easily implemented (for instance, such a programme takes a few lines in AXIOM code). Anyway, the complement of this semigroup requires the previous computation of a certain integral basis, what is equivalent to the desingularization of the affine part of the plane model, but what follows from such basis by means of a simple triangulation procedure. We will briefly illustrate these ideas with two examples.

Example 4.1. *Consider the affine plane curve* $F(X, Y) = Y^9 + Y^8 + XY^6 + X^2Y^3 + Y^2 + X^3$ *defined over* $\mathbb{F}_2$*, with only one branch at infinity* $P = (1 : 0 : 0)$*. The algorithm of approximate roots yields*

$$F_0 = X, \; \delta_0 = d_1 = 9, \; F_1 = Y$$

$$\delta_1 = deg_X Res_Y(F, Y) = 3, \; d_2 = gcd(9, 3) = 3$$

$$F_2 = app(3, F) = Y^3 + Y^2 + Y + X + 1$$

$$\delta_2 = deg_X Res_Y(F, F_2) = 8, \; d_3 = gcd(9, 3, 8) = 1$$

thus $h = 2$ *and* $S_P = \langle 9, 3, 8 \rangle$*.*

On the other hand, according to the lemma 3.3, take a $\mathbb{F}_2$*-basis for* A/B

$$h_1 = \frac{Y(1 + Y^6)}{X + Y^3} \qquad h_2 = \frac{Y(1 + Y^6)}{(X + Y^3)(Y^2 + Y + 1)}$$

$$h_3 = \frac{X^2 + Y^6}{Y^2 + Y + 1} \qquad h_4 = \frac{Y^2(1 + Y^3)(Y^2 + Y + 1)}{X + Y^3} \cdot$$

The values at P *of this functions are* $-v_P(h_1) = 13 \notin S_P$*,* $-v_P(h_2) = 7 \notin \Gamma_P^1$*,* $-v_P(h_3) = 10 \notin \Gamma_P^2$ *and* $-v_P(h_4) = 13 \in \Gamma_P^3$*. Then change* h_4 *by*

$$g_4 = h_4 + h_1 = \frac{Y(1 + Y^3)(Y^2 + Y + 1)}{X + Y^3}$$

and now $-v_P(g_4) = 10 \in \Gamma_P^3$*, so still one has to take the function*

$$g_4 = h_4 + h_1 + h_3 = \frac{Y(1 + Y^3)(Y^4 + Y^2 + 1) + (X + Y^3)^3}{(X + Y^3)(Y^2 + Y + 1)}$$

and now $-v_P(g_4) = 4 \notin \Gamma_P^3$*. Hence, the Weierstrass semigroup at* P *is*

$$\Gamma_P = \{0, 3, \mathbf{4}, 6, \mathbf{7}, 8, 9, \mathbf{10}, 11, 12, \mathbf{13}, 14, \dots\}$$

Unfortunately, there are examples where this method cannot be applied, and then the Brill-Noether method helps to compute Γ_P and the functions, even though it cannot compute in general the Feng-Rao distance.

Example 4.2. *Let* χ *be the Klein quartic over* $\mathbb{F}_2$ *given by the equation*

$$F(X, Y, Z) = X^3Y + Y^3Z + Z^3X = 0$$

whose adjunction divisor is $\mathcal{A} = 0$*, since* χ *is non-singular. We are going to compute now the Weierstrass semigroup at* $P = (0 : 0 : 1)$*, which is not the only one at infinity. Thus, by means of the Brill-Noether algorithm we compute a* $\mathbb{F}_2$*-basis of* $\mathcal{L}(7P)$

$$\{h_1 = 1, h_2 = \frac{Z}{Y}, h_3 = \frac{Z(Y^2 + YZ + Z^2)}{X^2Y}, h_4 = \frac{Z^2(Y + Z)}{X^2Y}, h_5 = \frac{Z^3}{X^2Y}\}$$

By using Hamburger-Noether expansions at P, one computes the pole order of these functions at such point

$$-v_P(h_1) = 0, -v_P(h_2) = 3, -v_P(h_3) = -v_P(h_4) = -v_P(h_5) = 7$$

Thus, we take $f_7 = h_5$ and replace $h_4 = h_4 + h_5 = \dfrac{Z^2}{X^2}$ and $h_3 = h_3 + h_5 = \dfrac{Z(Y + Z)}{X^2}$. Now the pole orders are

$$-v_P(h_1) = 0, -v_P(h_2) = 3, -v_P(h_3) = -v_P(h_4) = 6$$

and then we take $f_6 = h_4$. Thus, by replacing $h_3 = h_3 + h_4 = \dfrac{YZ}{X^2}$ we obtain now three different pole orders

$$-v_P(h_1) = 0, -v_P(h_2) = 3, -v_P(h_3) = 5$$

and we can stop. In particular, we have computed the Weierstrass semigroup, since we know the three [2] Weierstrass gaps $\{1, 2, 4\}$.

References

1. S.S. Abhyankar, *Lectures on expansion techniques in Algebraic Geometry*, Tata Institute of Fundamental Research, Bombay (1977).
2. S.S. Abhyankar, *Irreducibility criterion for germs of analytic functions of two complex variables*, Advances in Mathematics **74**, pp. 190-257 (1989).
3. G. Angermüller, *Die Wertehalbgruppe einer ebenen irreduziblen algebroiden Kurve*, Math. Zeit. **153**, pp. 267-282 (1977).
4. R. Apéry, *Sur les branches superlinéaires des courbes algébriques*, C.R. Acad. Sciences Paris **222**, pp. 1198-1200 (1946).
5. A. Campillo and J. Castellanos, *Curve singularities*, Univ. Valladolid, preprint (1997).
6. A. Campillo and J.I. Farrán, *Construction of AG codes from symbolic Hamburger-Noether expressions of plane curves*, Univ. Valladolid, preprint (1998).
7. J.I. Farrán, *Construcción y decodificación de códigos álgebro-geométricos a partir de curvas planas: algoritmos y aplicaciones*, Ph.D. thesis, Univ. Valladolid (1997).
8. G.L. Feng and T.R.N. Rao, *Decoding algebraic-geometric codes up to the designed minimum distance*, IEEE Trans. Inform. Theory **39**, pp. 37-45 (1993).
9. A. García, S.J. Kim and R.F. Lax, *Consecutive Weierstrass gaps and minimum distance of Goppa*, J. Pure Appl. Algebra **84**, pp. 199-207 (1993).
10. V.D. Goppa *Geometry and codes*, Kluwer Academic Publishers (1988).
11. G. Haché, *Construction effective des codes géométriques*, Ph.D. thesis, Univ. Paris 6 (1996).
12. G. Haché and D. Le Brigand, *Effective construction of algebraic geometry codes*, IEEE Trans. Inform. Theory **41**, pp. 1615-1628 (1995).
13. M. van Hoeij, *An algorithm for computing an integral basis in an algebraic function field*, Maple V Release 4 share library, preprint (1996).

---

[2] Notice that the genus of χ is $g = 3$.

14. C. Kirfel and R. Pellikaan, *The minimum distance of codes in an array coming from telescopic semigroups*, IEEE Trans. Inform. Theory **41**, pp. 1720-1732 (1995).

15. H. Pinkham, *Séminaire sur les singularités des surfaces (Demazure-Pinkham-Teissier)*, Cours donné au Centre de Math. de l'École Polytechnique (1977-1978).

16. S.C. Porter, B.-Z. Shen and R. Pellikaan, *On decoding geometric Goppa codes using an extra place*, IEEE Trans. Inform. Theory **38**, pp. 1663-1676 (1992).

17. S. Sakata, H.E. Jensen and T. Høholdt, *Generalized Berlekamp-Massey decoding of algebraic-geometric codes up to half the Feng-Rao bound*, IEEE Trans. Inform. Theory **41**, pp. 1762-1768 (1995).

18. B.M. Trager, *Integration of algebraic functions*, Ph.D. thesis, Dept. of EECS, Massachusetts Institute of Technology (1984).

19. M.A. Tsfasman, *Goppa codes that are better than Varshamov-Gilbert bound*, Prob. Peredachi Inform. **18**, pp. 3-6 (1982).

20. M.A. Tsfasman and S.G. Vlăduţ, *Algebraic-geometric codes*, Math. and its Appl., vol. 58, Kluwer Academic Pub., Amsterdam (1991).

On the Undetected Error Probability
of *m*-out-of-*n* Codes
on the Binary Symmetric Channel

Fang-Wei Fu[1], Torleiv Kløve[2], and Shu-Tao Xia[1]

[1] Department of Mathematics, Nankai University, Tianjin 300071, China.
[2] Department of Informatics, University of Bergen, N-5020 Bergen, Norway.

Abstract. The undetected error probability of *m*-out-of-*n* codes on the binary symmetric channel is studied in this paper. A new and simplified proof of the main results of Wang, Yang and Zhang in [6]-[11] is given.

1   Introduction

The well known *m*-out-of-*n* code Ω_n^m consists of all binary vectors of length n and weight m. The *m*-out-of-*n* codes have been widely used as the error-detecting codes in the digital communication systems with feedback, such as the automatic-repeat-request (ARQ) error-control system. The undetected error probability of an error-detecting code is one of the major parameters for evaluating the efficiency of ARQ error-control system. For a general introduction to the theory of the probability of undetected error for the codes, we refer the reader to [2] and its references.

Let C be a binary code of length n and size M. When the code C is used for error-detection on a binary symmetric channel with symbol error probability p, the undetected error probability is given by

$$P_{ud}(C, p) = \frac{1}{M} \sum_{\substack{\mathbf{a}, \mathbf{b} \in C \\ \mathbf{a} \neq \mathbf{b}}} (1-p)^{n - d_H(\mathbf{a}, \mathbf{b})} p^{d_H(\mathbf{a}, \mathbf{b})},$$

where $d_H(\mathbf{a}, \mathbf{b})$ is the Hamming distance between the vectors $\mathbf{a}$ and $\mathbf{b}$. If $P_{ud}(C, p)$ is an increasing function of p on the interval $[0, 1/2]$, the code C is called *proper* for error detection. If $P_{ud}(C, p) \leq P_{ud}(C, 1/2)$ for all $p \in [0, 1/2]$, the code C is called *good* for error detection. Clearly, proper codes are good codes for error detection.

In [9], Wang and Yang showed that the codes Ω_n^m with $n \leq 4$ are proper, and the codes Ω_n^m with $n \geq 9$ and $n \neq 2m$ are not good. They checked that the codes Ω_5^2, Ω_5^3, Ω_6^3, Ω_7^3, Ω_7^4, Ω_8^4 are proper, and all other codes Ω_n^m with $5 \leq n \leq 8$ are not good. They conjectured that the codes Ω_n^m with $n \geq 9$ and $n = 2m$ are not good. Subsequently, in paper [10], Wang and Zhang proved this conjecture. They showed that the codes Ω_n^m with $n \geq 18$ and $n = 2m$ are not good. They checked that the codes Ω_n^m with $9 \leq n \leq 16$ and $n = 2m$ are not good. For checking a specific code Ω_n^m being proper or not being proper (resp. being good or not being good), they first computed the distance distribution and $P_{ud}(\Omega_n^m, p)$, then examined whether $P_{ud}(\Omega_n^m, p)$ is an increasing function of p on the interval $[0, 1/2]$ (resp. $P_{ud}(C, p) \leq P_{ud}(C, 1/2)$ for all $p \in [0, 1/2]$). For some cases, they checked the codes Ω_n^m being proper or not being proper (resp. being good or not being good) by drawing the curves of $P_{ud}(\Omega_n^m, p)$ with computer.

We summarize the results by Wang et al. in the following theorem.

Theorem 1 (Wang, Yang, Zhang, [6]–[11])
i) *The codes* Ω_n^m *with* $n \leq 4$ *are proper for error detection.*
ii) *The codes* Ω_5^2, Ω_5^3, Ω_6^3, Ω_7^3, Ω_7^4, Ω_8^4 *are proper for error detection.*
iii) *All other codes* Ω_n^m *are not good for error detection.*

The goal of this paper is to present a new method to study the error detection capability of the codes Ω_n^m. Necessary and sufficient conditions for the codes Ω_n^m to be proper or good for error detection are given. Using these conditions, we give a new and simpler proof of Theorem 1. In the process, we derive some properties for the dual distance distribution of the codes Ω_n^m.

Note that the code Ω_n^{n-m} is the complement code of Ω_n^m, i.e. is obtained by changing 0 to 1 and 1 to 0 in every codeword of Ω_n^m. Hence, the codes Ω_n^m and Ω_n^{n-m} have the same distance distributions. This implies that $P_{ud}(\Omega_n^m, p) = P_{ud}(\Omega_n^{n-m}, p)$. Therefore, we only need to study the error detection capability of codes Ω_n^m with $2m \leq n$.

2   Preliminaries

Let V_n be the binary n dimensional vector space and $w_H(\cdot)$ be the Hamming weight of a vector. Let $\langle \cdot, \cdot \rangle$ be the scalar product between two vectors, i.e.,

$$\langle (x_1, \ldots, x_n), (y_1, \ldots, y_n) \rangle = \sum_{i=1}^{n} x_i y_i.$$

Let C be a binary code of length n and size M. The *distance distribution* of C is given by

$$D_i = \frac{1}{M} |\{(\mathbf{a}, \mathbf{b}) \mid \mathbf{a}, \mathbf{b} \in C, d_H(\mathbf{a}, \mathbf{b}) = i\}|, \quad i = 0, 1, \ldots, n.$$

The *distance enumerator* of C is given by

$$W_C(s) = \frac{1}{M} \sum_{\mathbf{a}, \mathbf{b} \in C} s^{d_H(\mathbf{a}, \mathbf{b})} = \sum_{i=0}^{n} D_i s^i.$$

The *dual distance distribution* of C is given by

$$\hat{D}_i = \frac{1}{M^2} \sum_{\substack{\mathbf{u} \in V_n \\ w_H(\mathbf{u}) = i}} \left[\sum_{\mathbf{a} \in C} (-1)^{\langle \mathbf{u}, \mathbf{a} \rangle} \right]^2, \quad i = 0, 1, \ldots, n. \tag{1}$$

The *dual distance enumerator* of C is given by

$$\hat{W}_C(s) = \sum_{i=0}^{n} \hat{D}_i s^i. \tag{2}$$

The MacWilliams-Delsarte identity (see [3, pp. 135-141]) gives a relationship between $W_C(s)$ and $\hat{W}_C(s)$.

$$\hat{W}_C(s) = \frac{1}{M}(1+s)^n W_C\left(\frac{1-s}{1+s}\right). \tag{3}$$

$$W_C(s) = \frac{M}{2^n}(1+s)^n \hat{W}_C\left(\frac{1-s}{1+s}\right). \tag{4}$$

We note that $W_C(s) = W_C(-s)$ if and only $D_i = 0$ for all odd i. Also $\hat{D}_i = \hat{D}_{n-i}$ for all i if and only if $\hat{W}_C(s) = s^n \hat{W}_C(1/s)$. Hence an immediate consequence of (3) is the following:

$$D_i = 0 \text{ for all odd } i \text{ if and only if } \hat{D}_i = \hat{D}_{n-i} \text{ for all } i. \tag{5}$$

Similarly, (4) implies that

$$\hat{D}_i = 0 \text{ for all odd } i \text{ if and only if } D_i = D_{n-i} \text{ for all } i. \tag{6}$$

It is easy to see that $P_{ud}(C, p)$ can be expressed by $W_C(s)$ and $\hat{W}_C(s)$ as follows.

$$P_{ud}(C, p) = \sum_{i=1}^{n} D_i p^i (1-p)^{n-i} \tag{7}$$

$$= (1-p)^n W_C\left(\frac{p}{1-p}\right) - (1-p)^n$$

$$= \frac{M}{2^n} \hat{W}_C(1-2p) - (1-p)^n \tag{8}$$

$$= \frac{1}{2^n} \sum_{i=0}^{n} \left\{ M\hat{D}_i - \binom{n}{i} \right\} (1-2p)^i. \tag{9}$$

Below we present some properties of the distance distribution and the dual distance distribution for the m-out-of-n codes Ω_n^m.

Lemma 1 (Wang and Yang [9]) *The distance distribution of the m-out-of-n codes Ω_n^m is given by*

$$D_{2i}(n, m) = \binom{m}{i}\binom{n-m}{i} \quad \text{for} \quad i = 0, 1, \cdots, m,$$

$$D_j(n, m) = 0 \quad \text{otherwise.}$$

Lemma 2 *The dual distance distribution of the m-out-of-n codes Ω_n^m is given by*

$$\hat{D}_i(n, m) = \frac{\binom{n}{i}[P_m(i; n)]^2}{\binom{n}{m}^2} \quad for \quad i = 0, 1, \cdots, n,$$

where $P_m(x; n)$ is the Krawtchouk polynomial (see [3, pp. 130]) defined by

$$P_m(x; n) = \sum_{j=0}^{m} (-1)^j \binom{x}{j} \binom{n-x}{m-j}. \tag{10}$$

In particular,

$$\hat{D}_1(n, m) = \frac{(n - 2m)^2}{n}, \tag{11}$$

and

$$\hat{D}_2(n, m) = \frac{[(n - 2m)^2 - n]^2}{2n(n - 1)}. \tag{12}$$

Proof: Let A be the binary $\binom{n}{m} \times n$ matrix, where the row vectors consist of all of the binary vectors of length n and weight m, that is, the codewords of Ω_n^m. Let $\mathbf{h}_1, \mathbf{h}_2, \ldots, \mathbf{h}_n$ be the column vectors of A. It is not hard to see from (1) that the dual distance distribution of the code Ω_n^m can be expressed by

$$\hat{D}_i(n, m) = \frac{1}{\binom{n}{m}^2} \sum_{1 \leq j_1 < j_2 < \cdots < j_i \leq n} \left[\binom{n}{m} - 2w_H(\mathbf{h}_{j_1} + \mathbf{h}_{j_2} + \cdots + \mathbf{h}_{j_i}) \right]^2 \tag{13}$$

for $i = 1, 2, \ldots, n$. Da Rocha [1], Tolhuizen and van Lint [5] studied the combinatorial code $C(n, m)$ generated by A^T (the transposed matrix of A). They showed that the weight of the sum of any i columns of A only depends on i, n and m and is given by

$$F(n, i, m) = \frac{1}{2} \left[\binom{n}{m} - P_m(i; n) \right], \tag{14}$$

where $P_m(x; n)$ is the Krawtchouk polynomial defined by (10). Combining (13) and (14) implies that

$$\hat{D}_i(n, m) = \frac{1}{\binom{n}{m}^2} \binom{n}{i} \left[\binom{n}{m} - 2F(n, i, m) \right]^2 = \frac{\binom{n}{i} P_m(i; n)^2}{\binom{n}{m}^2}. \tag{15}$$

∎

Remark. By the definition (1) of $\hat{D}_i(n, m)$ and the fact (see [3, pp. 135]) that

$$\sum_{\substack{\mathbf{a} \in V_n \\ w_H(\mathbf{a}) = m}} (-1)^{<\mathbf{u},\mathbf{a}>} = P_m(i; n),$$

for every $\mathbf{u} \in V_n$ with $w_H(\mathbf{u}) = i$, one can give another proof of Lemma 2. We adopt the above proof here, because below we will use some known properties of $F(n, i, m)$.

Lemma 3
i) $\hat{D}_i(n, m) = \hat{D}_{n-i}(n, m)$ *for* $i = 0, 1, \cdots, n$,
ii) $\hat{D}_0(n, m) = \hat{D}_n(n, m) = 1$,

$$\sum_{i=0}^{n} \hat{D}_i(n, m) = \frac{2^n}{\binom{n}{m}}, \qquad \sum_{i=0}^{n} i\hat{D}_i(n, m) = \frac{n2^{n-1}}{\binom{n}{m}},$$

iii) $\hat{D}_i(2m, m) = 0$ *for* i *odd*,
iv) *for* i *even and* $1 \le i < 2m$ *we have*

$$\hat{D}_i(2m, m) \le \binom{2m}{i}/(2m-1)^2,$$

v) *for* $2m < n$ *and* $1 \le i \le n-1$ *we have*

$$\hat{D}_i(n, m) \le \binom{n}{i}(n-2m)^2/n^2.$$

Proof: i) follows from Lemma 1 and (5).
ii) Since $D_1(n, m) = 0$, the result follows from the Pless identities (see e.g. [4][pp.69-70]).
iii) follows from Lemma 1 and (6).
iv) Tolhuizen and van Lint [5, Theorem 1' b] showed that if $1 \le i \le 2m - 1$ and i is even, then

$$2\binom{2m-2}{m-2} \le F(2m, i, m) \le 2\binom{2m-2}{m-1}.$$

Hence

$$\binom{2m}{m} - 2F(2m, i, m) \le \binom{2m}{m} - 4\binom{2m-2}{m-2} = \binom{2m}{m}/(2m-1)$$

and

$$\binom{2m}{m} - 2F(2m, i, m) \ge \binom{2m}{m} - 4\binom{2m-2}{m-1} = -\binom{2m}{m}/(2m-1).$$

Therefore,

$$\hat{D}_i(2m, m) \le \frac{\binom{2m}{i}}{(2m-1)^2}.$$

v) Tolhuizen and van Lint [5, Theorem 1] further showed that if $2m < n$ and $1 \le i \le n-1$, then

$$\binom{n-1}{m-1} \le F(n, i, m) \le \binom{n-1}{m}.$$

By (15), we have,

$$\hat{D}_i(n,m) \le \frac{1}{\binom{n}{m}^2}\binom{n}{i}\left[\binom{n-1}{m} - \binom{n-1}{m-1}\right]^2 = \frac{\binom{n}{i}(n-2m)^2}{n^2}.$$

∎

3 A New Proof of Theorem 1

From Lemma 1 and (7) we get, after differentiation,

$$P'_{ud}(\Omega_n^m, p) = \sum_{i=1}^{m}\binom{m}{i}\binom{n-m}{i}p^{2i-1}(1-p)^{n-2i-1}(2i-np). \qquad (16)$$

In particular, if $n \le 4$, then the right hand side of (16) is positive for all $p \in (0,1/2]$. This proves Theorem 1 i), as already observed by Wang and Yang [9].

From (9) we similarly get

$$P'_{ud}(C,p) = \frac{1}{2^{n-1}}\sum_{i=1}^{n}i\left[\binom{n}{i} - \binom{n}{m}\hat{D}_i(n,m)\right](1-2p)^{i-1}. \qquad (17)$$

Proof of Theorem 1 ii) and iii) in the case $n \ne 2m$

The main result of this section is the following characterization.

Lemma 4 *If $n \ne 2m$, then the following conditions are equivalent.*

i) *The code Ω_n^m is proper for error detection.*
ii) *The code Ω_n^m is good for error detection.*
iii) $\binom{n}{m} \le n^2/(n-2m)^2$.

Proof: The implication i) $\Longrightarrow$ ii) follows from the definitions of good and proper. ii) $\Longrightarrow$ iii). We have $P_{ud}(\Omega_n^m, p) \le P_{ud}(\Omega_n^m, 1/2)$ for all $p \in [0,1/2]$ since the code Ω_n^m is good for error detection. Hence, $P'_{ud}(\Omega_n^m, 1/2) \ge 0$. By (11) and (17), we have

$$P'_{ud}(\Omega_n^m, 1/2) = \frac{n}{2^{n-1}} - \frac{\binom{n}{m}}{2^{n-1}}\hat{D}_1(n,m) = \frac{1}{2^{n-1}}\left[n - \frac{\binom{n}{m}(n-2m)^2}{n}\right].$$

Therefore, $\binom{n}{m} \le n^2/(n-2m)^2$.
iii) $\Longrightarrow$ i). First we note that Lemma 3 ii) implies that

$$\sum_{i=0}^{n}i\left[\binom{n}{i} - \binom{n}{m}\hat{D}_i(n,m)\right] = n2^{n-1} - n2^{n-1} = 0,$$

and so

$$n\left[1 - \binom{n}{m}\hat{D}_n(n, m)\right] = -\sum_{i=0}^{n-1} i\left[\binom{n}{i} - \binom{n}{m}\hat{D}_i(n, m)\right].$$

It follows from (17) that

$$2^{n-1}P'_{ud}(\Omega_n^m, p) = \sum_{i=1}^{n-1} i\left[(1 - 2p)^{i-1} - (1 - 2p)^{n-1}\right]\left[\binom{n}{i} - \binom{n}{m}\hat{D}_i(n, m)\right].$$

From Lemma 3 v), we get that

$$\binom{n}{i} - \binom{n}{m}\hat{D}_i(n, m) \geq \binom{n}{i}\left[1 - \frac{\binom{n}{m}(n - 2m)^2}{n^2}\right] \geq 0$$

for $1 \leq i \leq n - 1$. Hence, $P'_{ud}(\Omega_n^m, p) \geq 0$ for all $p \in [0, 1/2]$ and so Ω_n^m is proper. ∎

Remark. By using some combinatorial identities, Wang and Yang [9] presented a quite long proof for the implication ii) $\Longrightarrow$ iii).

We now prove the case $2m < n$ of Theorem 1:
if $m = 1$ and $n \geq 5$, then

$$\binom{n}{m}(n - 2m)^2 = n(n - 2)^2 > n^2;$$

if $m = 2$ and $n \geq 6$, then

$$\binom{n}{m}(n - 2m)^2 = \frac{n(n - 1)}{2}(n - 4)^2 > n^2;$$

if $m \geq 3$ and $n \geq 9$, then

$$\binom{n}{m}(n - 2m)^2 \geq \binom{n}{m} \geq \binom{n}{3} > n^2.$$

In particular, if $n > 2m$ and $n \geq 9$, then

$$\binom{n}{m} > \frac{n^2}{(n - 2m)^2}.$$

Therefore, by Lemma 1, the codes Ω_n^m are not good for $n \geq 9$.

It is easy to verify that for $5 \leq n \leq 8$ and $2m < n$, only $(n, m) = (5, 2), (7, 3)$ satisfy the inequality $\binom{n}{m} \leq \frac{n^2}{(n-2m)^2}$. Hence, by Lemma 4, we know that the codes Ω_5^2, Ω_5^3, Ω_7^3, Ω_7^4 are proper and all other codes Ω_n^m with $5 \leq n \leq 8$ and $n \neq 2m$ are not good for error detection.

Proof of Theorem 1 ii) and iii) in the case $n = 2m$

Lemma 5 *If $m \geq 1$ and Ω_{2m}^m is good for error detection, then*

$$\binom{2m}{m} \leq m(2m-1)\left\{\left(\frac{m+1}{m}\right)^{2m} - 1\right\}. \tag{18}$$

In particular

$$\binom{2m}{m} < 7m(2m-1). \tag{19}$$

Proof: Since Ω_{2m}^m is good for error detection, we get, by (8), (12) and Lemma 3 ii), iii),

$$0 \leq 2^{2m}\left[P_{ud}(\Omega_{2m}^m, 1/2) - P_{ud}(\Omega_{2m}^m, p)\right]$$

$$= \binom{2m}{m} - 1 - \binom{2m}{m}\sum_{i=0}^{m}\hat{D}_{2i}(2m, m)(1-2p)^{2i} + 2^{2m}(1-p)^{2m}$$

$$\leq -1 - \binom{2m}{m}\hat{D}_2(2m, m)(1-2p)^2 + 2^{2m}(1-p)^{2m}$$

$$= (2-2p)^{2m} - 1 - \binom{2m}{m}\frac{m}{2m-1}(1-2p)^2.$$

for all $p \in [0, 1/2]$. Substituting $p = \frac{m-1}{2m}$, the inequality (18) follows. Further, (19) follows from the fact that

$$\left(\frac{m+1}{m}\right)^m < e \approx 2.71$$

and so

$$\left\{\left(\frac{m+1}{m}\right)^{2m} - 1\right\} < (e^2 - 1) \approx 6.39 < 7$$

for all m. ∎

Since

$$252 = \binom{10}{5} > 5 \cdot 9 \cdot \left(\left(\frac{6}{5}\right)^{10} - 1\right) \approx 233.6,$$

the inequality (18) does not hold for $m = 5$. Hence, Ω_{10}^5 is not good for error detection.

If $m \geq 6$, then

$$\binom{2m}{m} \geq \binom{2m}{6} = m(2m-1)\frac{(m-1)(2m-3)(m-2)(2m-5)}{3 \cdot 5 \cdot 6}$$

$$\geq m(2m-1)\frac{5 \cdot 9 \cdot 4 \cdot 7}{3 \cdot 5 \cdot 6}$$

$$= 14m(2m-1).$$

Hence inequality (19) does not hold and so Ω_{2m}^m is not good for error detection.

It remains to check Ω_{2m}^m for $m = 3, 4$, and this is most easily done by a suitable rewriting of $P'_{ud}(\Omega_{2m}^m, p)$ as a sum of terms which are immediately seen to be positive for $p \in (0, 1/2)$. By (16)

$$P'_{ud}(\Omega_6^3, p) = 9p(1-p)^3(2-6p) + 9p^3(1-p)(4-6p) + 6p^5$$
$$= 18p(1-p)(1-2p)\{(1-2p)^2 + p(1-p)\} + 6p^5.$$

Hence $P'_{ud}(\Omega_6^3, p) \geq 0$ for all $p \in [0, 1/2]$.

Similarly

$$P'_{ud}(\Omega_8^4, p) = 16p(1-p)(1-2p)\{2(1-2p)^3 + 5p^2(1-2p)^2 + p^3(2-3p)\} + 8p^7,$$

and so $P'_{ud}(\Omega_8^4, p) \geq 0$ for all $p \in [0, 1/2]$.

Hence, Ω_6^3 and Ω_8^4 are both proper for error detection.

Acknowledgements. This research work is supported in part by the Chinese Foundation of Institution of Higher Education for Doctoral Program, National Natural Science Foundation of China, and The Norwegian Research Council.

References

1. V. C. Da Rocha, "Combinatorial codes," *Electronics Letters*, vol. 21, no. 21, pp. 949-950, 1985.
2. T. Kløve and V. Korzhik, *Error Detecting Codes, General Theory and Their Application in Feedback Communication Systems.* Boston: Kluwer Acad. Press, 1995.
3. F.J. MacWilliams and N.J.A. Sloane, *The Theory of Error-Correcting Codes*, New York: North-Holland, 1977.
4. W. W. Peterson and E. J. Weldon, Jr., *Error-Correcting Codes*, Cambridge, Mass.: MIT Press, 1972.
5. J.H. van Lint and L. Tolhuizen, "On the minimum distance of combinatorial codes," *IEEE Trans. Inform. Theory*, vol. IT-36, no. 4, pp. 922-923, 1990.
6. X.M. Wang, "Existence of proper binary $(n, 2, w)$ constant weight codes and a conjecture," *Science in China*, Series A (in Chinese), vol. 17, no. 11, pp. 1225-1232, 1987.
7. X.M. Wang, "The undetected error probability of constant weight codes," *Acta Electronica Sinica* (in Chinese), vol. 17, no. 1, pp. 8-14, 1989.
8. X.M. Wang, "Further analysis of the performance on the error detection for constant weight codes," *Journal of China Institute of Communication* (in Chinese), vol. 13, no. 4, pp. 10-17, 1992.
9. X.M. Wang and Y.X. Yang, "On the undetected error probability of nonlinear binary constant weight codes," *IEEE Trans. Communication*, vol. COM-42, pp. 2390-2393, 1994.
10. X.M. Wang and Y.W. Zhang, "The existence problem of proper constant weight codes," *Acta Electronica Sinica* (in Chinese), vol. 23, no. 7, pp. 113-114, 1995.
11. Y.X. Yang, "Proof of Wang's conjecture," *Chinese Science Bulletin* (in Chinese), vol. 34, No. 1, pp. 78-80, 1989.

Skew Pyramids of Function Fields
Are Asymptotically Bad

Arnaldo Garcia[1] and Henning Stichtenoth[2]

[1] Instituto de Matématica Pura e Aplicada IMPA
Estrada Dona Castorina 110, 22460-320 Rio de Janeiro RJ, Brazil
garcia@impa.br
[2] Mathematik und Informatik, Universität GH Essen
FB 6, D-45117 Essen, Germany
stichtenoth@uni-essen.de

A tower of function fields over a finite field $\mathbb{F}_q$ is a sequence $\mathcal{F} = (F_1, F_2, F_3, \dots)$ of function fields $F_i/\mathbb{F}_q$ satisfying the following conditions:

  i) $F_1 \subseteq F_2 \subseteq F_3 \subseteq \dots$.
 ii) for all $n \geq 1$, F_{n+1}/F_n is a separable extension of degree > 1.
iii) the genus $g(F_j) > 1$, for some j.

In a series if papers [2–4] we studied asymptotic properties of such towers. In particular we are interested in the limit

$$\lambda(\mathcal{F}) = \lim_{n \to \infty} N(F_n)/g(F_n),$$

where $N(F_n)$ (resp. $g(F_n)$) denotes the number of places of degree one (resp. the genus) of $F_n/\mathbb{F}_q$. Note that this limit always exists [3, p. 253].

The tower $\mathcal{F}$ is said to be asymptotically good if $\lambda(\mathcal{F}) > 0$, and asymptotically bad if $\lambda(\mathcal{F}) = 0$. This terminology comes from coding theory: asymptotically good towers of function fields can be used for constructing asymptotically good sequences of codes [6].

In general it seems to be difficult to find asymptotically good towers. It is known that certain class field towers [5] and modular towers [1,6] are good. Explicit constructions of good towers were given in [1–4]. These explicit examples are all of the following type: $F_n = \mathbb{F}_q(x_1, \dots, x_n)$ where x_n satisfies an equation $f(x_n, x_{n-1}) = 0$, with some irreducible polynomial $f(X, Y) \in \mathbb{F}_q[X, Y]$ (independent of n). In this note we want to point out why many attempts to construct in a similar manner other explicit examples failed, cf. [3, Ex. 4.5].

<u>Theorem</u> *Assume that the tower $\mathcal{F} = (F_1, F_2, \dots)$ is given by $F_n = \mathbb{F}_q(x_1, \dots, x_n)$ where (for all $n \geq 2$) x_n satisfies an irreducible equation over F_{n-1} of the form*

$$f(x_n, x_{n-1}) = 0,$$

with a polynomial $f(X,Y) \in \mathbb{F}_q[X,Y]$. Suppose that f is separable both in X and Y, and that $\deg_X f \neq \deg_Y f$. Then the tower $\mathcal{F}$ is asymptotically bad.

<u>Proof.</u> We have the following "pyramid" of subextensions of F_n:

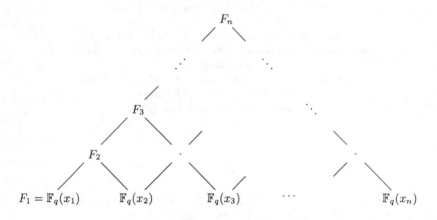

Let $a := \deg_X f$ and $b := \deg_Y f$. It follows from our assumptions that

$$[F_n : \mathbb{F}_q(x_1)] = a^{n-1} \quad \text{and} \quad [F_n : \mathbb{F}_q(x_n)] = b^{n-1}.$$

We assume that $a > b$ (the case $a < b$ is similar), and we choose r such that $g(F_r) \geq 2$. For all $n \geq r$, the Hurwitz genus formula yields

$$g(F_n) - 1 \geq [F_n : F_r](g(F_r) - 1) = a^{n-r}(g(F_r) - 1)). \tag{1}$$

On the other hand, considering the extension $F_n/\mathbb{F}_q(x_n)$ we find that

$$N(F_n) \leq (q+1) \cdot [F_n : \mathbb{F}_q(x_n)] = (q+1) \cdot b^{n-1}. \tag{2}$$

Hence we have for $n \geq r$

$$\frac{N(F_n)}{g(F_n)} \leq \frac{(q+1) \cdot b^{n-1}}{(g(F_r) - 1) \cdot a^{n-r}} = \frac{(q+1) \cdot a^r}{(g(F_r) - 1) \cdot b} \cdot \left(\frac{b}{a}\right)^n. \tag{3}$$

As $b < a$, the right hand side of (3) tends to 0 for $n \to \infty$. $\square$

As an application of our theorem we consider the following example: Let $q = l^s$, where l is a prime power and $s \geq 2$. Denote by

$$N(Z) := Z^{l^{s-1}} \cdot Z^{l^{s-2}} \cdot \ldots \cdot Z^l \cdot Z$$

resp.

$$T(Z) := Z^{l^{s-1}} + Z^{l^{s-2}} + \ldots + Z^l + Z$$

the norm (resp. trace) polynomial of $\mathbb{F}_q/\mathbb{F}_l$. Define the tower $\mathcal{F} = (F_1, F_2, F_3, ...)$ over $\mathbb{F}_q$ by $F_n = \mathbb{F}_q(x_1, \dots, x_n)$ where

$$T(x_n) = \frac{N(x_{n-1})}{T(x_{n-1})}.$$

For $s = 2$ this tower is asymptotically good, see [3]. However it is asymptotically bad for all $s \geq 3$. This follows immediately from our theorem: the corresponding polynomial $f(X, Y)$ is here

$$f(X, Y) = X^{-1}(T(Y) \cdot T(X) - N(X)),$$

with $\deg_X f = l^{s-1} + l^{s-2} + \dots + l$ and $\deg_Y f = l^{s-1}$.

References

1. Elkies, N., Explicit modular towers Preprint. Harvard Univ., 1997.
2. Garcia, A. and Stichtenoth, H., A tower of Artin-Schreier extensions of function fields attaining the Drinfeld-Vladut bound. Invent. Math. **121** (1995), 211–222.
3. Garcia, A. and Stichtenoth, H., On the asymptotic behaviour of some towers of function fields over finite fields. J. Number Theory **61** (1996), 248–273.
4. Garcia, A., Stichtenoth, H. and Thomas, M., On towers and composita of towers of function fields over finite fields. Finite Fields and their Appl. **3** (1997), 257–274.
5. Serre, J.-P., Sur le nombre des points d'une courbe algébrique sur un corps fini. C. R. Acad. Sci. Paris. Ser. I Math. **269** (1983), 397–402.
6. Tsfasman, M. A. and Vladut. S., "Algebraic-Geometric Codes", Kluwer, Dordrecht, 1991.

A Public Key Cryptosystem Based on Sparse Polynomials

D. Grant, K. Krastev, D. Lieman, and I. Shparlinski

[1] Department of Mathematics, University of Colorado
Boulder, CO 80309, USA
grant@boulder.colorado.edu
[2] School of MPCE, Macquarie University
Sydney, NSW 2109, Australia
kate@mpce.mq.edu.au
[3] Department of Mathematics, University of Missouri
Columbia, MO 65211, USA
lieman@math.missouri.edu
[4] School of MPCE, Macquarie University
Sydney, NSW 2109, Australia
igor@mpce.mq.edu.au

Abstract. This paper introduces a new type of cryptosystem which is based on sparse polynomials over finite fields. We evaluate its theoretic characteristics and give some security analysis. Some preliminary timings are presented as well, which compare quite favourably with published optimized RSA timings. We believe that similar ideas can be used in some other settings as well.

1 Overview

In this paper we present a new idea for the construction of one-way functions.

The hard problem underlying our one-way functions can be stated as follows: *Given a system of sparse polynomial equations of high degree over certain large rings, it is hard to find a solution to this system.*

On the other hand, because the polynomials involved are sparse, their values at any point can be computed quite efficiently. We have conducted tests of our cryptosystem with parameter choices equivalent to four different levels of security, including the two most popular RSA levels of security and a 2^{80} level of security. Even with no serious attempt at optimization, our cryptosystem can encrypt and decrypt a message at speeds roughly equal to that of optimized RSA. In addition, key generation in our scheme is several orders of magnitude faster than in RSA.

We remark that several other cryptosystems based on polynomials have been developed, see [7, 9, 10] for example, but all of them exploit quite different ideas.

Throuout the paper $\log x$ and $\ln x$ denote the binary logarithm and the natural logarithm of $x > 0$, respectively.

2   Construction of a Cryptosystem

Here we describe one of several possible variants of this cryptosystem, which we construct from polynomials over finite fields.

Following the established tradition, we call the communicating parties *Alice* and *Bob*.

The algorithm ENROOT (encryption with roots) can be described as follows:

Algorithm ENROOT

Step 1
   Alice and *Bob* choose a large finite field $\mathbb{F}_q$, and positive integers k, s_i and t_i, $i = 1, \ldots, k$. This information is *public*.
Step 2
   Alice puts $e_1 = 1$ and selects a random element $\vartheta \in \mathbb{F}_q$ and $k - 1$ exponents $e_2, \ldots, e_k \in \mathbb{Z}/(q-1)$, which are all *secret*.
Step 3
   Alice selects k random polynomials $h_i \in \mathbb{F}_q[X_1, \ldots, X_k]$ of degree at most $q - 1$, containing at most $t_i - 1$ monomials, and makes the polynomials

$$f_i(X_1, \ldots, X_k) = h_i(X_1, \ldots, X_k) - h_i(a_1, \ldots, a_k), \qquad i = 1, \ldots, k,$$

   public, where
$$a_i = \vartheta^{e_i}, \qquad i = 1, \ldots, k.$$

Step 4
   To send a message $m \in \mathbb{F}_q$, *Bob* selects k random polynomials

$$g_i \in \mathbb{F}_q[X_1, \ldots, X_k], \qquad i = 1, \ldots, k,$$

   of degree at most $q - 1$, with each containing at most s_i monomials and having non-zero constant coefficients. *Bob* then computes the reduction Ψ of the polynomial $f_1 g_1 + \ldots + f_k g_k$ modulo the ideal generated by

$$X_1^q - X_1, \ldots, X_k^q - X_k,$$

   and sends the polynomial $\Phi = m + \Psi$.
Step 5
   To decrypt the message, *Alice* merely computes $\Phi(a_1, \ldots, a_k) = m$.

It is obvious that the computational cost of this algorithm is polynomial. More precisely, let us denote

$$M(r) = r \log r \log \log r.$$

It is known that the bit cost of multiplication and addition of r-bit integers as well as the bit cost of multiplication and addition over $\mathbb{F}_q$, where the prime power q is r-bits long, can be estimated by $O(M(r))$, see [1, 3, 12].

Put

$$T = \sum_{i=1}^{k} t_i, \qquad S = \sum_{i=1}^{k} s_i, \qquad R = \sum_{i=1}^{k} t_i s_i.$$

Theorem. *Let a prime power q be r-bits long. The Algorithm* ENROOT *has the following characteristics:*

- ○ *the complexity of generating the public key, that is, the set of polynomials $f_1, \ldots, f_k$, is $O((k+r)TM(r))$ bit operations plus the cost of generating $O(krT)$ random bits;*
- ○ *the size of the public key is $O(krT)$ bits;*
- ○ *the complexity of encryption, that is, generating the polynomial Φ, is $O(kRM(r))$ bit operations plus the cost of generating $O(krS)$ random bits;*
- ○ *the size of the encrypted message is $O(krR)$ bits;*
- ○ *the complexity of decryption, that is, finding the plain text message $m \in \mathbb{F}_q$, is $O((k+r)RM(r))$ bit operations.*

Proof.   First of all, we remark that the value of any monomial $X_1^{n_1} \ldots X_k^{n_k}$, with exponents $0 \leq n_1, \ldots, n_k \leq q-1$, can be computed at $(a_1, \ldots, a_k)$ with $O((k+r)M(r))$ bit operations, by using repeated squaring. Indeed, first of all one may compute

$$E \equiv \sum_{i=1}^{k} e_k n_k \pmod{q-1}, \quad 1 \leq E \leq q-1,$$

with $kM(r)$ bit operations. After this the computation of

$$a_1^{n_1} \ldots a_k^{n_k} = \vartheta^E$$

can be done with $O(rM(r))$ bit operations.

To generate ϑ and the exponents $e_2, \ldots, e_k$ we need to generate $O(rk)$ random bits.

To generate the coefficients of the polynomials $f_1, \ldots, f_k$, we need to generate $T - k$ random elements of $\mathbb{F}_q$. This requires $O(rT)$ random bits. We also need to generate $T - k$ random k-tuples $(n_1, \ldots, n_k)$, with $0 \leq n_1, \ldots, n_k \leq q-1$, giving the exponents of the $T - k$ non-constant monomials involved in these polynomials. This requires $O(krT)$ random bits and, as it follows from the above remark, $O((k+r)TM(r))$ bit operations.

Similar analysis applies to the cost of generating $g_1, \ldots, g_k$.

Next, the cost of computing the sum of the products $f_i g_i$, $i = 1, \ldots, k$ is $O(kRM(r))$, which consists of the cost of computing $O(R)$ products over $\mathbb{F}_q$ and the cost of computing kR sums of $O(r)$ bit integers (to compute th exponents for each monomial in the product). The cost of reduction modulo the ideal generated by $X_1^q - X_1, \ldots, X_k^q - X_k$ involves only $O(kR)$ subtractions of $O(r)$ bit integers.

Noting that Φ contains at most $O(R)$ monomials and that each of them can be computed at $(a_1, \ldots, a_k)$ with $O\left((k+r)M(r)\right)$ bit operations, we obtain the desired result. $\qquad\square$

We remark that the implied constants in these estimates can be easily evaluated.

3   Security Considerations

One possible attack on this cryptosystem is to try to find a solution to the system of equations

$$f_i(x_1, \ldots, x_k) = 0, \qquad i = 1, \ldots, k. \tag{1}$$

All known algorithms to solve systems of polynomial equations of total degree n require (regardless of sparsity) time polynomial in n, see [6, 12], but the degree of the polynomials in (1) is very large in our settings, namely it can be of order q. Thus this attack is totally infeasible, taking into account that n is exponentially large in our setting.

Another possible attack is to guess a solution. However, one expects that a system of k sparse polynomial equations in k variables of high degree over $\mathbb{F}_q$ has few zeroes over $\mathbb{F}_q$. Thus the probability that such a random guess gives a solution is, apparently, very small. The best known estimate on this problem when $k = 1$ is given in [2, 5] and it confirms that sparse polynomials over $\mathbb{F}_q$ have very few zeros in $\mathbb{F}_q$. Thus this brute force attack should take about $0.5q^k$ trials "on average".

Of course, it is very tempting to select $k = 1$. Unfortunately it seems that in this case there are more intelligent attacks, one of which is based upon considering the difference set of the powers of monomials of the polynomial Φ.

Indeed, if

$$f(X) = \sum_{i=1}^{t} A_i X^{n_i} \qquad \text{and} \qquad g(X) = \sum_{j=1}^{s} B_i X^{m_j}$$

are the polynomials selected by *Alice* and *Bob*, respectively, with $n_1 = m_1 = 0$, then $\Phi(X)$ contains st monomials $C_{ij} X^{r_{ij}}$, where

$$r_{ij} \equiv n_i + m_j \pmod{q-1}, \qquad i = 1, \ldots, t; \ j = 1, \ldots, s.$$

In particular, for any pair $j_1, j_2 = 1, \ldots, s$, we have

$$r_{ij_1} - r_{ij_2} \equiv m_{j_1} - m_{j_2} \pmod{q-1},$$

for any $i = 1, \ldots, t$.

Therefore, finding the repeated elements in the difference set

$$\Delta = \{r_{i_1 j_1} - r_{i_2 j_2} \pmod{q-1} \ : \ i_1, i_2 = 1, \ldots, t; \ j_1, j_2 = 1, \ldots, s\},$$

which is considered as a subset of the residue ring $\mathbb{Z}/(q-1)\mathbb{Z}$, may reveal some information about the polynomial g.

In addition, if $k = 1$, one may also compute the greatest common divisor of $f(X)$ with $X^q - X$. This yields a product of the linear factors of f. If f has few roots, it may be easy to find a root of this new polynomial, which will have much smaller degree than f. Although it is not clear how to do this in time that would be polynomial in the sparsity t (rather than in the degree of f, which is of order q) and $\log q$, potentially this may be a threat.

On the other hand, even for $k = 2$, these attacks seem to fail. Indeed, the first attack may help to get some information about the total set of monomials in all the polynomials $g_1, \ldots, g_k$, but does not provide any information about the individual polynomials because it is not clear which monomial comes from which product $f_i g_i$, $i = 1, \ldots, k$. In order to try all possible partitions into k groups of $s_i t_i$ monomials, $i = 1, \ldots, k$, one should examine

$$N = \frac{R!}{(s_1 t_1)! \ldots (s_k t_k)!} \tag{2}$$

combinations. In particular, in the most interesting case when all s_i are of approximately the same size and so are t_i, that is, if $s_i \sim s$, $t_i \sim t$, $i = 1, \ldots, k$, then

$$\log N \sim R \log k.$$

Thus the number N of combinations to consider grows exponentially with respect to all parameters, provided that $k \geq 2$.

The second attack fails as well, because the notion of the greatest common divisor of multivariate polynomials is not defined, and taking resolvents to reduce to one variable is too costly.

Moreover, it may be that if the polynomials $h_1, \ldots, h_k$ contain the same monomials (or monomials which differ by the same degrees), then the cryptosystem is more secure, and it may also help to reduce the computational cost of the encryption and decryption.

We have also constructed several lattice attacks to recover the private key, but these attacks are based on lattices of dimension equal to the cardinality of the base field. They are thus completely impractical provided the size of the base field is large, as in the sample parameters below.

4   Parameter Choices and Runtimes

We have tested ENROOT with four parameter choices which provide different levels of security. In all our experiments we use $k = 3$ and work over the prime field $\mathbb{F}_q$ with $q = 2^{31} - 1$. Thus for these values of parameters the brute force attack of searching for a common root of the polynomials f_i, $i = 1, \ldots, k$, takes about 2^{92} trials.

Our implementation uses the NTL library [13] quite substantially. Replacing some of the general purpose programs of this library by some more specialized and better tuned to our applications programs should provide an essential speeding up of the process.

We tested the following combination parameters $\mathbf{s}_i = (s_{i1}, s_{i2}, s_{i3})$ and $\mathbf{t}_i = (t_{i1}, t_{i2}, t_{i3})$

$$\mathbf{s}_1 = (4, 4, 4), \qquad \mathbf{s}_2 = (4, 4, 4), \qquad \mathbf{s}_3 = (4, 4, 5), \qquad \mathbf{s}_4 = (4, 5, 5);$$
$$\mathbf{t}_1 = (3, 3, 4), \qquad \mathbf{t}_2 = (4, 4, 4), \qquad \mathbf{t}_3 = (4, 4, 4), \qquad \mathbf{t}_4 = (4, 4, 4).$$

From (2) we estimate the corresponding security levels with respect to our best idea of attack

$$N_1^{ER} = 2^{57}, \qquad N_2^{ER} = 2^{70}, \qquad N_3^{ER} = 2^{76}, \qquad N_4^{ER} = 2^{82}.$$

With these parameter choices, the total time required to execute a complete cycle of loading the cryptosystem; choosing a private key; constructing a public key; encrypting a message and decrypting that message is given below

$$T_1^{ER} = 0.009 \text{ sec}, \quad T_2^{ER} = 0.010 \text{ sec}, \quad T_3^{ER} = 0.011 \text{ sec}, \quad T_4^{ER} = 0.013 \text{ sec}.$$

These times are on a 600 MHz DEC AlphaStation.

The results are compared with corresponding results for RSA (the RSA time is scaled from the runtimes announced in [8] on a 255 MHz DEC AlphaStation - these times may be compared directly to those in the previous paragraph). Note that the RSA times include only encryption and decryption, and do not include substantial key generation times (as much as 1 second!).

To estimate the level of security of RSA we use the formula

$$T = \exp\left(1.639 \ln^{1/3} M \ln^{2/3} \ln M\right)$$

from [4] for the expected complexity of factoring of an integer M by the number field sieve.

For the key lengths (in bits)

$$K_1^{RSA} = 512, \qquad K_2^{RSA} = 768, \qquad K_3^{RSA} = 1024$$

and the security levels

$$N_1^{RSA} = 2^{55} \qquad N_2^{RSA} = 2^{65} \qquad N_3^{RSA} = 2^{74}.$$

we have corresponding times

$$T_1^{RSA} = 0.004 \text{ sec}, \qquad T_2^{RSA} = 0.011 \text{ sec}, \qquad T_3^{RSA} = 0.019 \text{ sec}.$$

Moreover, the key generation time for ENROOT is several orders of magnitude faster than for RSA. Please note that the highest security level tested for RSA is the same (roughly) as our medium security level, and that the ENROOT times do include key generation!

5   Concluding Remarks

Clearly, our cryptosystem is naturally suited to private key sharing among multiple parties.

The initial set and decryption can probably be accelerated in ENROOT if one uses more sophisticate algorithms to evaluate sparse polynomials, see [11, 14].

We remark that this entire cryptosystem is based on a special case of the following problem: *Let $\mathcal{R}$ be a commutative ring with identity. Given a set of elements $f_1, \ldots f_k$ in an $\mathcal{R}$-algebra S, find an $\mathcal{R}$-algebra homomorphism $\varphi : S \to \mathcal{R}$ such that $\varphi(f_i) = 0$ for all $i = 1, \ldots, k$.*

Even more generally, the problem could be stated as: *Given a morphism of schemes $f : X \to Y$, find a section $s : Y \to X$ for f.*

One inherent weakness of our cryptosystem is its high message expansion cost. Perhaps working with noncommutative rings or rings which are not principal ideal domains will allow the possibility of more secure or more efficient implementations of the above algorithm.

Acknowledgments

The authors would like to thank Michael Larsen for a number of fruitful discussions, Michael Johnson for support and Richard Miller for help with computation.

A part of this work was done during visits by I. S. to the University of Missouri and by D. L. to Macquarie University and the University of Colorado, whose hospitality and support are gratefully acknowledged.

Work supported in part, for D. L. by the National Science Foundation and a Big 12 Faculty Fellowship from the University of Missouri and for I. S. by the Australian Research Council.

References

1. A.V. Aho, J.E. Hopcroft and J.D. Ullman, *The design and analysis of computer algorithms*, Addison-Wesley, Reading, MA, 1975.
2. R. Canetti, J. Friedlander, S. Konyagin, M. Larsen, D. Lieman and I. E. Shparlinski, 'On the statistical properties of the Diffie–Hellman distribution', *Israel J. Math*, to appear.
3. D. G. Cantor and E. Kaltofen, 'On fast multiplication of polynomials over arbitrary algebras', *Acta Inform.*, **28** (1991), 693–701.
4. D. Coppersmith, 'Modifications to the number field sieve', *J. Cryptology*, **6** (1993), 169–180.
5. J. Friedlander, M. Larsen, D. Lieman and I. E. Shparlinski, 'On correlation of binary M-sequences', *Designs, Codes and Cryptography*, to appear.
6. M.-D. A. Huang and Y.-C. Wong, 'Solving systems of polynomial congruences modulo a large prime', *Proc. 37 IEEE Symp. on Found. of Comp. Sci.*, 1996, 115–124.
7. N. Koblitz, *Algebraic aspects of cryptography*, Springer-Verlag, Berlin, 1998.

8. NTRU Cryptosystems, Inc., 'The NTRU public key cryptosystem: Operating characteristics and comparison with RSA, ElGamal, and ECC cryptosystems', http://www.ntru.com/tutorials/operatingchar.htm, 1998.

9. J. Patarin, Asymmetric cryptography with a hidden monomial, *Lect. Notes in Comp. Sci.*, Springer-Verlag, Berlin, **1109** (1996), 45–60

10. J. Patarin, L. Goubin and N. Courtois, 'Improved algorithm for isomorphism of polynomials', *Lect. Notes in Comp. Sci.*, Springer-Verlag, Berlin, **1403** (1998), 184–200.

11. N. Pippenger, 'On the evaluation of powers and monomials', *SIAM J. Comp.*, **9** (1980), 230–250.

12. I. E. Shparlinski, *Finite fields: Theory and computation*, Kluwer Acad. Publ., Dordrecht, 1999.

13. V. Shoup, 'NTL: A library for doing number theory (version 3.1b)', http://www.cs.wisc.edu/~shoup/ntl/, 1998.

14. A. C.-C. Yao, 'On the evaluation of powers', *SIAM J. Comp.*, **5** (1976), 100–103.

Higher Weights of Grassmann Codes

Sudhir R. Ghorpade[1,*] and Gilles Lachaud[2]

[1] Department of Mathematics
Indian Institute of Technology, Bombay
Powai, Mumbai 400076, India
srg@math.iitb.ernet.in

[2] Équipe "Arithmétique et Théorie de l'Information"
Institut de Mathématiques de Luminy
Luminy Case 907, 13288 Marseille, Cedex 9, France
lachaud@iml.univ-mrs.fr

Abstract. Using a combinatorial approach to studying the hyperplane
sections of Grassmannians, we give two new proofs of a result of Nogin
concerning the higher weights of Grassmann codes. As a consequence,
we obtain a bound on the number of higher dimensional subcodes of the
Grassmann code having the minimum Hamming norm. We also discuss
a generalization of Grassmann codes.

1   Introduction

Let C be linear $[n, k]_q$–code, that is, a k-dimensional subspace of the n-dimensional vector space $\mathbb{F}_q^n$ over the finite field $\mathbb{F}_q$ with q elements. Given any (linear) subspace D of $\mathbb{F}_q^n$, one defines (cf. [14]) the *Hamming norm* of D as

$$\|D\| = |\{i \in \{1, \dots, n\} : \text{ there exists } v \in D \text{ with } v_i \neq 0\}|.$$

Given $r \geq 1$, the r-th *higher weight* of the code C is defined by

$$d_r = d_r(C) = \min\{\|D\| : D \text{ is a subspace of } C \text{ with } \dim D = r\}.$$

Note that $d_1 = d_1(C)$ is the classical Hamming weight or the minimum distance of C. The notion of higher weights was first introduced by V. Wei in [15] and, besides being a natural generalization of Hamming weights, it has found a number of applications in Cryptography and Coding Theory. For a survey of this topic and a detailed bibliography, we refer to [14].

In this paper, we consider the so called Grassmann codes, which were introduced, and studied in a series of papers (cf. [8]–[12]) by C. Ryan and K. Ryan in the binary case ($q = 2$) and later by D. Nogin [7] in the general case. These codes may be viewed as a generalization of Reed-Muller codes. In [7], Nogin has shown that the higher weights d_r of the Grassmann code $C(\ell, m)$ satisfy the

---

* Partially supported by a 'Career Award' grant from AICTE, New Delhi and an
IRCC grant from IIT Bombay.

Griesmer-Wei bound for $r \leq \max\{\ell, m - \ell\} + 1$. One of the main aims of this paper is to give an alternate, short proof of this result of Nogin. Our approach is based on using the Plücker embedding of the Grassmannian and a classical and elementary result concerning the same. This approach has turned out to be quite fruitful elsewhere in the study of MDS codes (cf. [1]) and it seems conceivable that this may also be helpful in studying a possibly far-reaching generalization of the Grassmann codes that is proposed in Section 5 of this paper. Besides the abovementioned alternate, short proof of Nogin's result, we also indicate how, using a certain combinatorial structure theorem proved in [1], we can obtain a yet another proof. This second proof of Nogin's result on higher weights of Grassmann codes $C(\ell, m)$ is arguably more 'natural' in the sense that, unlike in Nogin's proof and our first proof, one does not have to pass to the dual Grassmannian in order to assume something like $m - \ell \geq \ell$. In fact, this 'natural' proof makes it transparent why a quantity such as "$\max\{\ell, m - \ell\} + 1$" should appear, and, moreover, one can also obtain as a corollary, a bound on the number of r-dimensional subcodes of $C(\ell, m)$ with minimum Hamming norm, for $r \leq \max\{\ell, m - \ell\} + 1$.

This paper is organized as follows. In Section 2 below, we collect some preliminaries concerning projective systems and Grassmannians. In Section 3, we recall a basic fact about Grassmannians and deduce some results for the number of $\mathbb{F}_q$-rational points of sections of Grassmannians by coordinate hyperplanes. These are applied to give two proofs of Nogin's result on higher weights in Section 4. Finally, in Section 5, we discuss a possible generalization of the Grassmann codes, make a conjecture about the minimum distance of these general codes, and prove a partial result towards this conjecture by using the techniques of Section 3. It may be remarked that the methods used in Sections 3 and 4 are similar to those in [1]. Also, a few results (Corollary 1, for example) are analogous to those in [1] but the proofs here are somewhat different. At any rate, it seemed desirable that for an alternate proof which is claimed to be 'short', the exposition here should be reasonably self-contained.

A part of this work was done when the first author was visiting the Institut de Mathématiques de Luminy (IML) in Marseille for a few months during 1997 and 1998. He would like to use this opportunity to thank the IML for its hospitality and the CNRS of France for its support. Thanks are also due to Michael Tsfasman for a number of helpful discussions, and to Hao Chen for his comments on a preliminary version of this paper.

2   Preliminaries

For studying, or even defining, the Grassmann codes, it is convenient to use the language of projective systems due to Tsfasman and Vlăduţ. Thus, we begin by briefly recalling some basic aspects of this terminology. For details, one may refer to [13], [14].

An $[n, k]_q$-*projective system* is a collection of n not necessarily distinct points in the $(k - 1)$-dimensional projective space $\mathbb{P}^{k-1}_{\mathbb{F}_q}$ over $\mathbb{F}_q$. It is called *nondegenerate* if these n points are not contained in any hyperplane. Incidentally, a linear

$[n, k]_q$–code $C \subseteq \mathbb{F}_q^n$ is called *nondegenerate* if C is not contained in any coordinate hyperplane. One has a natural notion of equivalence of projective systems as well as of linear codes. There is a natural one-to-one correspondence between the equivalence classes of nondegenerate $[n, k]_q$–projective systems and the equivalence classes of nondegenerate linear $[n, k]_q$–codes. Under this correspondence, if X is an $[n, k]_q$–projective system and C is the corresponding code, then the nonzero codewords of C correspond to hyperplanes in $\mathbb{P}_{\mathbb{F}_q}^{k-1}$ and more generally, subcodes $D \subset C$ of dimension r correspond to (projective) subspaces of codimension r in $\mathbb{P}_{\mathbb{F}_q}^{k-1}$. Consequently, the higher weights of C are given by

$$d_r(C)$$
$$= \min \left\{ |X \cap \Pi^c| : \Pi \text{ is a projective subspace of codimension } r \text{ in } \mathbb{P}^{k-1} \right\}$$
$$= n - \max \left\{ |X \cap \Pi| : \Pi \text{ is a projective subspace of codimension } r \text{ in } \mathbb{P}^{k-1} \right\} .$$

Here Π^c denotes the complement (in $\mathbb{P}_{\mathbb{F}_q}^{k-1}$) of Π.

Nice examples of projective systems can be obtained by considering (the set of $\mathbb{F}_q$-rational points of) projective algebraic varieties defined over $\mathbb{F}_q$. In particular, the Grassmannian over $\mathbb{F}_q$:

$$G_{\ell,m} = G_{\ell,m}(\mathbb{F}_q) = \{W : W \text{ is a subspace of } \mathbb{F}_q^m \text{ with } \dim W = \ell\}$$

defines an $[n, k]_q$–projective system, with

$$n := |G_{\ell,m}(\mathbb{F}_q)| = \begin{bmatrix} m \\ \ell \end{bmatrix}_q := \frac{(q^m - 1)(q^m - q) \dots (q^m - q^{\ell-1})}{(q^\ell - 1)(q^\ell - q) \dots (q^\ell - q^{\ell-1})} \quad \text{and} \quad k := \binom{m}{\ell}.$$

Note that the above formula for $|G_{\ell,m}(\mathbb{F}_q)|$ is a well-known fact whose proof can be found, for example, in [2]. Here, we use the natural nondegenerate embedding of $G_{\ell,m}(\mathbb{F}_q)$ in $\mathbb{P}_{\mathbb{F}_q}^{k-1}$, called the *Plücker embedding*. This is concretely obtained as follows. Let

$$I(\ell, m) = \{\alpha = (\alpha_1, \dots, \alpha_\ell) \in \mathbb{Z}^\ell : 1 \leq \alpha_1 < \dots < \alpha_\ell \leq m\}$$

be an indexing set [ordered, say, lexicographically] for the points of $\mathbb{P}^{k-1}(\mathbb{F}_q)$. Given any $\alpha \in I(\ell, m)$ and any $\ell \times m$ matrix $A = (a_{ij})$, let

$$p_\alpha(A) = \alpha\text{-th minor of } A := \det \left(a_{i\alpha_j} \right)_{1 \leq i, j \leq \ell}.$$

Now, for any $W \in G_{\ell,m}(\mathbb{F}_q)$, we can find a $\ell \times m$ matrix A_W whose rows give a basis of W, and then

$$p(W) = (p_\alpha(A_W))_{\alpha \in I(\ell, m)} \in \mathbb{P}_{\mathbb{F}_q}^{k-1}$$

is called the *Plücker coordinate* of W. It is easy to see that this depends only on W and not on the choice of A_W. Moreover, the map $W \mapsto p(W)$ of $G_{\ell,m}(\mathbb{F}_q) \to \mathbb{P}_{\mathbb{F}_q}^{k-1}$ is injective and its image equals the zero locus of certain quadratic polynomials (cf. [4], [5] for details); this map, then, is the *Plücker embedding*. Henceforth, we shall identify W with $p(W)$. It may be remarked that in defining the Plücker embedding, we have tacitly used representations of elements of $\mathbb{F}_q^m$ in terms of the natural basis of $\mathbb{F}_q^m$; if instead, we use another basis to define the 'coordinates', then we have a linear isomorphism (i.e., a collineation) of the ambient space $\mathbb{P}_{\mathbb{F}_q}^{k-1}$, which induces an isomorphism between the two different embeddings of $G_{\ell,m}(\mathbb{F}_q)$.

The nondegenerate linear $[n, k]_q$–code corresponding to the projective system defined by $G_{\ell,m}(\mathbb{F}_q)$ (with its Plücker embedding) is denoted by $C(\ell, m)$ and is called the *Grassmann code*. It may be noted that if $\ell = 1$, then $G_{\ell,m}(\mathbb{F}_q) = \mathbb{P}_{\mathbb{F}_q}^{m-1}$, and thus in this case $C(\ell, m)$ is essentially the Reed-Muller code. On the other hand, if $l = m$, then $G_{\ell,m}(\mathbb{F}_q)$ reduces to a single point. To avoid trivialities, hereafter we shall tacitly assume that $1 \leq \ell < m$.

3   Linear Sections of Grassmannians

We begin by stating a classical result on Grassmannians and then prove some auxiliary results concerning sections of the Grassmannian by linear subspaces that are intersections of coordinate hyperplanes. First, we need some notation.

Given any $\alpha \in I(\ell, m)$, we let

$$B_\alpha = \{p \in G_{\ell,m}(\mathbb{F}_q) : p_\alpha = 1\}$$

and

$$C_\alpha = \{\mathbf{t} = (t_{ij}) : \mathbf{t} \text{ an } \ell \times m \text{ matrix over } \mathbb{F}_q \text{ with } t_{i\alpha_j} = \delta_{ij} \text{ for } 1 \leq i, j \leq \ell\},$$

where δ_{ij} is the usual Kronecker delta. Given any $\gamma_1, \ldots, \gamma_\ell \in \{1, \ldots, m\}$ and any $p \in \mathbb{P}^{k-1}(\mathbb{F}_q)$, we let

$$p_{\gamma_1 \ldots \gamma_\ell} = \begin{cases} 0 & \text{if } \gamma_i = \gamma_j \text{ for some } i \neq j \\ p_\alpha & \text{if there exists a permutation } \sigma \text{ of } \{1, \ldots, \ell\} \\ & \text{such that } \left(\gamma_{\sigma(1)}, \ldots, \gamma_{\sigma(\ell)}\right) = \alpha \in I(\ell, m). \end{cases}$$

The following result appears essentially as Proposition 2 in [5], and is also implicit in [4]. For more motivation and references about this result, see [1].

Lemma 1 (Basic Cell Lemma). *Fix any $\alpha \in I(\ell, m)$. Then the polynomial maps (or morphisms) $\Phi : B_\alpha \to C_\alpha$ and $\Psi : C_\alpha \to B_\alpha$ defined by*

$$\Phi(p) = (t_{ij}(p)), \text{ where } t_{ij}(p) := p_{\alpha_1 \ldots \alpha_{i-1} j \alpha_{i+1} \ldots \alpha_\ell}, \text{ for } 1 \leq i \leq \ell, \ 1 \leq j \leq m$$

and

$$\Psi(\mathbf{t}) = (p_\beta), \text{ where } p_\beta := \beta\text{--th minor of } \mathbf{t} = \det \left(t_{i\beta_j}\right)_{1 \leq i,j \leq \ell}, \text{ for } \beta \in I(\ell, m),$$

are bijective and inverses of each other.

Now we introduce the following notation for intersections of the Grassmannian by coordinate hyperplanes. Given any subset Λ of $I(\ell, m)$, we let

$$E_\Lambda = \{p \in G_{\ell,m}(\mathbb{F}_q) : \; p_\alpha = 0 \text{ for all } \alpha \in \Lambda\}.$$

For a "small" subset Λ such as $\{\alpha\}$, $\{\alpha, \beta\}$, $\{\alpha, \beta, \gamma\}$, the corresponding E_Λ would be simply denoted by E_α, $E_{\alpha\beta}$, $E_{\alpha\beta\gamma}$, respectively. By standard conventions, E_Λ equals $G_{\ell,m}(\mathbb{F}_q)$ if Λ is empty.

Given any $\alpha \in I(\ell, m)$, by $\bar{\alpha}$ we shall denote the corresponding set, i.e., $\bar{\alpha} = \{\alpha_1, \ldots, \alpha_\ell\}$. A family $\Lambda \subseteq I(\ell, m)$ will be called *close* if $|\bar{\alpha} \cap \bar{\beta}| = k - 1$ for all $\alpha, \beta \in \Lambda$, $\alpha \neq \beta$.

Proposition 1. *Let $\Lambda \subseteq I(\ell, m)$ be a close family of cardinality r. Let $\delta = \ell(m - \ell)$. Then*

$$|E_\Lambda| = \begin{bmatrix} m \\ \ell \end{bmatrix}_q - q^\delta - q^{\delta-1} - \cdots - q^{\delta-r+1}.$$

Proof. We proceed by induction on r. The case when $r = 0$ (i.e., when Λ is empty) is trivial. Assume that $r \geq 1$ and that the result holds for smaller values of r. Fix any $\alpha \in \Lambda$. Since Λ is close, for any $\beta \in \Lambda$ with $\beta \neq \alpha$, there exist unique integers $u = u(\beta)$ and $v = v(\beta)$ such that $1 \leq u \leq \ell$, $1 \leq v \leq m$, $v \notin \bar{\alpha}$ and $\beta = (\alpha_1, \ldots, \alpha_{u-1}, v, \alpha_{u+1}, \ldots, \alpha_\ell)$. Clearly the pairs $(u(\beta), v(\beta))$ are distinct for distinct elements $\beta \in \Lambda$. Moreover, for any $p \in B_\alpha$, and any $\beta \in \Lambda$ with $\beta \neq \alpha$, the β-th minor of $\mathbf{t} = \Phi(p)$ is precisely equal to t_{uv}. Thus it follows from Lemma 1 that the set $A_{r-1} = \{p \in G_{\ell,m} : p_\alpha \neq 0 \text{ and } p_\beta = 0 \text{ for all } \beta \in \Lambda \setminus \{\alpha\}\}$ is in bijection with the zero locus of $r - 1$ distinct coordinates in $\mathbb{F}_q^\delta$, and hence $|A_{r-1}| = q^{\delta-r+1}$. Now note that $E_\Lambda = E_{\Lambda \setminus \{\alpha\}} \setminus A_{r-1}$, and apply the induction hypothesis.

Corollary 1. *Given any $\alpha \in I(\ell, m)$, we have $|E_\alpha| = \begin{bmatrix} m \\ \ell \end{bmatrix}_q - q^{\ell(m-\ell)}$.*

Proof. Any singleton subset of $I(\ell, m)$ is close. So we can apply Proposition 1.

4   Computation of Higher Weights

In this section, we shall give two proofs of Nogin's result on the determination of certain higher weights d_r of the Grassmann code $C(\ell, m)$. As in Proposition 1, we let $\delta = \ell(m - \ell)$. The following result is proved in [7, Thm. 4.1].

Proposition 2. $d_1(C(\ell, m)) = q^\delta$.

Remarks. 1. Notice that a weaker version of the above result, namely, the inequality, $d_1 \leq q^\delta$ is an immediate consequence of Corollary 1.

2. Nogin's proof of Proposition 2 uses the language of Exterior Algebra, and he shows that if $\omega \in \Lambda^{m-\ell} \mathbb{F}_q^m$ is "completely decomposable" then the corresponding hyperplane H_ω is of minimum weight, that is, $|H_\omega \cap G_{\ell,m}(\mathbb{F}_q)| = q^\delta$.

It may be noted that these completely decomposable hyperplanes are essentially the same as the coordinate hyperplanes $E_\alpha = \{p \in G_{\ell,m}(\mathbb{F}_q) : p_\alpha = 0\}$. Indeed, if a completely decomposable element $\omega = v_1 \wedge \cdots \wedge v_{m-\ell} \in \Lambda^{m-\ell}\mathbb{F}_q^m$ is nonzero, then we can extend $\{v_1, \ldots, v_{m-\ell}\}$ to a basis $\{v_1, \ldots, v_m\}$ of $\mathbb{F}_q^m$. If we use this basis to define the Plücker coordinates, then for any $\omega' \in \Lambda^\ell\mathbb{F}_q^m$, the condition $\omega' \wedge \omega = 0$ (that defines H_ω) corresponds precisely to the condition $p_\alpha = 0$ (that defines E_α) for a uniquely determined α.

3. A change of basis of the ambient space can be viewed as an instance of the natural action of the general linear group $GL_m(\mathbb{F}_q)$ on the Grassmannian $G_{\ell,m}(\mathbb{F}_q)$. (See, for example, [1] for an explicit description of this group action). This action is transitive and thus an 'orbit' of p_α will consist of $|G_{\ell,m}(\mathbb{F}_q)| = \begin{bmatrix} m \\ \ell \end{bmatrix}_q$ elements. Moreover, the α-th Plücker coordinate p_α is determined only up to multiplication by nonzero elements of $\mathbb{F}_q$. This shows that the number of hyperplanes H in $\mathbb{P}_{\mathbb{F}_q}^{k-1}$ such that $|H \cap G_{\ell,m}(\mathbb{F}_q)| = \begin{bmatrix} m \\ \ell \end{bmatrix}_q - q^\delta$, or equivalently, the number of minimum weight codewords of the Grassmann code $C(\ell, m)$, is $\geq (q-1)\begin{bmatrix} m \\ \ell \end{bmatrix}_q$. This inequality may be viewed as a weak version of Corollary 4.5 of [7].

We are now ready to state and prove Nogin's main result on higher weights of Grassmann codes.

Theorem 1. *For $1 \leq r \leq \max\{\ell, m - \ell\} + 1$, we have*

$$d_r(C(\ell, m)) = q^\delta + q^{\delta-1} + \cdots + q^{\delta-r+1} = q^\delta\left(1 + \frac{1}{q} + \cdots + \frac{1}{q^{r-1}}\right).$$

Proof. It is well-known (cf. [15]) that the higher weights of any linear code satisfy the Griesmer-Wei bound:

$$d_r \geq \sum_{i=0}^{r-1} \left\lceil \frac{d_1}{q^i} \right\rceil.$$

Thus, for the Grassmann code $C(\ell, m)$, we find using Proposition 2 that

$$d_r \geq q^\delta + q^{\delta-1} + \cdots + q^{\delta-r+1}.$$

Now, since $G_{\ell,m}$ and $G_{m-\ell,m}$ are isomorphic, we may assume without loss of generality that $(m - \ell) \geq \ell$. Choose any subset λ of $\{1, 2, \ldots, m\}$ of cardinality $\ell - 1$. Given any positive integer r such that $r \leq m - \ell + 1$, we can choose r distinct integers from the complement of λ in $\{1, 2, \ldots, m\}$, and add them to λ to generate a close subset Λ of $I(\ell, m)$ cardinality r. Therefore, by Proposition 1 and the description, given in Section 2, of higher weights in terms of projective systems, it follows that

$$d_r \leq q^\delta + q^{\delta-1} + \cdots + q^{\delta-r+1}.$$

This completes the proof.

For a more 'natural' proof of Proposition 1, we shall make use of a 'Structure Theorem for Close Families' that is proved in [1]. It suffices to note only the following consequence of this Structure Theorem, which is also proved in [1].

Proposition 3. *For $r \geq 2$, the cardinality $c_r = c_r(\ell, m)$ of the set of close families $\Lambda \subseteq I(\ell, m)$ with $|\Lambda| = r$, is given by*

$$c_r = \binom{m}{r}\left[\binom{m-r}{\ell-1} + \Delta_{2,r}\binom{m-r}{m-\ell-1}\right],$$

where $\Delta_{2,r}$ is the complement of Kronecker delta, which is 0 if $r = 2$ and 1 otherwise.

Henceforth, for $r \geq 1$, we let $c_r = c_r(\ell, m)$ be the integer defined above if $r \geq 2$ and $c_r(\ell, m) = \binom{m}{\ell}$ if $r = 1$. Note that this definition is consistent with the set theoretic description in Proposition 3.

Corollary 2. *Assume that $1 \leq \ell < m$ and $r \geq 1$. Then*

$$c_r(\ell, m) > 0 \Longleftrightarrow r \leq \max\{\ell, m - \ell\} + 1.$$

Proof. Since $l \geq 1$, we have

$$\binom{m-r}{\ell-1} > 0 \Longleftrightarrow \ell - 1 \leq m - r \Longleftrightarrow r \leq m - \ell + 1.$$

In particular, since $l < m$, we have $c_2(\ell, m) > 0$. Further, since $l < m$, we have

$$\binom{m-r}{m-\ell-1} > 0 \Longleftrightarrow m - \ell - 1 \leq m - r \Longleftrightarrow r \leq \ell + 1.$$

So the desired result follows from Proposition 3 if $r \geq 2$. The case of $r = 1$ is obvious.

Proof (Second Proof of Theorem 1). Suppose $1 \leq r \leq \max\{\ell, m - \ell\} + 1$. By Corollary 2, there exists a close family $\Lambda \subseteq I(\ell, m)$ of cardinality r. Now Proposition 1 implies that

$$d_r \leq q^\delta + q^{\delta-1} + \cdots + q^{\delta-r+1}.$$

The other inequality is obtained, as in the previous proof of Theorem 1, from the Griesmer-Wei bound.

Corollary 3. *For $1 \leq r \leq \max\{\ell, m - \ell\} + 1$, the number of r-dimensional subcodes of $C(\ell, m)$ of minimum Hamming norm is at least $c_r(\ell, m)$.*

Proof. The r-dimensional subcodes D of $C(\ell, m)$ of minimum Hamming norm, viz., $\|D\| = d_r = d_r(C(\ell, m))$, are in one-to-one correspondence with projective subspaces Π of $\mathbb{P}_{\mathbb{F}_q}^{k-1}$ of codimension r for which $|G_{\ell,m}(\mathbb{F}_q) \cap \Pi| = n - d_r$. Every close family $\Lambda \subseteq I(\ell, m)$ of cardinality r defines such a projective subspace $\Pi(\Lambda)$, and evidently, $\Pi(\Lambda) \neq \Pi(\Lambda')$ if $\Lambda' \subseteq I(\ell, m)$ is a close family of cardinality r with $\Lambda \neq \Lambda'$. Thus the desired result follows from Proposition 3.

5   A Generalization of the Grassmann code

In this section, we discuss what appears to be a natural and, from the viewpoint of Algebraic Geometry, obvious generalization of Grassmann codes. This is obtained by considering the so called Schubert subvarieties of the Grassmannian. These subvarieties may be described as follows.

Fix any $\alpha = (\alpha_1, \ldots, \alpha_\ell) \in I(\ell, m)$. Let $A_1 \subset \cdots \subset A_\ell$ be the canonical partial flag of subspaces of $\mathbb{F}_q^m$ associated to α, given by

$$A_i = \mathrm{span}\{e_1, \ldots, e_{\alpha_i}\}, \quad \text{for} \quad i = 1, \ldots, \ell$$

where $\{e_1, \ldots, e_m\}$ denotes the natural basis of $\mathbb{F}_q^m$. Define the *Schubert variety* in $G_{\ell,m}(\mathbb{F}_q)$ associated to α (or more pedantically, to the above partial flag) by

$$\Omega_\alpha = \Omega_\alpha(\mathbb{F}_q) = \{W \in G_{\ell,m}(\mathbb{F}_q) : \dim(W \cap A_i) \geq i \text{ for } i = 1, \ldots, \ell\}.$$

It is well-known (cf. [4], [5]) that Ω_α is a projective algebraic variety. Indeed, if we define for $\beta = (\beta_1, \ldots, \beta_\ell)$ and $\beta' = (\beta'_1, \ldots, \beta'_\ell)$ in $I(\ell, m)$,

$$\beta \leq \beta' \iff \beta_i \leq \beta'_i \quad \text{for all } i = 1, \ldots, \ell,$$

then we obtain a partial order on $I(\ell, m)$, and as a consequence of the Basic Cell Lemma 1, it can be shown (see, for example, [5, Prop. 3]) that

$$\Omega_\alpha(\mathbb{F}_q) = \{p \in G_{\ell,m}(\mathbb{F}_q) : p_\beta = 0 \text{ for all } \beta \not\leq \alpha\},$$

and moreover that for any $\beta \leq \alpha$, there exists some $p \in \Omega_\alpha(\mathbb{F}_q)$ such that $p_\beta \neq 0$. In fact, the (defining) ideal $I(\Omega_\alpha)$ of Ω_α is generated by the generic Plücker coordinates p_β, for $\beta \not\leq \alpha$ and certain quadratic polynomials (cf. [4] for details). It follows that the Plücker coordinates p_β, for $\beta \leq \alpha$, do not satisfy any linear relation. Thus if we let

$$k_\alpha = |\{\beta \in I(\ell, m) : \beta \leq \alpha\}| \quad \text{and} \quad n_\alpha = |\Omega_\alpha(\mathbb{F}_q)|$$

then $\Omega_\alpha(\mathbb{F}_q)$ gives rise to a nondegenerate $[n_\alpha, k_\alpha]_q$-projective system, via the induced Plücker embedding, and therefore, a linear $[n_\alpha, k_\alpha]_q$-code. This code may be called a *Schubert code* and denoted by $C_\alpha(\ell, m)$.

Notice that if $\alpha = (m - \ell + 1, m - \ell + 2, \ldots, m)$, then $\beta \leq \alpha$ for all $\beta \in I(\ell, m)$, and thus $\Omega_\alpha(\mathbb{F}_q) = G_{\ell,m}(\mathbb{F}_q)$. Thus in this case $C_\alpha(\ell, m)$ equals the Grassmann code $C(\ell, m)$.

The following conjecture seems plausible.

Conjecture 1 (Ghorpade). The minimum distance $d_1(C_\alpha(\ell, m))$ of the Schubert code $C_\alpha(\ell, m)$ equals q^{δ_α}, where

$$\delta_\alpha = \sum_{i=1}^{\ell} (\alpha_i - i) = \alpha_1 + \cdots + \alpha_\ell - \frac{\ell(\ell+1)}{2}.$$

Notice that if $\alpha = (m - \ell + 1, m - \ell + 2, \ldots, m)$, then $\delta_\alpha = \ell(m - \ell)$, and in this case, the conjectured equality follows from Proposition 2. In the general case, the following inequality can be proved as an application of the Basic Cell Lemma.

Proposition 4. $d_1\left(C_\alpha(\ell, m)\right) \leq q^{\delta_\alpha}$.

Proof. Identify $\mathbb{P}^{k_\alpha - 1}$ as a subset of $\mathbb{P}^{k-1}$ in a natural way. Let $p = p(W) \in B_\alpha \cap \Omega_\alpha(\mathbb{F}_q)$. Then we can choose a basis of W such that the corresponding $\ell \times m$ matrix $A_W = (a_{ij})$ has the property that $a_{ij} = 0$ for $j > \alpha_i$. Now if $\Phi : B_\alpha \to C_\alpha$ is as in Lemma 1 and $\mathbf{t} = \Phi(p)$, then using Laplace development for determinants, it is easy to see that $\mathbf{t}_{ij} = 0$ whenever $j > \alpha_i$. Conversely, if $\mathbf{t} \in C_\alpha$ has this property, then it is evident that the subspace corresponding to $p = \Psi(\mathbf{t}) \in G_{\ell,m}(\mathbb{F}_q)$ satisfies the intersection conditions defining $\Omega_\alpha(\mathbb{F}_q)$. It follows that $B_\alpha \cap \Omega_\alpha(\mathbb{F}_q)$ is in one-to-one correspondence with the set

$$\{\mathbf{t} \in C_\alpha : \mathbf{t}_{ij} = 0, \text{ for } 1 \leq i \leq \ell \text{ and } 1 \leq j \leq m \text{ with } j > \alpha_i\},$$

which is evidently in bijection with $\mathbb{F}_q^{\delta_\alpha}$. Consequently, for the coordinate hyperplane $H_\alpha = \{p \in \mathbb{P}_{\mathbb{F}_q}^{k_\alpha - 1} : p_\alpha = 0\}$, we have $|\Omega_\alpha(\mathbb{F}_q) \cap H_\alpha| = n_\alpha - q^{\delta_\alpha}$. This proves the desired inequality.

Remarks. 1. It may be noted that unlike in the case of Grassmannians, the intersection of the Schubert variety $\Omega_\alpha(\mathbb{F}_q)$ with an arbitrary coordinate hyperplane $H_\beta = \{p \in \mathbb{P}_{\mathbb{F}_q}^{k_\alpha - 1} : p_\beta = 0\}$ is not the complement of an affine space unless $\beta = \alpha$. Indeed, this intersection is the complement of a 'determinantal variety' if $\beta \leq \alpha$, and it is $\Omega_\alpha(\mathbb{F}_q)$ if $\beta \not\leq \alpha$. For some details concerning this, one may refer to [6].

2. The Grassmann variety $G_{\ell,m}$ admits a natural generalization to homogeneous varieties of the form G/P, where G is a nice (say, semisimple) algebraic group and P a maximal parabolic (or more generally, any parabolic) subgroup. Moreover, Schubert subvarieties can also be defined in this general context as the Zariski closures of the corresponding Schubert cells, and these are indexed by the quotients W/W_P of the Weyl group of G by the Weyl group of P. This would of course lead to a further generalization of the Grassmann code.

3. However, before proceeding with further generalizations as indicated above, it may be worthwhile to note that already in the classical case which we have considered here, the Schubert code $C_\alpha(\ell, m)$ is far from well understood. For example, even the basic parameters n_α and k_α are not so easy to describe explicitly, not to mention the entire weight hierarchy. Recently, M. Tsfasman and the first author have looked at these and other related problems, and are able to answer some of them. The details may appear elsewhere.

4. Professor Hao Chen, who came across a preliminary version of this paper, has recently informed us that he is able to prove Conjecture 1 for Schubert codes in $G_{2,m}(\mathbb{F}_q)$, that is, in the case of $\ell = 2$.

References

1. Ghorpade, S. R., Lachaud, G.: Hyperplane sections of Grassmannians and the number of MDS linear codes . Preprint, 1999.
2. Goldman, J., Rota, G.-C.: The number of subspaces of a vector space. In: "Recent Progress in Combinatorics" (W. T. Tutte Ed.), pp. 75–84, Academic Press, New York, 1969.
3. Hirschfeld, J. W. P., Tsfasman, M. A., Vlăduţ, S. G.: The weight hierarchy of higher-dimensional Hermitian codes. IEEE Trans. Inform. Theory **40** (1994) 275–278.
4. Hodge, W. V. D., Pedoe, D.: "Methods of Algebraic Geometry", Vol. II, Cambridge Univ. Press, Cambridge, 1952.
5. Kleiman, S. L., Laksov, D.: Schubert Calculus. Amer. Math. Monthly **79** (1972) 1061–1082.
6. Musili, C.: Applications of standard monomial theory. In: "Proceedings of the Hyderabad Conference on Algebraic Groups", (S. Ramanan, Ed.), pp. 381–406, Manoj Prakashan, Madras (Distributed outside India by the American Math. Society, Providence), 1991.
7. Nogin, D. Yu.: Codes associated to Grassmannians. In: "Arithmetic, Geometry and Coding theory" (R. Pellikan, M. Perret, S. G. Vlăduţ, Eds.), pp. 145–154, Walter de Gruyter, Berlin/New York, 1996.
8. Ryan, C. T.: An application of Grassmannian varieties to coding theory. Congr. Numer. **57** (1987) 257–271.
9. Ryan, C. T. Projective codes based on Grassmann varieties. *Congr. Numer.* **57** (1987) 273–279.
10. Ryan, C. T.: The weight distribution of a code associated to intersection properties of the Grassmannian variety $G(3,6)$. *Congr. Numer.* **61** (1988) 183–198.
11. Ryan, C. T., Ryan, K. M.: An application of geometry to the calculation of weight enumerators. *Congr. Numer.* **67** (1988) 77–90.
12. Ryan, C. T., Ryan, K. M.: The minimum weight of the Grassmann codes $C(k,n)$. Discrete Appl. Math. **28** (1990) 149–156.
13. Tsfasman M. A., Vlăduţ, S. G.: "Algebraic Geometric Codes", Kluwer, Amsterdam, 1991.
14. Tsfasman, M. A., Vlăduţ, S. G.: Geometric approach to higher weights. IEEE Trans. Inform. Theory **41** (1995) 1564–1588.
15. Wei, V. K.: Generalized Hamming weights for linear codes. IEEE Trans. Inform. Theory **37** (1991) 1412–1418.

Toric Surfaces and Error-correcting Codes

Johan P. Hansen[1]

Matematisk Institut, Århus Universitet, 8000 Århus C, Denmark
matjph@imf.au.dk

Abstract. From an integral convex polytope in $\mathbb{R}^2$ we give an explicit description of an error-correcting code over the finite field $\mathbb{F}_q$ of length $(q-1)^2$. The codes are obtained from toric surfaces and the results are proved using the cohomology and intersection theory of such surfaces. The parameters of three such families of toric codes are determined.

1   Toric codes

The theory can be extended to higher dimensions, here we will consider only the case of dimension 2.

Let $M \simeq \mathbb{Z}^2$ be a free $\mathbb{Z}$-module of rank 2 over the integers $\mathbb{Z}$. Let $\square$ be an integral convex polytope in $M_{\mathbb{R}} = M \otimes_{\mathbb{Z}} \mathbb{R}$, i.e. a compact convex polyhedron such that the vertices belong to M.

Let q be a prime power and let $\xi \in \mathbb{F}_q$ be a primitive element. For any i such that $0 \leq i \leq q-1$ and any j such that $0 \leq j \leq q-1$, we let $P_{ij} = (\xi^i, \xi^j) \in \mathbb{F}_q^* \times \mathbb{F}_q^*$. Let m_1, m_2 be a $\mathbb{Z}$-basis for M. For any $m = \lambda_1 m_1 + \lambda_2 m_2 \in M \cap \square$, we let $\mathbf{e}(m)(P_{ij}) = (\xi^i)^{\lambda_1}(\xi^j)^{\lambda_2}$.

Definition 1. *The toric code $C_\square$ associated to $\square$ is the linear code of length $n = (q-1)^2$ generated by the vectors*

$$\{(\mathbf{e}(m)(P_{ij}))_{i=0,\ldots,q-1;j=0,\ldots,q-1} \mid m \in M \cap \square\}. \tag{1}$$

We present a general method to obtain the dimension and a lower bound for the minimal distance. In particular we obtain the following three results. The second code is a subcode of the Reed Muller code on $\mathbb{P}^2$.

Theorem 1. *Let d be a positive integer and let $\square$ be the polytope in $M_{\mathbb{R}}$ with vertices $(0,0), (d,d), (0,2d)$, see figure 1. Assume that $2d < q-1$. The toric code $C_\square$ has length equal to*

$$(q-1)^2, \tag{2}$$

dimension equal to

$$\#(M \cap \square) = (d+1)^2 \tag{3}$$

(the number of lattice points in $\square$) and minimal distance greater or equal to

$$(q-1)^2 - 2d(q-1). \tag{4}$$

Theorem 2. *Let d be a positive integer and let $\square$ be the polytope in $M_{\mathbb{R}}$ with vertices $(0,0), (d,0), (0,d)$, see figure 2. Assume that $d < q - 1$. The toric code $C_\square$ has length equal to*

$$(q-1)^2, \tag{5}$$

dimension equal to

$$\#(M \cap \square) = \frac{(d+1)(d+2)}{2} \tag{6}$$

(the number of lattice points in $\square$) and minimal distance greater or equal to

$$(q-1)^2 - d(q-1). \tag{7}$$

Theorem 3. *Let d, e be positive integers and let $\square$ be the polytope in $M_{\mathbb{R}}$ with vertices $(0,0), (d,0), (d,e), (0,e)$, see figure 3. Assume that $d < q-1$ and that $e < q - 1$. The toric code $C_\square$ has length equal to*

$$(q-1)^2, \tag{8}$$

dimension equal to

$$\#(M \cap \square) = (d+1)(e+1) \tag{9}$$

(the number of lattice points in $\square$) and minimal distance greater or equal to

$$(q-1)^2 - (d(q-1) + (q-1-d)e). \tag{10}$$

2   The method of toric varieties

The toric codes are obtained from evaluating certain rational functions in rational points on toric varieties. For the general theory of toric varieties we refer to [1] and [3]. Here we will be using toric surfaces and we recollect their theory.

In 2.2 we present the method using toric varieties, their cohomology and intersection theory to obtain bounds for the number of rational zeroes of a rational function. In 2.3 this is used to prove the theorems on dimension and minimal distance of the codes $C_\square$ presented above.

2.1   Toric surfaces and their cohomology

Let M be an integer lattice $M \simeq \mathbb{Z}^2$. Let $N = \mathrm{Hom}_{\mathbb{Z}}(M, \mathbb{Z})$ be the dual lattice with canonical $\mathbb{Z}$ - bilinear pairing $< \ , \ >: M \times N \to \mathbb{Z}$ Let $M_{\mathbb{R}} = M \otimes_{\mathbb{Z}} \mathbb{R}$ and $N_{\mathbb{R}} = N \otimes_{\mathbb{Z}} \mathbb{R}$ with canonical $\mathbb{R}$ - bilinear pairing $< \ , \ >: M_{\mathbb{R}} \times N_{\mathbb{R}} \to \mathbb{R}$.

Given a 2-dimensional integral convex polytope $\square$ in $M_{\mathbb{R}}$. The support function $h_\square : N_{\mathbb{R}} \to \mathbb{R}$ is defined as $h_\square(n) := \inf\{< m, n > \mid m \in \square\}$ and $\square$ can be reconstructed:

$$\square_h = \{m \in M \mid \ < m, n > \geq h(n) \quad \forall n \in N\}. \tag{11}$$

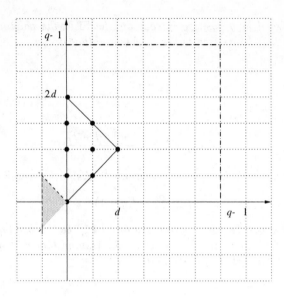

Fig. 1. The polytope $\square$ of Theorem 1 is the triangle with vertices $(0,0)$, (d,d), $(0,2d)$.

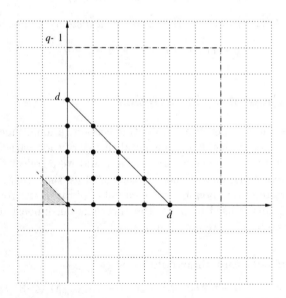

Fig. 2. The polytope $\square$ of Theorem 2 is the triangle with vertices $(0,0)$, $(d,0)$, $(0,d)$.

The support function $h_\square$ is piecewise linear in the sense that $N_\mathbb{R}$ is the union of a non-empty finite collection of strongly convex polyhedral cones in $N_\mathbb{R}$ such that $h_\square$ is linear on each cone. A fan is a collection Δ of strongly convex polyhedral cones in $N_\mathbb{R}$ such that every face of $\sigma \in \Delta$ is contained in Δ and $\sigma \cap \sigma' \in \Delta$ for all $\sigma, \sigma' \in \Delta$.

The *normal fan* Δ is the coarsest fan such that $h_\square$ is linear on each $\sigma \in \Delta$, i.e. for all $\sigma \in \Delta$ there exists $l_\sigma \in M$ such that

$$h_\square(n) = < l_\sigma, n > \quad \forall n \in \sigma. \tag{12}$$

The 1-dimensional cones $\rho \in \Delta$ are generated by unique primitive elements $n(\rho) \in N \cap \rho$ such that $\rho = \mathbb{R}_{\geq 0} n(\rho)$.

Upon refinement of the normal fan, we can assume that two successive pairs of $n(\rho)$'s generate the lattice and we obtain *the refined normal fan*.

Example 1. Consider the polytope of Theorem 1, see figure 1 . The refined normal fan is show in figure 4. We have that $n(\rho_1) = \begin{pmatrix} 1 \\ 0 \end{pmatrix}$, $n(\rho_2) = \begin{pmatrix} -1 \\ 1 \end{pmatrix}$, $n(\rho_3) = \begin{pmatrix} -1 \\ 0 \end{pmatrix}$ and $n(\rho_4) = \begin{pmatrix} -1 \\ -1 \end{pmatrix}$. Let σ_1 be the cone generated by $n(\rho_1)$ and $n(\rho_2)$, σ_2 be the cone generated by $n(\rho_2)$ and $n(\rho_3)$, σ_3 the cone generated by $n(\rho_3)$ and $n(\rho_4)$ and σ_4 the cone generated by $n(\rho_4)$ and $n(\rho_1)$.

The support function is:

$$h_\square \begin{pmatrix} n_1 \\ n_2 \end{pmatrix} = \begin{cases} \begin{pmatrix} 0 \\ 0 \end{pmatrix} \cdot \begin{pmatrix} n_1 \\ n_2 \end{pmatrix} & \text{if } \begin{pmatrix} n_1 \\ n_2 \end{pmatrix} \in \sigma_1, \\ \begin{pmatrix} d \\ d \end{pmatrix} \cdot \begin{pmatrix} n_1 \\ n_2 \end{pmatrix} & \text{if } \begin{pmatrix} n_1 \\ n_2 \end{pmatrix} \in \sigma_2, \\ \begin{pmatrix} d \\ d \end{pmatrix} \cdot \begin{pmatrix} n_1 \\ n_2 \end{pmatrix} & \text{if } \begin{pmatrix} n_1 \\ n_2 \end{pmatrix} \in \sigma_3, \\ \begin{pmatrix} 0 \\ 2d \end{pmatrix} \cdot \begin{pmatrix} n_1 \\ n_2 \end{pmatrix} & \text{if } \begin{pmatrix} n_1 \\ n_2 \end{pmatrix} \in \sigma_4. \end{cases} \tag{13}$$

Example 2. Consider the polytope of Theorem 2, see figure 2 . The refined normal fan is show in figure 5. We have that $n(\rho_1) = \begin{pmatrix} 1 \\ 0 \end{pmatrix}$, $n(\rho_2) = \begin{pmatrix} 0 \\ 1 \end{pmatrix}$, $n(\rho_3) = \begin{pmatrix} -1 \\ -1 \end{pmatrix}$. Let σ_1 be the cone generated by $n(\rho_1)$ and $n(\rho_2)$, σ_2 be the cone generated by $n(\rho_2)$ and $n(\rho_3)$ and σ_3 the cone generated by $n(\rho_3)$ and $n(\rho_1)$.

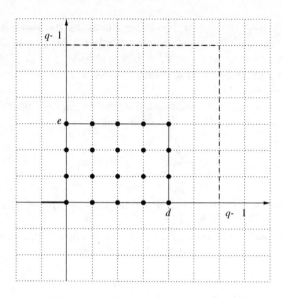

Fig. 3. The polytope $\square$ of Theorem 3 is the square with vertices $(0,0)$, $(d,0)$, (d,e), $(0,e)$.

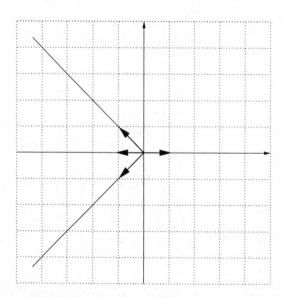

Fig. 4. The normal fan af the polytope in figure 1

The support function is:

$$
h_\square \begin{pmatrix} n_1 \\ n_2 \end{pmatrix} = \begin{cases} \begin{pmatrix} 0 \\ 0 \end{pmatrix} \cdot \begin{pmatrix} n_1 \\ n_2 \end{pmatrix} & \text{if } \begin{pmatrix} n_1 \\ n_2 \end{pmatrix} \in \sigma_1, \\[2mm] \begin{pmatrix} d \\ 0 \end{pmatrix} \cdot \begin{pmatrix} n_1 \\ n_2 \end{pmatrix} & \text{if } \begin{pmatrix} n_1 \\ n_2 \end{pmatrix} \in \sigma_2, \\[2mm] \begin{pmatrix} 0 \\ d \end{pmatrix} \cdot \begin{pmatrix} n_1 \\ n_2 \end{pmatrix} & \text{if } \begin{pmatrix} n_1 \\ n_2 \end{pmatrix} \in \sigma_3. \end{cases} \tag{14}
$$

Example 3. Consider the polytope of Theorem 3, see figure 3 . The refined normal fan is show in figure 6. We have that $n(\rho_1) = \begin{pmatrix} 1 \\ 0 \end{pmatrix}$, $n(\rho_2) = \begin{pmatrix} 0 \\ 1 \end{pmatrix}$, $n(\rho_3) = \begin{pmatrix} -1 \\ 0 \end{pmatrix}$ and $n(\rho_4) = \begin{pmatrix} 0 \\ -1 \end{pmatrix}$. Let σ_1 be the cone generated by $n(\rho_1)$ and $n(\rho_2)$, σ_2 be the cone generated by $n(\rho_2)$ and $n(\rho_3)$, σ_3 the cone generated by $n(\rho_3)$ and $n(\rho_4)$ and σ_4 the cone generated by $n(\rho_4)$ and $n(\rho_1)$. The support function is:

$$
h_\square \begin{pmatrix} n_1 \\ n_2 \end{pmatrix} = \begin{cases} \begin{pmatrix} 0 \\ 0 \end{pmatrix} \cdot \begin{pmatrix} n_1 \\ n_2 \end{pmatrix} & \text{if } \begin{pmatrix} n_1 \\ n_2 \end{pmatrix} \in \sigma_1, \\[2mm] \begin{pmatrix} d \\ 0 \end{pmatrix} \cdot \begin{pmatrix} n_1 \\ n_2 \end{pmatrix} & \text{if } \begin{pmatrix} n_1 \\ n_2 \end{pmatrix} \in \sigma_2, \\[2mm] \begin{pmatrix} d \\ e \end{pmatrix} \cdot \begin{pmatrix} n_1 \\ n_2 \end{pmatrix} & \text{if } \begin{pmatrix} n_1 \\ n_2 \end{pmatrix} \in \sigma_3, \\[2mm] \begin{pmatrix} 0 \\ e \end{pmatrix} \cdot \begin{pmatrix} n_1 \\ n_2 \end{pmatrix} & \text{if } \begin{pmatrix} n_1 \\ n_2 \end{pmatrix} \in \sigma_4. \end{cases} \tag{15}
$$

The 2-dimensional *algebraic torus* $T_N \simeq \overline{\mathbb{F}}_q^* \times \overline{\mathbb{F}}_q^*$ is defined by $T_N := \operatorname{Hom}_{\mathbb{Z}}(M, \overline{\mathbb{F}}_q^*)$. The multiplicative character $\mathbf{e}(m)$, $m \in M$ is the homomorphism $\mathbf{e}(m) : T \to \overline{\mathbb{F}}_q^*$ defined by $\mathbf{e}(m)(t) = t(m)$ for $t \in T_N$. Specifically, if $\{n_1, n_2\}$ and $\{m_1, m_2\}$ are dual $\mathbb{Z}$-bases of N and M and we denote $u_j := \mathbf{e}(m_j)$, $j = 1, 2$, then we have an isomorphism $T_N \simeq \overline{\mathbb{F}}_q^* \times \overline{\mathbb{F}}_q^*$ sending t to $(u_1(t), u_2(t))$. For $m = \lambda_1 m_1 + \lambda_2 m_2$ we have

$$
\mathbf{e}(m)(t) = u_1(t)^{\lambda_1} u_2(t)^{\lambda_2} \tag{16}
$$

The *toric surface* $X_\square$ associated to the refined normal fan Δ of $\square$ is irreducible, non-singular and complete

$$
X_\square = \cup_{\sigma \in \Delta} U_\sigma \tag{17}
$$

where U_σ is the $\overline{\mathbb{F}}_q$-valued points of the affine scheme $\operatorname{Spec}(\overline{\mathbb{F}}_q[S_\sigma])$, i.e.

$$
U_\sigma = \{u \in S_\sigma \to \overline{\mathbb{F}}_q | u(0) = 1, \ u(m + m') = u(m)u(m') \ \forall n, m' \in S_\sigma\} \tag{18}
$$

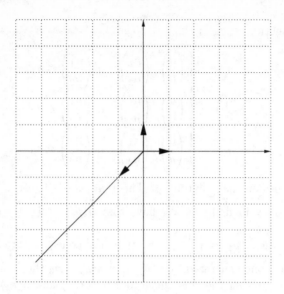

Fig. 5. The normal fan af the polytope in figure 2

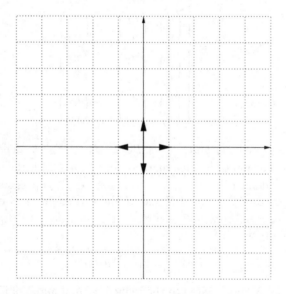

Fig. 6. The normal fan af the polytope in figure 3

If $\sigma, \tau \in \Delta$ and τ is a face of σ, then U_τ is an open subset of U_σ. Obviously $S_0 = M$ and $U_0 = T_N$ such that the algebraic torus T_N is an open subset of $X_\square$. T_N acts algebraically on $X_\square$. On $u \in U_\sigma$ the action of $t \in T_N$ is obtained as

$$(tu)(m) := t(m)u(m)   m \in S_\sigma \tag{19}$$

such that $tu \in U_\sigma$ and U_σ is T_N-stable. The orbits of this action is in one-to-one correspondance with Δ. For each $\sigma \in \Delta$ let

$$\operatorname{orb}(\sigma) := \{u : M \cap \sigma \to \overline{\mathbb{F}}_q{}^* | u \text{ is a group homomorphism}\}.$$

Then $\operatorname{orb}(\sigma)$ is a T_N orbit in $X_\square$. Define $V(\sigma)$ to be the closure of $\operatorname{orb}(\sigma)$ in $X_\square$.

A Δ-linear support function h gives rise to the Cartier divisor D_h. Let $\Delta(1)$ be the 1-dimensional cones in Δ then

$$D_h := - \sum_{\rho \in \Delta(1)} h(n(\rho)) V(\rho) \tag{20}$$

In particular

$$D_m = \operatorname{div}(\mathbf{e}(-m))   m \in M \tag{21}$$

Following [3] Lemma 2.3 we have the lemma.

Lemma 1. *Let h be a Δ-linear support function with associated Cartier divisor D_h and convex polytope $\square_h$ defined in (11). The vector space $\mathrm{H}^0(X, O_X(D_h))$ of global sections of $O_X(D_h)$, i.e. rational functions f on $X_\square$ such that $\operatorname{div}(f) + D_h \geq 0$ has dimension $\#(M \cap \square_h)$ and has $\{\mathbf{e}(m)| m \in M \cap \square_h\}$ as a basis.*

Remark 1. In Example 1

$$D_h := - \sum_{\rho \in \Delta(1)} h(n(\rho)) V(\rho) = d V(\rho_3) + 2d V(\rho_4) \tag{22}$$

and

$$\dim \mathrm{H}^0(X, O_X(D_h)) = (d+1)^2 \tag{23}$$

In Example 2

$$D_h := - \sum_{\rho \in \Delta(1)} h(n(\rho)) V(\rho) = d V(\rho_3) \tag{24}$$

and

$$\dim \mathrm{H}^0(X, O_X(D_h)) = \frac{(d+1)(d+2)}{2} \tag{25}$$

In Example 3

$$D_h := - \sum_{\rho \in \Delta(1)} h(n(\rho)) V(\rho) = d V(\rho_3) + e V(\rho_4) \tag{26}$$

and

$$\dim \mathrm{H}^0(X, O_X(D_h)) = (d+1)(e+1). \tag{27}$$

2.2   Intersection theory and the number of rational zeroes of a rational function

For a Δ-linear support function h and a 1-dimensional cone $\rho \in \Delta(1)$ we will determine the intersection number $(D_h; V(\rho))$ between the Cartier divisor D_h and $V(\rho)) = \mathbb{P}^1$. This is number is obtained in [3], Lemma 2.11. The cone ρ is the common face fo two 2-dimensional cones $\sigma', \sigma'' \in \Delta(2)$. Choose primitive elements $n', n'' \in N$ such that

$$n' + n'' \in \mathbb{R}\rho$$
$$\sigma' + \mathbb{R}\rho = \mathbb{R}_{\geq 0}n' + \mathbb{R}\rho$$
$$\sigma'' + \mathbb{R}\rho = \mathbb{R}_{\geq 0}n'' + \mathbb{R}\rho$$

Lemma 2. *For any $l_\rho \in M$ such that h coincides with l_ρ on ρ, let $\overline{h} = h - l_\rho$. Then*

$$(D_h; V(\rho)) = -(\overline{h}(n') + \overline{h}(n'')) \tag{28}$$

Also we will need the self-intersection number of a Cartier divisor D_h. This is number is obtained in [3], Prop. 2.10.

Lemma 3. *Let D_h be a Cartier divisor and let $\square_h$ be the polytope associated to h , see (11). Then*

$$(D_h; D_h) = 2\,\mathrm{vol}_2(\square_h), \tag{29}$$

where vol_2 is the normalized Lebesgue-measure.

2.3   Determination of parameters

We start by exhibiting the toric codes as evaluation codes.
    For each $t \in T \simeq \overline{\mathbb{F}}_q{}^* \times \overline{\mathbb{F}}_q{}^*$, we can evaluate

$$\mathrm{H}^0(X, O_X(D_h)) \to \overline{\mathbb{F}}_q{}^*$$
$$f \mapsto f(t)$$

Taking all points in $T(\mathbb{F}_q)$ we obain the code $C_\square$:

$$\mathrm{H}^0(X, O_X(D_h))^{\mathrm{Frob}} \to C_\square \subset (\mathbb{F}_q^*)^{T(\mathbb{F}_q^*)}$$
$$f \mapsto (f(t))_{t \in T(\mathbb{F}_q)}$$

and the generators of the code is obtained as the image of the basis:

$$\mathbf{e}(m) \mapsto (\mathbf{e}(m)(t))_{t \in T(\mathbb{F}_q)} \tag{30}$$

as in (1).

    Let $m_1 = (1,0)$. The $\mathbb{F}_q$-rational points of $T \simeq \overline{\mathbb{F}}_q{}^* \times \overline{\mathbb{F}}_q{}^*$ belong to the $q-1$ lines on $X_\square$ given by $\prod_{\eta \in \mathbb{F}_q}(\mathbf{e}(m_1) - \eta) = 0$. Let $0 \neq f \in \mathrm{H}^0(X, O_X(D_h))$ and assume that f is identically zero along precisely a of these lines. As $\mathbf{e}(m_1) - \eta$ and $\mathbf{e}(m_1)$ have the same divisors of poles, they have equivalent divisors of zeroes, so

$$(\operatorname{div}(\mathbf{e}(m_1) - \eta))_0 \sim (\operatorname{div}(\mathbf{e}(m_1)))_0. \tag{31}$$

Therefore

$$\operatorname{div}(f) + D_h - a(\operatorname{div}(\mathbf{e}(m_1)))_0 \geq 0 \tag{32}$$

or equivalently

$$f \in \mathrm{H}^0(X, O_X(D_h - a(\operatorname{div}(\mathbf{e}(m_1)))_0). \tag{33}$$

In the cases of Theorem 1, Theorem 2 and Theorem 3 this implies that $a \leq d$ according to Lemma 1. On any of the other $q - 1 - a$ lines the number of zeroes of f is according to [2] at most the intersection number:

$$(D_h - a(\operatorname{div}(\mathbf{e}(m_1)))_0; (\operatorname{div}(\mathbf{e}(m_1)))_0) \tag{34}$$

which is equal to

$$(D_h; (\operatorname{div}(\mathbf{e}(m_1)))_0) - a((\operatorname{div}(\mathbf{e}(m_1)))_0; (\operatorname{div}(\mathbf{e}(m_1)))_0). \tag{35}$$

The number can be calculated using Lemma 2 and Lemma 3. In the situation of Theorem 1 (the volume-element is shown as gray in figure 1) this number is $2d - a \cdot 2 \cdot (\frac{1}{2} \cdot 1 \cdot 2) = 2d - 2a$. In the situation of Theorem 2 (the volume-element is shown as gray in figure 2) it is $d - a \cdot 2 \cdot (\frac{1}{2} \cdot 1 \cdot 1) = d - a$. In the situation of Theorem 3 (the volume-element is the line segment shown in bold in figure 3) it is e. As $0 \leq a \leq d$ the total number of $\mathbb{F}_q$- rational zeroes for f in the three cases is at most: In Theorem 1:

$$a(q - 1) + (q - 1 - a)(2d - 2a) \leq (q - 1)2d \tag{36}$$

In Theorem 2:

$$a(q - 1) + (q - 1 - a)(d - a) \leq d(q - 1) \tag{37}$$

In Theorem 3:

$$a(q - 1) + (q - 1 - a)e \leq d(q - 1) + (q - 1 - d)e \tag{38}$$

This implies that the evaluation maps

$$\mathrm{H}^0(X, O_X(D_h))^{\mathrm{Frob}} \to C_\square \subset (\mathbb{F}_q^*)^{T(\mathbb{F}_q^*)}$$
$$f \mapsto (f(t))_{t \in T(\mathbb{F}_q)}$$

are injective in all three situations and using Remark 1 that the parameters of the toric codes are as claimed.

References

1. W. Fulton, "Introduction to Toric Varieties," *Annals of Mathematics Studies; no. 131*, Princeton University Press, 1993.
2. S. H. Hansen, "Error-correcting codes from higher-dimensional varieties," *Preprint Series No. 4, 1998, University of Aarhus.*
3. T. Oda, "Convex Bodies and Algebraic Geometry, An Introduction to the Theory of Toric Varieties," *Ergebnisse der Mathematik und ihrer Grenzgebiete, 3. Folge, Band 15*, Springer Verlag, 1985.

Decoding Spherical Codes Generated by Binary Partitions of Symmetric Pointsets

John K. Karlof[1] and Guodong Liu[2]

[1] Department of Mathematics and Statistics
The University of North Carolina at Wilmington
Wilmington, NC 28403
[2] Intelligent Information Systems
Durham, NC 27713

Abstract. Recently, Ericson and Zinoviev presented a clever, new construction for spherical codes for the Gaussian channel using ideas of code concatenation and set partitioning. This family of new spherical codes is generated from sets of binary codes using equally spaced symmetric pointsets on the real line. The family contains some of the best known spherical codes in terms of minimum distance. However, no efficient decoding algorithm is known for this new construction. In this paper, we present a new decoding algorithm for this family of spherical codes which is more efficient than maximum likelihood decoding.

1 Introduction

We consider a communications model in which messages are transmitted over a noisy channel to a receiver. We assume the messages come from an n dimensional spherical code, i.e. a finite set $X \subset \Re^n$ such that $\|x\|$ is constant $\forall x \in X$. We also assume the noise is Gaussian. Thus if x is the sent message then $y = x + z$ is the received signal where $z = (z_1, \ldots, z_n)$ is a random sample from a Gaussian distribution of mean zero. The receiver must make an estimate of x. The process of estimating is called decoding and the model is referred to as a Gaussian channel. If all the codewords of a spherical code are equally likely, then the decoding algorithm that minimizes the average message error probability is maximum likelihood decoding. In maximum likelihood decoding, the received message, y, is decoded as the codeword x that minimizes the Euclidean distances $\{d(y, z) : z \in X\}$. Good spherical codes should have i) a large Euclidean minimum distance and ii) an efficient decoding scheme that is as close as possible to maximum likelihood decoding.

In the past, permutation groups [2, 10, 11], groups of orthogonal matrices [1, 3, 5, 6, 9, 12], and shells of lattices [4] have all been used to generate spherical codes. Recently, Ericson and Zinoviev [7] presented a new construction of a family of spherical codes. This code construction begins with an equally spaced symmetric alphabet from the real line of size K. Then, a tree of binary codes, whose construction depends on the value of K, is used to construct the spherical code. This new family of codes contains some of the best known spherical codes in terms of minimum distance.

In this paper, we present a new efficient decoding algorithm for Ericson and Zinoviev's code construction. This decoding scheme is an m step algorithm. At each step the algorithm attempts to identify the binary codeword in the tree that was used to generate the sent spherical codeword. We incorporate Forney's [8] idea of error and erasure decoding and Zinoviev and Litsyn's [14] idea of distance decoding in our new decoding algorithm. We prove the error correcting ability of the new algorithm and through simulation compare its performance to maximum likelihood decoding. At low signal to noise ratios, it is 99% equivalent to maximum likelihood but takes just 2% of the computational time.

Section 2 contains a description of Ericson and Zinoviev's construction. In section 3, we present our decoding algorithm for these new spherical codes. We include descriptions of the code construction and our decoding algorithm only for the case of even alphabet size. The case of odd alphabet size is similar, but its inclusion in this paper would considerably lengthen the paper. In section 4 we discuss the error-correcting ability of our decoding algorithm. In section 5, we give a performance comparison between our decoding algorithm and maximum likelihood decoding based on simulation of these two decoding algorithms.

2   Ericson and Zinoviev's Coding Construction

In [7], a clever construction of spherical codes, some with optimal minimum distance, for Gaussian channel is presented. We include those same results in a modified form for even alphabet size.

The code construction begins with choosing K even and the code alphabet

$$L_K = \left\{ \pm\frac{1}{2}, \pm\frac{3}{2}, \ldots, \pm\frac{K-1}{2} \right\}$$

Let $\overline{L}_K = \{0, 1, \ldots, \frac{K}{2} - 1\}$ and form a tree with node labels, $\Gamma = \overline{L}_K \cup \{\lambda, *\}$, using the following rules. (Our nomenclature is taken from [13]).

1. The root of the tree is λ and λ is adjacent only to $*$. Every internal node has exactly two children except for λ. We will say that node λ is at level -1, $*$ is at level 0, the children of $*$ are at level 1, etc.
2. The children of $*$ are labeled 0 and 1 with 0 being the left child.
3. For succeeding levels, say level k, the left child of a node at level $k-1$ is labeled the same as its parent and the right child is chosen from $\overline{L}_K$ so that the sum of the labels of the two children is $2^k - 1$. If that is impossible, the node at level $k-1$ is a leaf.

Example 1.

$$\begin{aligned}
K &= 10 & L_{10} &= \{\pm\tfrac{1}{2}, \pm\tfrac{3}{2}, \ldots, \pm\tfrac{9}{2}\} \\
\overline{L}_{10} &= \{0, 1, \ldots, 4\} & \Gamma &= \{0, 1, \ldots, 4, \lambda, *\}
\end{aligned}$$

We choose a binary code for each internal node of the tree. Codes at level k will be designated C_γ^k where γ is the label of the corresponding node on the tree. An arbitrary code, C_λ^{-1} of length n is chosen for node λ. A code, C_*^0 of length n and constant weight w_* is chosen for node $*$. Suppose internal node γ at level $k-1, (k \geq 1)$ has internal node left child γl and internal node right child γr and code C_γ^{k-1} of length n_γ^{k-1} and constant weight w_γ^{k-1} has been chosen for node γ. Then code $C_{\gamma l}^k$ of length $n_{\gamma l}^k = n_\gamma^{k-1} - w_\gamma^{k-1}$ and constant weight $w_{\gamma l}^k$ is chosen for node γl and code $C_{\gamma r}^k$ of length $n_{\gamma r}^k = w_{\gamma r}^{k-1}$ and constant weight $w_{\gamma l}^k$ is chosen for node γr. If internal node γ at level $k-1$ has only one child that is internal (it will always be the right child), then code $C_{\gamma r}^k$ of length $n_{\gamma r}^k = w_{\gamma r}^{k-1}$ and constant weight $w_{\gamma l}^k$ is chosen for node γr. The only restrictions on the codes chosen is that they be constant weight and of the given length.

Example 2. (Previous example continued.) The tree for the example is presented below. The circles represent the nodes with the node labels inside them. To the left of each internal node is the name of the binary code associated with that node and on the right n is the length of the code and w is the constant weight.

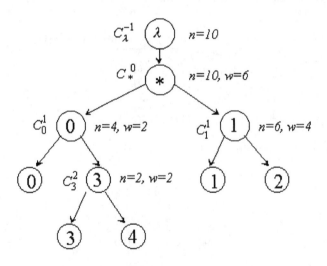

Fig. 1. Binary Code Tree for $K = 10$

The tree of binary codes and alphabet L_K is used to form a spherical code, X, of length n for the Gaussian channel. For each collection of codewords $\{c_\gamma^i \in C_\gamma^i \mid C_\gamma^i$ is a code in the tree$\}$, we form a codeword $x \in X$ in the following manner. Suppose the tree has $m+1$ levels of internal vertices. We form a $m+1$ by n matrix where the rows are labeled by the levels of the tree and the ith row consists of the codewords chosen from the codes at that level in the tree. Suppose the codewords have been arranged in the row labeled $i - 1, i \geq 1$. We arrange the codewords in row i in the following manner. Suppose codeword $c^i_{\gamma l} \in C^i_{\gamma l}$ and codeword $c^i_{\gamma r} \in C^i_{\gamma r}$ where $C^i_{\gamma l}$ and $C^i_{\gamma r}$ are the left and right children respectively of level $i-1$ code C^{i-1}_γ. Then the components of $c^i_{\gamma l}$ are placed under the $0's$ of c^{i-1}_γ and the components of $c^i_{\gamma r}$ are placed under the $1's$ of c^{i-1}_γ. If any of the children of C^{i-1}_γ are leaves, then $0's$ are placed under the corresponding components of c^{i-1}_γ.

The binary sequences that are the columns of the matrix correspond to the components of x. Suppose $\tilde{b} = \tilde{b}_0, \tilde{b}_1, \ldots, \tilde{b}_m$ is a binary sequence in the jth column of the matrix. We form the whole number $B = \sum_{i=0}^{m-1} b_i 2^i$ where $b_{m-1} = \tilde{b}_m$ and $b_i = b_{i+1} \oplus \tilde{b}_{i+1}, i = 0, \ldots, m-2$ and $\oplus$ represents binary addition. Then $x_j = \pm B\frac{1}{2}$. The $+$ or $-$ is given by the following table.

$\tilde{b}_0$	B is odd	B is even
1	$+$	$-$
0	$-$	$+$

Example 3. (Previous example continued.) Suppose the following codewords were chosen from the binary codes in the tree.

		symbol in matrix
$c^{-1}_\lambda = (0110100111) \in C^{-1}_\lambda$		
$c^0_* = (1011000111) \in C^0_*$		
$c^1_0 = (0110) \in C^1_0$	$-$
$c^1_1 = (110101) \in C^1_1$	$\sim$
$c^2_3 = (11) \in C^2_3$	$\wedge$

These codewords determine the following matrix.

$$\begin{bmatrix} 0 & 1 & 1 & 0 & 1 & 0 & 0 & 1 & 1 & 1 \\ 1 & 0 & 1 & 1 & 0 & 0 & 0 & 1 & 1 & 1 \\ \tilde{1} & \tilde{0} & \tilde{1} & \tilde{0} & \tilde{1} & \tilde{1} & \tilde{0} & \tilde{1} & \tilde{0} & \tilde{1} \\ 0 & 0 & 0 & 0 & \hat{1} & \hat{1} & 0 & 0 & 0 & 0 \end{bmatrix}$$

The spherical codeword $x \in X$ determined by this set of binary codewords is
$x = (\frac{5}{2}, -\frac{1}{2}, -\frac{5}{2}, -\frac{3}{2}, -\frac{9}{2}, \frac{9}{2}, \frac{1}{2}, -\frac{5}{2}, \frac{3}{2}, -\frac{5}{2})$

The following result relating the minimum distance of the spherical code X to the minimum distances of the binary codes $\{C^k_\gamma | k \geq -1\}$ appears in [7].

Theorem 1. *Let X be the spherical code generated by Ericson and Zinoviev's construction using the binary codes $\{C^k_\gamma | k \geq -1\}$. Let d^k_γ be the minimum Hamming distance of the code C^k_γ and let d^2 be the (unnormalized) minimum squared distance of X. Then $d^2 \geq min\{d^k_\gamma \cdot 4^{k+1} | k \geq -1\}$.*

3 Decoding Algorithm

3.1 Forming the Subalphabets

The first step is to perform binary partitions of the alphabet L_K which we now simply denote L. Our partitions have the same properties of partitions of the set $Z + \frac{1}{2}$ in [7]:

1. The distance within the subsets of subsequent partitions are increased by a factor of 2 in comparison with the distance in the original sets.
2. Odd symmetry is retained.

We use a binary tree to describe our partitions:

1. Each node of the binary tree has a subalphabet of the alphabet L associated with it.
2. Leaves in the tree have a single element associated with them. If a subalphabet with cardinality 2 or more is associated with a node, then this node must be an internal node.
3. Each node in the binary tree is a leaf or has exactly two children.
4. The root of the binary tree is at level 1. The children of the root are at level 2, etc.
5. The full alphabet L is associated with the root at level 1. The left child of the root L is labeled by the subalphabet L_0. The right child of the root L is labeled by the subalphabet L_1.
6. At level 1, for K even, $L \subset \frac{1}{2}(2Z+1)$, it is partitioned as $L = L_0 \cup L_1$, where $L_0 \subset \frac{1}{2}(4Z+1)$ and $L_1 \subset \frac{1}{2}(4Z-1)$.
7. Suppose subalphabet $L_{u_1 u_2 \ldots u_{i-1}}$, where $u_1, u_2, \ldots, u_{i-1}$ are binary numbers, is associated with an internal node at level $i-1$, the subalphabet associated with the left child is $L_{u_1 u_2 \ldots u_{i-1} 0}$ and the subalphabet associated with the right child is $L_{u_1 u_2 \ldots u_{i-1} 1}$. These subalphabets are chosen as follows:
   (a) If subalphabet $L_{u_1 u_2 \ldots u_{i-1}} \subset \frac{1}{2}(2^i Z + j)$, where $i > 1$ and $0 < j < 2^{i-1}$, then it is partitioned as $L_{u_1 u_2 \ldots u_{i-1}} = L_{u_1 u_2 \ldots u_{i-1} 0} \cup L_{u_1 u_2 \ldots u_{i-1} 1}$, where $L_{u_1 u_2 \ldots u_{i-1} 0} \subset \frac{1}{2}(2^{i+1} Z + j)$ and $L_{u_1 u_2 \ldots u_{i-1} 1} \subset \frac{1}{2}(2^{i+1} Z + j - 2^i)$.
   (b) If subalphabet $L_{u_1 u_2 \ldots u_{i-1}} \subset \frac{1}{2}(2^i Z - j)$, where $0 < j < 2^{i-1}$, then it is partitioned as $L_{u_1 u_2 \ldots u_{i-1}} = L_{u_1 u_2 \ldots u_{i-1} 0} \cup L_{u_1 u_2 \ldots u_{i-1} 1}$, where $L_{u_1 u_2 \ldots u_{i-1} 0} \subset \frac{1}{2}(2^{i+1} Z - j)$ and $L_{u_1 u_2 \ldots u_{i-1} 0} \subset \frac{1}{2}(2^{i+1} Z - j + 2^i)$.

3.2 Decoding Algorithm

Let $x = (x_1, x_2, \ldots, x_n) \in X$, where $x_1, x_2, \ldots, x_n \in L$, be the word obtained by Ericson and Zinoviev's construction from the code words $c^1, c^2, \ldots, c^s$ of $C^1, C^2, \ldots, C^s$, respectively. Suppose d_i = minimum Hamming distance of C^i and ρ_i = squared minimum distance of the subalphabets at level i. Let $y = (y_1, y_2, \ldots, y_n), y_j \in R$ be the received word corrupted by noise. The new decoding algorithm consists of s steps, where each step finds $c^i, i = 1, \ldots, s$. At each step, the decoding algorithm is divided into an inner code decoding algorithm and an outer code decoding algorithm. Assume that $c^1, c^2, \ldots, c^{i-1}$ are already found prior to step i, we find c^i during step i. Now we describe the inner coder decoder and outer code decoder in step i.

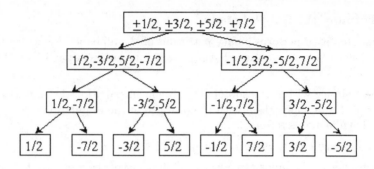

Fig. 2. Binary Partition Tree for $K = 8$

Inner Code Decoder The inner code decoding algorithm is as follows:

1. For $j = 1, 2, \ldots, n$, y_j is decoded in the subalphabet $L_{c_j^1, c_j^2, \ldots, c_j^{i-1}}$, which for simplicity is denoted by L_j^{i-1}, where $i > 1$. When $i = 1$, y_i is decoded in the whole alphabet L which is denoted by L_j^0. The output of the inner code decoder is $\hat{c}_j^i$. We define a distance parameter ρ_j^i associated with it.

2. If the cardinality of L_j^{i-1} is 1, then the node associated with L_j^{i-1} in the binary tree is a leaf. In this case, $\hat{c}_j^i = 0$ and $\rho_j^i = 0$.

3. If the cardinality of L_j^{i-1} is 2 or more, then the node associated with L_j^{i-1} in the binary tree is an internal node with both a left child and a right child. The decoder chooses an element z_j^i from L_j^{i-1} which has the minimum distance to y_j, i.e. $|y_j - z_j^i| \le |y_j - \bar{z}_j^i|, \forall \bar{z}_j^i \in L_j^{i-1}$. We define the distance parameter ρ_j^i by the following rules: If $\rho(y_j, z_j^i) = |y_j - z_j^i|^2 < \rho_i/4$, then set $\rho_j^i = \rho(y_j, z_j^i)$. Otherwise, set $\rho_j^i = \rho_i/4$ and the result of the decoding is an erasure. Now we will determine the number $\hat{c}_j^i$ of $L_{c_j^1, c_j^2, \ldots, \hat{c}_j^i}$ using the procedure in section 2. The result of the inner code decoder for $j = 1, \ldots, n$ at step i is a vector $\hat{c}^i = (\hat{c}_1^i, \hat{c}_2^i, \ldots, \hat{c}_n^i)$ and a tuple of n numbers $(\rho_1^i, \rho_2^i, \ldots, \rho_n^i)$.

Outer Code Decoder Without loss of generality, we may assume that the numbers $\rho_1^i, \ldots, \rho_n^i$ are arranged in nonincreasing order. Let r_i be the number of erasures claimed by inner coder decoder, the outer code decoder is defined as follows: Begin with $m = d_i - 1$, then $m = d_i - 3, \ldots$, until $m = r_i$ or $m = r_i + 1$. We decode the word $\hat{c}^i$ by an "error- and-erasure decoding algorithm". At trial m, we find a candidate codeword $c^i(m) \in C^i$.

Definition 1. *In trial m, the decoding of $\hat{c}^i_j$ produces an error if $c^i_j(m) \neq \hat{c}^i_j$ and $\hat{c}^i_j$ is not an erasure claimed by the outer code decoder. Let $t =$ number of errors.*

First, erase the first m symbols $\hat{c}^i_j, j = 1, \ldots, m$. Then we choose codewords one by one from the code C^i to compare with $\hat{c}^i$ until we exaust the code C^i or find a candidate $c^i(m)$ from C^i which satisfies $m + 2t < d_i$. Here t is the number of errors as defined in definition 3.1. We define a number $F(c^i(m))$ associated with $c^i(m)$,

$$F(c^i(m)) = \sum_{j=1}^{n} f(c^i_j(m))$$

where

$$f(c^i_j(m)) = \begin{cases} \rho^i_j & \text{if } c^i_j(m) = \hat{c}^i_j \text{ or } \hat{c}^i_j \text{ is erased by inner code decoder} \\ \left(\sqrt{\rho_i} - \sqrt{\rho^i_j}\right)^2 & \text{otherwise, } \hat{c}^i_j \text{ is an error.} \end{cases}$$

Then we check $F(c^i(m)) < d_i\rho_i/4$. If so, then $c^i(m)$ is the output of the decoder in step i. If not, then we take the next smaller m and repeat the outer code decoding of step i. If all candidates fail to satisfy $F(c^i(m)) < d_i\rho_i/4$, then we say that the received word is beyond the correcting ability of our decoding algorithm. The receiver will ask the transitter to re-send this codeword.

Proposition 1. *For a fixed erasure number m and outer code C^i, if there is a codeword c^i from code C^i which satisfies $m + 2t < d_i$, where t is the number of errors, then it is unique.*

Proof: Assume the contrary. Then there are two codewords c^{i1} and c^{i2} in C^i such that $m + 2t_1 < d_i$ and $m + 2t_2 < d_i$. Then $t_1 < \frac{d_i-m}{2}$ and $t_2 < \frac{d_i-m}{2}$. Since the minimum Hamming distance of code C^i is d_i, the Hamming distance between codewords c^{i1} and c^{i2} is $d_H(c^{i1}, c^{i2}) = \sum_{j=1}^{n} c^{i1}_j \oplus c^{i2}_j \geq d_i$. The symbol $\oplus$ represents mod 2 addition. So in the rest of the $n - m$ unerased positions, there are at least $d_i - m$ positions where c^{i1} and c^{i2} are different. Let J be the set of unerased positions where c^{i1} and c^{i2} are different. Since $t_1 \geq \sum_{j \in J} c^{i1}_j \oplus \hat{c}^i_j$ and $t_2 \geq \sum_{j \in J} c^{i2}_j \oplus \hat{c}^i_j$, we have

$$t_1 + t_2 \geq \sum_{j \in J} c^{i1}_j \oplus \hat{c}^i_j + \sum_{j \in J} c^{i2}_j \oplus \hat{c}^i_j$$

$$= \sum_{j \in J} c^{i1}_j \oplus \hat{c}^i_j + c^{i2}_j \oplus \hat{c}^i_j$$

$$= \sum_{j \in J} 1$$

$$\geq d_i - m$$

contradicting $t_1 < \frac{d_i-m}{2}$ and $t_2 < \frac{d_i-m}{2}$.
Q.E.D.

4   Error-correcting Ability of the Decoding Algorithm

In [14], several results on the error-correcting ability of the distance decoding algorithm for generalized concatenated codes are presented. In this section, we prove similar results, which are tailored to fit our decoding algorithm.

Lemma 1. *The inequalities*

$$\rho^i_{j'} + \left(\sqrt{\rho_i} - \sqrt{\rho^i_j}\right)^2 \geq \rho_i/2 \tag{1}$$

and

$$\left(\sqrt{\rho_i/2} - \sqrt{2\rho^i_j}\right)^2 + \rho^i_{j'} - \rho^i_j \geq 0 \tag{2}$$

are equivalent.

Proof: Note that

$$\left(\sqrt{\rho_i} - \sqrt{\rho^i_j}\right)^2 = \rho_i + \rho^i_j - 2\sqrt{\rho_i}\sqrt{\rho^i_j}$$
$$= \rho_i + \rho^i_j - 2\sqrt{\rho_i/2}\sqrt{2\rho^i_j}.$$

So

$$\rho^i_{j'} + \left(\sqrt{\rho_i} - \sqrt{\rho^i_j}\right)^2 - \rho_i/2 = \rho_i + \rho^i_j - 2\sqrt{\rho_i/2}\sqrt{2\rho^i_j} + \rho^i_{j'} - \rho_i/2$$
$$= \rho_i/2 + 2\rho^i_j - 2\sqrt{\rho_i/2}\sqrt{2\rho^i_j} + \rho^i_{j'} - \rho^i_j$$
$$= \left(\sqrt{\rho_i/2} - \sqrt{2\rho^i_j}\right)^2 + \rho^i_{j'} - \rho^i_j.$$

Hence the inequalities (1) and (2) are equivalent.
Q.E.D.

Proposition 2. *If there is a codeword $c^i \in C^i$ for which the parameter $F(c^i) < \frac{d_i \rho_i}{4}$, then it is unique.*

Proof: Assume the contrary. There are two codewords c^{i1} and c^{i2} for which

$$F(c^{i1}) < \frac{d_i \rho_i}{4} \tag{3}$$

and

$$F(c^{i2}) < \frac{d_i \rho_i}{4} \tag{4}$$

Let J be the set of positions where c^{i1} and c^{i2} are different, then $|J| \geq d_i$. Let $j \in J$. If $\hat{c}^i_j$ is an erasure claimed by the inner code decoder, then $f(c^{i1}_j) = f(c^{i2}_j) = \rho_i/4$. Hence $f(c^{i1}_j) + f(c^{i2}_j) = \rho_i/2$. If $\hat{c}^i_j$ is not an erasure claimed by

the inner coder decoder, then for one of the codewords(say, c^{i1}), we have $\hat{c}_j^{i1}$ $= c_j^{i1}$, so $f(c_j^{i1}) = \rho_j^i$, but since $j \in J$, $c_j^i \neq c_j^{i2}$, so $\hat{c}_j^i \neq c_j^{i2}$. We then have $f(c_j^{i2}) = \left(\sqrt{\rho_i} - \sqrt{\rho_j^i}\right)^2$. Hence $f(c_j^{i1}) + f(c_j^{i2}) = \rho_j^i + \left(\sqrt{\rho_i} - \sqrt{\rho_j^i}\right)^2 \geq \rho_i/2$. Therefore $F(c^{i1}) + F(c^{i2}) \geq \sum_{j\in J} f(c_j^{i1}) + f(c_j^{i2}) \geq \frac{d_i\rho_i}{2}$, contradicting (3) and (4).
Q.E.D.

Proposition 3. *Let x be the transmitted codeword constructed by the binary codewords $c^1, c^2, \ldots, c^i, \ldots, c^s$ and y the received word corrupted by noise. If*

$$\rho(x,y) < d_i\rho_i/4 \tag{5}$$

then

$$F(c^i) < d_i\rho_i/4. \tag{6}$$

Proof: If $c_j^i = \hat{c}_j^i$ and $\hat{c}_j^i$ is not an erasure claimed by the inner code decoder, then $\rho(x_j, y_j) = \rho_j^i = f(c_j^i)$. If $c_j^i \neq \hat{c}_j^i$ and $\hat{c}_j^i$ is not an erasure claimed by the inner code decoder, then $\rho(x_j, y_j) \geq \left(\sqrt{\rho_i} - \sqrt{\rho_j^i}\right)^2 = f(c_j^i)$. If $\hat{c}_j^i$ is an erasure claimed by the inner code decoder, then $\rho(x_j, y_j) \geq \rho_i/4 = f(c_j^i)$. Hence $\rho(x,y) = \sum_{j=1}^n \rho(x_j, y_j) \geq \sum_{j=1}^n f(c_j^i) = F(c^i)$. Since $\rho(x,y) < d_i\rho_i/4$,then $F(c^i) < d_i\rho_i/4$.
Q.E.D.

Proposition 4. *Let x be the transmitted codeword constructed by the binary codewords $c^1, c^2, \ldots, c^i, \ldots, c^s$ and let y be the received word corrupted by noise. Assume that the first $i-1$ code vectors $c^1, c^2, \ldots, c^{i-1}$ have been found correctly. If*

$$\rho(x,y) < d_i\rho_i/4 \tag{7}$$

then at least one error-erasure decoding trial will successfully decode to c^i, i.e.

$$m + 2t < d_i$$

for some trial where m is the number of erasures claimed by the outer code decoder and t is the number of errors when decoding to c^i in this trial.

Proof: We prove this proposition by assuming the contrary, which means that all trials fail to decode to c^i when m is chosen as $d_i - 1$, $d_i - 3$, $\ldots$, until m is r_i or $r_i + 1$ where r_i is the number of erasures claimed by the inner code decoder. This is only possible when the $n - m$ unerased symbols of $\hat{c}^i$ contain at least t_m errors where $m + 2t_m \geq d_i$ where d_i is the minimum Hamming distance of the outer code C^i [8].
If $d_i = r_i + 1$ (mod 2), assume that one of every two consective components in the word $\hat{c}^i$ with the number from m to $d_i + 1$ is an error. Since the sequence of numbers ρ_j^i is nonincreassing, where $j = 1, \ldots, m$, the inequality (2) holds

only when $j' < j$. We assume the worst case where errors occur in positions $m+1, m+3, \ldots, d_i - 2, d_i$, then

$$\rho(x,y) = \sum_{j=1}^{n} \rho(x_j, y_j)$$

$$\geq \sum_{j=1}^{r_i} \rho(x_j, y_j) + \sum_{j=r_i+1}^{d_i+1} \rho(x_j, y_j)$$

$$\geq \frac{r_i \rho_i}{4} + \sum_{j=r_i+1, r_i+3, \ldots}^{d_i+1} [(\sqrt{\rho_i} - \sqrt{\rho_j^i})^2 + \rho_{j+1}^i]$$

$$\geq \frac{r_i \rho_i}{4} + \sum_{j=r_i+1, r_i+3, \ldots}^{d_i} \{[(\sqrt{\rho_i/2} - \sqrt{2\rho_j^i})^2 + \rho_{j+1}^i - \rho_j^i] + \rho_i/2\}$$

$$= \frac{r_i \rho_i}{4} + \frac{\rho_i}{2}\frac{d_i - r_i + 1}{2} + \sum_{j=r_i+1, r_i+3, \ldots}^{d_i} [(\sqrt{\rho_i/2} - \sqrt{2\rho_j^i})^2 + \rho_{j+1}^i - \rho_j^i]$$

$$\geq \frac{\rho_i}{4}(d_i + 1) + \sum_{j=r_i+1, r_i+3, \ldots}^{d_i} (\rho_{j+1}^i - \rho_j^i)$$

$$= \frac{\rho_i}{4}d_i + \frac{\rho_i}{4} - \rho_{r_i+1}^i + \rho_{r_i+2}^i - \cdots - \rho_{d_i-1}^i + \rho_{d_i}^i.$$

Since ρ_j^i is nonincreasing and $\rho_j^i \leq \frac{\rho_i}{4}$, where $j = r_i + 1, \ldots, d_i$, then

$$\frac{\rho_i}{4} - \rho_{r_i+1}^i + \rho_{r_i+2}^i - \cdots - \rho_{d_i-1}^i + \rho_{d_i}^i \geq 0$$

Hence $\rho(x,y) \geq d_i \rho_i/4$, contradicting (7).

Now let $d_i = r_i \pmod 2$. In this case, make an additional erasure, i.e., erase the symbol $\hat{c}_{r_i+1}^i$, and set $r_i' = r_i + 1$. then, as in the previous case,

$$\rho(x,y) \geq \sum_{j=1}^{r_i'} \rho(x_j, y_j) + \sum_{j=r_i'+1}^{d_i+1} \rho(x_j, y_j)$$

$$\geq r_i'\frac{\rho_i}{4} + (\rho_{r_i'}^i - \frac{\rho_i}{4}) + \frac{\rho_i}{4}(d_i + 1 - r_i') - \rho_{r_i'+1}^i$$

$$= \frac{\rho_i}{4}d_i + \rho_{r_i'}^i - \rho_{r_i'+1}^i$$

$$\geq d_i \rho_i/4.$$

also contradicting (7).
Q.E.D.

Theorem 2. *Let x be the trasmitted codeword constructed by the binary codewords $c^1, c^2, \ldots, c^i, \ldots, c^s$ and y the received word corrupted by noise. Assume that the first code vectors $c^1, c^2, \ldots, c^{i-1}$ have been found correctly, if*

$$\rho(x, y) < d_i \rho_i / 4$$

then the decoding algorithm will correctly decode to codeword c^i.

Proof: In proposition 5.3, we proved that if $\rho(x, y) < d_i \rho_i / 4$, we must be able to successfully decode to the codeword c^i in at least one error-erasure trial, i.e. $m + 2t < d_i$. In propositions 5.1 and 5.2, we proved that if $\rho(x, y) < d_i \rho_i / 4$, then c^i is the one and the only one codeword c^i which satisfies the inequality:

$$F(c^i) < d_i \rho_i / 4.$$

Hence we conclude that, if $\rho(x, y) < d_i \rho_i / 4$, then the decoding algorithm will correctly decode the codeword c^i.
Q.E.D.

5   Simulation Results and Performance Analysis

In this section, we present simulation results run on a solaris computer and compare our new decoding algorithm to maximum likelihood decoding. We present comparisons for a spherical codes constructed from alphabets of size four and size eight. For both of the spherical codes constructed we "send" each codeword over the Gaussian channel and decode it using both maximum likelihood decoding and our new decoder. The following tables contain the simulation results. The symbols used in the tables are defined as below:

M : cardinality of the code.
σ : standard deviation of Gaussian noise.
MLD : maximum likelihood decoder.
KLD : new decoder.
CPU : total time for decoding of the entire code in minutes:seconds.
equivalence : percent of time KLD decoding is equal to MLD decoding.

code	length	weight	M	type	
C_0^1	10	arbitrary	64	binary code	
C_0^2	10	6	210	binary code	
X	10		13440	spherical code	

	MLD		KLD		
σ	CPU time	correctness	CPU time	correctness	equivalence
0.14	31:32	100%	0:28	100%	100%
0.16	31:36	100%	0:28	99.96%	99.96%
0.18	31:33	100%	0:28	99.51%	99.51%
0.20	31:35	100%	0:29	97.73%	97.73%
0.22	31:29	99.98%	0:30	93.69%	93.69%
0.23	31:11	99.97%	0:30	90.52%	90.52%

Table 4: Decoding Comparison for Code with Alphabet Size 4

code	length	weight	M	type	
C_0^1	8	arbitrary	16	binary code	
C_0^2	8	4	70	binary code	
C_0^3	4	1	4		
C_1^3	4	3	4	binary code	
X	8		17920	spherical code	

	MLD		KLD		
σ	CPU time	correctness	CPU time	correctness	equivalence
0.12	47:03	100%	0:24	100%	100%
0.14	47:11	100%	0:24	99.97%	99.97%
0.16	47:08	100%	0:25	99.53%	99.53%
0.18	47:07	100%	0:29	97.87%	97.87%
0.20	47:02	99.97%	0:27	93.78%	93.78%

Table 5: Decoding Comparison for Code with Alphabet Size 8

6   Conclusions

In this paper, we presented a new decoding algorithm for a family of spherical codes developed by Ericson and Zinoviev. This new family of codes contains some of the best known spherical codes in terms of minimum distance. Simulation results show that the new algorithm takes approximately 2% of the computational time of maximum likelihood decoding and is almost equivalent to MLE decoding at low noise levels. However, at higher noise levels the performance of the new algorithm drops faster than MLE decoding. We believe that this could be improved by choosing binary codes with greater minimum distance for the upper level binary codes in the spherical code construction. Improving the accuracy of the new algorithm at higher noise levels is a subject for future research. Also, since only the binary codewords need to stored, the new algorithm requires less storage space than the MLE decoder.

References

1. E. Biglieri and M. Elia, "On the existence of group codes for the Gaussian channel," IEEE Trans. Inform. Theory, vol. IT-18, pp. 399-402, May 1972.
2. E. Biglieri and M. Elia, "Optimum permutation modulation codes and their asymptotic performance," IEEE Trans. Inform.Theory, vol. IT-22, no. 6, pp. 751-753, Nov. 1976.
3. I. F. Blake, "Distance properties of group codes for the Gaussian channel," SIAM J. Appl. Math., vol.23, no.3, 1972.
4. H. H. Conway and N. J. A. Sloane, Sphere Packings, Lattices, and Groups, New York: Springer Verlag, 1988.
5. C. P. Downey and J. K. Karlof, "On the existence of [M, n] Group codes for the Gaussian Channel with M and n Odd," IEEE Trans. Inform. Theory, vol.IT-23, pp. 500-503, July 1977.

6. C. P. Downey and J. K. Karlof, "Optimal [M,3] Group Codes for the Gaussian Channel", IEEE Trans, Inform. Theory, vol. IT-24, pp. 760-761, Nov. 1978.

7. T. Ericson and V. Zinoviev, "Spherical Codes Generated by Binary Partitions of Symmetric Pointsets," IEEE Trans. Inform. Theory, vol. 41, no.1, Jan 1995.

8. G. D. Forney, "Generalized Minimum Distance Decoding," IEEE Trans, Inform. Theory, vol.12, April 1966.

9. I. Ingemarsson, "Group Codes for the Gaussian Channel," in Lecture Notes in Control and Inform. Sciences, vol.128, M. Thoma and A. Wyner, Eds. New York: Springer- Verlag, 1989.

10. J. K. Karlof, "Permutation Codes for the Gaussian Channel," IEEE Trans. Inform. Theory, vol. 35, no.4, pp. 726-732, July 1989.

11. D. Slepian, "Permutation Modulation," Proc. IEEE, vol.53, pp.228-236, Mar. 1965.

12. D. Slepian, "group codes for the Gaussian channel," Bell Syst. Tech. J., vol. 47, no. 4, pp. 575-602, Apr. 1968.

13. A. Tucker, Applied Combinatorics, New York: John Wiley and Sons, 1995.

14. V. Zinoviev, S. Litsyn, and Portnoi, "Concatenated Codes in Euclidean Space," Probl. Inform. Transm., vol.25, no.3, 1989.

Worst-Case Analysis of an Algorithm for Computing the Greatest Common Divisor of n Inputs

Charles Lam[1], Jeffrey Shallit[2,*], and Scott Vanstone[3]

[1] Department of Combinatorics and Optimization, University of Waterloo
Waterloo, Ontario, Canada N2L 3G1
cy3lam@crypto2.uwaterloo.ca
[2] Department of Computer Science, University of Waterloo
Waterloo, Ontario, Canada N2L 3G1
shallit@graceland.uwaterloo.ca
[3] Department of Combinatorics and Optimization, University of Waterloo
Waterloo, Ontario, Canada N2L 3G1
savanstone@math.uwaterloo.ca

Abstract. We give an exact worst-case analysis of an algorithm for computing the greatest common divisor of n inputs. The algorithm is extracted from a 1995 algorithm of de Rooij for fixed-base exponentiation.

1 Introduction

In this paper we analyze an algorithm for computing the greatest common divisor (gcd) of n inputs. The algorithm we study was motivated by a method for doing efficient exponentiation with precomputation proposed by de Rooij [3]. A rigorous analysis of the expected number of multiplications required by de Rooij's algorithm has not been done. The results of this paper constitute a first step towards this goal.

The paper is organized as follows. Section 2 describes de Rooij's method for exponentiation, and from this an algorithm to compute the gcd of n inputs is extracted. In Section 3 we give an asymptotic upper bound on the complexity of the new algorithm using the familiar "naive bit complexity" model; see, for example, [1, Chap. 3].

Next we give an exact worst-case analysis of the algorithm. As is traditional with gcd algorithms, in this case our analysis is based on the number of division steps performed. For the worst-case analysis of the ordinary Euclidean algorithm on two inputs, see, for example, [4,5] and [1, §4.2].

Section 5 contains some remarks on the implications of the results of the paper for the problem of determining the number of multiplications required by the de Rooij exponentiation method.

---

* Research supported in part by a grant from NSERC.

2   The fixed-base Euclidean method for exponentiation

Let $\{b_0, b_1, \ldots, b_t\}$ be a set of integers for some $t \geq 1$ such that any exponent $e \geq 1$ (suitably bounded) can be written as a linear combination

$$e = \sum_{0 \leq i \leq t} e_i b_i$$

where $0 \leq e_i < h$ for some fixed positive integer h. The fixed-base Euclidean method for exponentiation is a procedure for computing α^e where α is an element of a group G. This technique was first proposed by de Rooij [3]. The term "fixed-base Euclidean" and the following description are taken from [6, §14.6.3].

Let $\{x_0, x_1, \ldots, x_t\}$ be a set of non-negative integers with $t \geq 1$. Let M be an integer in the interval $[0, t]$ such that $x_M \geq x_i$ for $0 \leq i \leq t$. Let N be an integer in the interval $[0, t]$, $N \neq M$, such that $x_N \geq x_i$ for $0 \leq i \leq t$ and $i \neq M$.

FBE
input: $\{\alpha^{b_0}, \alpha^{b_1}, \ldots, \alpha^{b_t}\}$ and $e = \sum_{0 \leq i \leq t} e_i b_i$
output: α^e
1. For i from 0 to t do the following:
    Set $\alpha_i \leftarrow \alpha^{b_i}$ and $x_i \leftarrow e_i$.
2. Determine the indices M and N for $\{x_0, x_1, \ldots, x_t\}$.
3. While $x_N \neq 0$ do the following:
    3.1. Set $q \leftarrow \lfloor x_M / x_N \rfloor$, $\alpha_N \leftarrow (\alpha_M)^q \cdot \alpha_N$, and $x_M \leftarrow x_M \bmod x_N$.
    3.2. Determine the indices M and N for $\{x_0, x_1, \ldots, x_t\}$.
4. Return $\alpha_M^{x_M}$.

The number of group operations required to compute α^e is a function of the number of divisions and size of the quotients q performed in steps 3 and 3.1 respectively. From the preceding algorithm one can extract an algorithm for determining the greatest common divisor of n inputs. The following terminology is useful.

Let L be a list of $k \geq 2$ non-negative integers. Extract-Max(L) is a routine which removes and returns the largest element of L. If there is more than one largest element, any choice is permitted. Max(L) returns the largest element of L, but does not alter L. Insert(L, a) inserts a into list L; the particular position of the new element is unspecified.

Here is the algorithm extracted from de Rooij's exponentiation method:

GCD
input: list L
output: $\gcd_{x \in L} x$
1. Set $u \leftarrow$ Extract-Max(L) and $v \leftarrow$ Max(L).
2. While ($v \neq 0$) do the following:

 2.1. Insert($L, u \bmod v$).
 2.2. Set $u \leftarrow$ Extract-Max(L) and $v \leftarrow$ Max(L).
3. Return u.

Correctness of the algorithm is left to the reader.

In the next section, we analyze the worst-case behaviour of the algorithm.

3   Asymptotic worst-case analysis

It is not difficult to prove that GCD uses a polynomial number of division steps. For an integer $r \geq 0$ define

$$\lg r = \begin{cases} 1, & \text{if } r = 0; \\ 1 + \lfloor \log_2 r \rfloor, & \text{if } r > 0. \end{cases}$$

Thus $\lg r$ counts the number of bits required to represent r.

Theorem 1. *Let $a_1 \geq a_2 \geq \cdots \geq a_k$. The number of division steps performed by algorithm* GCD *on input $(a_1, a_2, \ldots, a_k)$ is bounded by*

$$k + (\lg a_2) + \sum_{2 \leq i \leq k} \lg a_i = O(\sum_{2 \leq i \leq k} \lg a_i).$$

Proof. First, do one division step. If a_1 and a_2 are the two largest elements, the algorithm replaces a_1 with $a_1 a_2 \pmod{<}a_2$.

Now let u, v be the two largest elements of the new list with $u \geq v$. If $v = 0$, then the algorithm performs no division steps. Otherwise write $u = qv + r$ with $0 \leq r < v$. If $q \geq 2$, then $r < v \leq u/2$. If $q = 1$, then $v > u/2$, and $r = u - v < u/2$. Hence $r < u/2$ in both cases. Provided $(u, v) \neq (1, 1)$, this means that $\lg r < \lg u$, and the algorithm continues with r in place of u. Thus, exclusive of the number of division steps where $(u, v) = (1, 1)$, after the first step the algorithm uses at most $(\lg a_2) + \sum_{2 \leq i \leq k} \lg a_i$ division steps. When the pair $(u, v) = (1, 1)$ is encountered, it is replaced by $(0, 1)$, and this can occur at most $k - 1$ times before the algorithm halts. The total number of division steps is therefore $\leq 1 + (k - 1) + (\lg a_2) + \sum_{2 \leq i \leq k} \lg a_i$.

Now we discuss the naive bit complexity of the algorithm. In this model [1, Chap. 3], we can compare an integer u to an integer v (and decide if $u > v, u = v$, or $u < v$), in $O(\max(\lg u, \lg v))$ bit operations. We can write $u = qv + r$, with $0 \leq r < v$, in $O((\lg q)(\lg v))$ bit operations.

Theorem 2. *Let $k \geq 2$, and let $(a_1, a_2, \ldots, a_k)$ be a k-tuple of integers with $a_1 \geq a_2 \geq \cdots \geq a_k$. Then the cost of the division steps in the* GCD *algorithm on input* $\gcd(a_1, a_2, \ldots, a_k)$ *is $O((\lg a_2)(\sum_{1 \leq i \leq k} \lg a_i))$ bit operations.*

Proof. Let the division steps performed by the algorithm be $u_i = q_i v_i + r_i$ for $1 \leq i < t$. Let $A_1, A_2, \ldots, A_t$ be the successive k-tuples computed by the algorithm, and let B_i for $1 \leq i \leq t$ be the product of all the non-zero elements of A_i.

Then if $r_i \neq 0$,

$$B_i = B_{i+1} u_i / r_i = B_{i+1}(q_i v_i + r_i)/r_i > q_i B_{i+1},$$

while if $r_i = 0$ then

$$B_i = B_{i+1} u_i \geq q_i B_{i+1},$$

for $1 \leq i \leq t - 1$.

It now follows that $B_1 \geq q_1 q_2 \cdots q_{t-1} B_t$, and so $q_1 q_2 \cdots q_{t-1} \leq B_1$. Furthermore, $a_2 = v_1 \geq v_2 \geq \cdots v_{t-1}$. It now follows that the total cost for all the division steps is, up to constant factors,

$$\sum_{0 \leq i < t} (\lg q_i)(\lg v_i) \leq (\lg a_2) \sum_{0 \leq i < t} (1 + \log_2 q_i)$$
$$= (\lg a_2)(t + \log_2 q_1 q_2 \cdots q_{t-1})$$
$$\leq (\lg a_2)(t + \log_2 B_1)$$
$$= O((\lg a_2)(\sum_{1 \leq i \leq k} \lg a_i)),$$

where we have used Theorem 1. to bound t.

In this paper, we are concerned with bounding the cost of the arithmetic operations, but data manipulation is also a legitimate concern. One possible data structure to store the k-tuples is a heap, which efficiently supports the basic operations Max, Extract-Max, and Insert using $O(\log k)$ comparisons of numbers in L. See, for example, [2]. Using a heap, the cost for the data manipulation would then be $O((\log k)(\lg a_1)(\sum_{2 \leq i \leq k} \lg a_i))$.

4   Exact worst-case analysis

In this section we obtain the exact worst-case behavior of the GCD algorithm. That is, for integers n, we are interested in the "smallest" k-tuple that causes the algorithm to perform n division steps.

Of course, since the inputs are k-tuples, the choice of what constitutes "smallest" is somewhat subjective. For example, both $A_1 = (14, 10, 7, 5)$ and $A_2 = (17, 10, 5, 4)$ cause the algorithm to perform 9 division steps, but both 4-tuples have a sum of 36.

Our definition of smallest is as follows. Let $A = (a_1, a_2, \ldots, a_k)$ and $B = (b_1, b_2, \ldots, b_k)$ denote two k-tuples of integers with $a_1 \geq a_2 \geq \cdots \geq a_k$ and $b_1 \geq b_2 \geq \cdots \geq b_k$. We say A is *lexicographically less* than B (and write $A < B$) if there exists an index i, $1 \leq i \leq k$, such that $a_j = b_j$ for $j < i$ and $a_i < b_i$. In this section we find the lexicographically least k-tuple that causes the algorithm to perform n division steps.

First, we introduce a family of linear recurrences.

Definition 3. *Let k, n be integers with $k \geq 2$ and $n \geq 0$. Define*

$$A_k(n) = \begin{cases} 0, & \text{if } n = 0; \\ 1, & \text{if } 1 \leq n < k; \\ A_k(n-1) + A_k(n-k), & \text{if } n \geq k. \end{cases}$$

Here are the first few terms for the first few sequences:

$k\backslash n$	0	1	2	3	4	5	6	7	8	9	10	11	12	13	14
2	0	1	1	2	3	5	8	13	21	34	55	89	144	233	377
3	0	1	1	1	2	3	4	6	9	13	19	28	41	60	88
4	0	1	1	1	1	2	3	4	5	7	10	14	19	26	36
5	0	1	1	1	1	1	2	3	4	5	6	8	11	15	20
6	0	1	1	1	1	1	1	2	3	4	5	6	7	9	12

Note that in the case $k = 2$, we have $A_2(n) = F_n$, the ordinary Fibonacci sequence.

Lemma 4. (a) *For $k \geq 2$ and $i \geq 0$, we have $A_k(i+1) \geq A_k(i)$.*
(b) *For $k \geq 2$ and $1 \leq i \leq k$, we have $A_k(i) = 1$.*
(c) *For $k \geq 2$ and $0 \leq i \leq k$, we have $A_k(k+i) = i+1$.*
(d) *For $k \geq 2$ and $n \geq k+2$ $A_k(n-k) = A_k(n) \bmod A_k(n-1)$.*
(e) *For $k \geq 2$ and $n \geq k$ we have $A_k(n-k) \geq A_k(n) \bmod A_k(n-1)$.*

Proof. Left to the reader.

Lemma 5. *For $k \geq 2$ and $n \geq 1$ we have $A_k(n) \geq A_{k+1}(n+1)$.*

Proof. By induction on n. The base case is $1 \leq n \leq k$. In this case $A_k(n) = A_{k+1}(n+1) = 1$ by Lemma 4. (b).
For the induction step, assume the result is true for n. We prove it for $n+1$.

$$\begin{aligned} A_k(n+1) &= A_k(n) + A_k(n-k+1) & \text{(by definition of } A_k) \\ &\geq A_{k+1}(n+1) + A_{k+1}(n-k+2) & \text{(by induction)} \\ &\geq A_{k+1}(n+1) + A_{k+1}(n-k+1) & \text{(by Lemma 4.(a))} \\ &= A_{k+1}(n+2). & \text{(by definition of } A_k) \end{aligned}$$

Now we are able to state and prove our fundamental lemma.

Lemma 6. *Suppose k, n are integers with $n \geq k \geq 2$. Suppose $GCD(a_1, a_2, \ldots, a_k)$ uses n division steps and $a_1 \geq a_2 \geq \cdots \geq a_k \geq 0$. Then the following inequalities hold:*

(a) $a_1 \geq A_k(n+2)$;
(b) $a_2 \geq A_k(n+1)$; *and*
(c) $a_1 + \sum_{k-i \leq j \leq k} a_j \geq A_k(n+i+3)$ *for $0 \leq i \leq k-3$.*

Proof. By induction on $n + k$. The base case is $k = 2$. In this case it is well known (see, for example, [1, §4.3]) that $a_1 \geq F_{n+2} = A_2(n + 2)$ and $a_2 \geq F_{n+1} = A_2(n + 1)$. In this case (c) is vacuously true.

Now assume the result is true for all n, k with $k, n \geq 2$ and $n + k < q$. We prove it for $n + k = q$. We may assume $k \geq 3$. There are several cases.

Case 1: $k = n$: If $a_1 \leq 2$, every iteration of the algorithm must replace the largest element with 0. This can happen at most $k - 1$ times before the second largest element becomes 0 and the algorithm halts. The same conclusion follows if $a_2 \leq 1$. Hence if $GCD(a_1, a_2, \ldots, a_k) = k$, we must have $a_1 \geq 3 = A_k(k + 2)$ and $a_2 \geq 2 = A_k(k + 1)$. This proves (a) and (b).

It remains to prove (c), i.e., to show that $a_1 + \sum_{k-i \leq j \leq k} a_j \geq A_k(k + i + 3)$ for $0 \leq i \leq k - 3$. There are two subcases, corresponding to $a_k = 0$ and $a_k \geq 1$.

Case 1.1: $a_k = 0$: Then $GCD(a_1, a_2, \ldots, a_{k-1})$ also takes k division steps. By induction, then, we deduce that

(a) $a_1 \geq A_{k-1}(k + 2)$; and
(c) $a_1 + \sum_{k-i-1 \leq j \leq k-1} a_j \geq A_{k-1}(k + i + 3)$ for $0 \leq i \leq k - 4$.

Now add $a_k = 0$ to these inequalities; we get

(1.1.a) $a_1 + a_k \geq A_{k-1}(k + 2)$; and
(1.1.c) $a_1 + \left(\sum_{k-i-1 \leq j \leq k-1} a_j \right) + a_k \geq A_{k-1}(k + i + 3)$ for $0 \leq i \leq k - 4$.

Combining these two inequalities, and reindexing (with $i' = i - 1$) we get, for $0 \leq i' \leq k - 3$,

$$a_1 + \sum_{k-i' \leq j \leq k} a_j \geq A_{k-1}(k + i' + 2)$$
$$\geq A_k(k + i' + 3),$$

where we have used Lemma 5.. This proves (c).

Case 1.2: If $a_k \geq 1$, then since $a_1 \geq a_2 \geq \cdots \geq a_k$, we have $\sum_{k-i \leq j \leq k} a_j \geq i + 1$. Hence

$$a_1 + \sum_{k-i \leq j \leq k} a_j \geq a_1 + i + 1 \geq i + 4 = A_k(k + i + 3)$$

for $0 \leq i \leq k - 3$ by Lemma 4. (c).

Case 2: $n > k \geq 3$: We perform one step of the algorithm, writing $a_1 = qa_2 + r$, $0 \leq r < a_2$, and replacing a_1 with r. We know that $GCD(r, a_2, a_3, \ldots, a_k)$ takes $n - 1$ division steps, so we may apply the induction hypothesis provided we know the magnitude of r compared with $a_2, a_3, \ldots, a_k$. There are a number of cases to consider. (Note that the case $r \geq a_2$ is impossible.)

Case 2.1: $a_2 \geq r \geq a_3$. In this case the induction hypothesis applied to $GCD(a_2, r, a_3, \ldots, a_k)$ gives the following inequalities:

(a) $a_2 \geq A_k(n+1)$;
(b) $r \geq A_k(n)$;
(c) $a_2 + \sum_{k-i\leq j\leq k} a_j \geq A_k(n+i+2)$ for $0 \leq i \leq k-3$.

From these hypotheses the following conclusions are deduced:

(**2.1.a**) $a_2 \geq A_k(n+1)$;
(**2.1.b**) $r = a_1 - qa_2 \geq A_k(n) \Longrightarrow a_1 \geq A_k(n) + qa_2 \geq A_k(n) + A_k(n+1) \geq A_k(n+2)$;
(**2.1.c**) We have

$$a_1 + \sum_{k-i\leq j\leq k} a_j \geq a_1 - qa_2 + a_2 + \sum_{k-i\leq j\leq k} a_j$$
$$\geq A_k(n) + A_k(n+i+2)$$
$$\geq A_k(n+i+3)$$

for $0 \leq i \leq k-3$.

Case 2.2: There is an integer t, $1 \leq t \leq k-2$, such that $a_2 \geq a_3 \geq \cdots \geq a_{k-t-1} \geq r \geq a_{k-t} \geq \cdots \geq a_k$.
The induction hypotheses give:

(a) $a_2 \geq A_k(n+1)$;
(b) $a_3 \geq A_k(n)$;
(**c.1**) $a_2 + \sum_{k-i\leq j\leq k} a_j \geq A_k(n+i+2)$ for $0 \leq i \leq t$;
(**c.2**) $a_2 + r + \sum_{k-i\leq j\leq k} a_j \geq A_k(n+i+3)$ for $t \leq i \leq k-4$;

From these hypotheses the following conclusions are deduced:

(**2.2.a**) $a_2 \geq A_k(n+1)$;
(**2.2.b**) $a_1 \geq a_1 - qa_2 + a_2 = r + a_2 \geq a_k + a_2 \geq A_k(n+2)$, where the last inequality follows by (c.1) with $i = 0$.
(**2.2.c**) There are three subcases:

Case (2.2.c.1): $0 \leq i \leq t-1$. By (c.1) we have $a_2 + \sum_{k-i\leq j\leq k} a_j \geq A_k(n+i+2)$. Hence

$$a_1 + \sum_{k-i\leq j\leq k} a_j \geq a_1 - qa_2 + a_2 + \sum_{k-i\leq j\leq k} a_j$$
$$\geq r + a_2 + \sum_{k-i\leq j\leq k} a_j$$
$$\geq a_{k-t} + a_2 + \sum_{k-i\leq j\leq k} a_j$$
$$\geq a_2 + \sum_{k-(i+1)\leq j\leq k} a_j$$
$$\geq A_k(n+i+3).$$

Case (2.2.c.2): $i = t$. We have

$$a_1 + \sum_{k-t \le j \le k} a_j \ge a_1 - qa_2 + a_2 + \sum_{k-t \le j \le k} a_j$$

$$\ge r + a_2 + \sum_{k-t \le j \le k} a_j$$

$$\ge A_k(n + t + 3)$$

by (c.2).

Case (2.2.c.3): $t + 1 \le i \le k - 4$. We have

$$a_1 + \sum_{k-i \le j \le k} a_j \ge a_1 - qa_2 + a_2 + \sum_{k-t \le j \le k} a_j$$

$$\ge r + a_2 + \sum_{k-t \le j \le k} a_j$$

$$\ge A_k(n + i + 3)$$

by (c.2).

Case 2.3. $a_2 \ge a_3 \ge \cdots \ge a_k \ge r$.
The induction hypotheses give:

(a) $a_2 \ge A_k(n + 1)$;
(b) $a_3 \ge A_k(n)$;
(c.1) $a_2 + r \ge A_k(n + 2)$;
(c.2) $a_2 + r + \sum_{k-i \le j \le k} a_j \ge A_k(n + i + 3)$ for $0 \le i \le k - 4$.

From these hypotheses we deduce the following conclusions:

(2.3.a) $a_1 = qa_2 + r \ge a_2 + r \ge A_k(n + 2)$, by (c.1).
(2.3.b) $a_2 \ge A_k(n + 1)$, by (a).
(2.3.c) $a_1 + \sum_{k-i \le j \le k} a_j \ge a_2 + r + \sum_{k-i \le j \le k} a_j \ge A_k(n + i + 3)$, by (c.1) and (c.2).

This completes the proof of our lemma.

As a corollary we immediately get the following result:

Corollary 7. *Let* $(a_1, a_2, \ldots, a_k)$ *be a k-tuple that causes the algorithm* GCD *to perform n division steps, with* $n \ge k \ge 2$. *Then*

$$\sum_{1 \le i \le k} a_i \ge A_k(n + k + 1).$$

Proof. Adding together the inequalities in Lemma 6. (b) and (c) (when $i = k-3$), we get

$$\sum_{1 \leq i \leq k} a_i = A_k(n+1) + A_k(n+k) = A_k(n+k+1).$$

We now observe that Lemma 6. (b) in fact holds for a wider range of n.

Lemma 8. *Let $k \geq 2$ be an integer. Suppose $\mathrm{GCD}(a_1, a_2, \ldots, a_k)$ uses n division steps and $a_1 \geq a_2 \geq \cdots \geq a_k \geq 0$. Then $a_2 \geq A_k(n+1)$ for all $n \geq 1$.*

Proof. For $n \geq k$ the result has been established above in Lemma 6. (b). Hence it suffices to prove it for $1 \leq n < k$. Consider the k-tuple

$$A = (\overbrace{1, 1, \ldots, 1,}^{n+1} \overbrace{0, 0, \ldots, 0}^{k-n-1}).$$

Clearly A causes the algorithm to perform exactly n division steps, and any lexicographically smaller k-tuple (subject to the condition that the entries are in decreasing order) must have more zero entries. Hence A is the lexicographically least k-tuple, and it follows that $a_2 = 1$. But by Lemma 4. (b), $A_k(n+1) = 1$ for $1 \leq n < k$.

We now prove our main result.

Theorem 9. *Let $n \geq k \geq 2$. Then*

(a) *on input $A = (A_k(n+2), A_k(n+1), \ldots, A_k(n+3-k))$ the algorithm GCD performs n division steps, and*

(b) *A is the lexicographically least k-tuple $(a_1, a_2, \ldots, a_k)$ causing the algorithm to perform n division steps, with $a_1 \geq a_2 \geq \cdots \geq a_k$.*

Proof. We prove (a) by induction on n. The base case is $n = k$. In this case $A = (A_k(k+2), A_k(k+1), \ldots, A_k(3))$, so $A = (3, 2, 1, \ldots, 1)$ by Lemma 4.. It is easy to verify that GCD performs n division steps on input A.

Now, the induction step. Assume that on input $A = (A_k(n+2), A_k(n+1), \ldots, A_k(n+3-k))$, the algorithm GCD performs n division steps. Now consider $A' = (A_k(n+3), A_k(n+2), \ldots, A_k(n+4-k))$. In the first step, the algorithm replaces $A_k(n+3)$ with $A_k(n+3) \bmod A_k(n+2)$. But $A_k(n+3) = A_k(n+2) + A_k(n+3-k)$. Now $A_k(n+3-k) < A_k(n+2)$ for $n \geq k-1$, by Lemma 4.. Hence $A_k(n+3) \bmod A_k(n+2) = A_k(n+3-k)$. It follows one division step of the algorithm transforms A' to A. But on input A, the algorithm performs n division steps. Hence on input A', the algorithm performs $n+1$ division steps, as desired. This proves (a).

To prove (b), let $n \geq k \geq 2$ and let $B = (b_1, b_2, \ldots, b_k)$ be a k-tuple with $B \leq A$ and $b_1 \geq b_2 \geq \cdots \geq b_k$ such that $\mathrm{GCD}(b_1, b_2, \ldots, b_k)$ performs n division steps on input B. Then, by Lemma 6., parts (a) and (b) we have $b_1 \geq A_k(n+2)$ and $b_2 \geq A_k(n+1)$. If $b_1 > A_k(n+2)$, or if $b_1 = A_k(n+2)$ and $b_2 > A_k(n+1)$,

B would be lexicographically greater than A. Hence we have $b_1 = A_k(n+2)$ and $b_2 = A_k(n+1)$.

Now assume $k \geq 3$. By Lemma 6. (c), we know $b_1 + b_k \geq A_k(n+3)$. Since we have established that $b_1 = A_k(n+2)$, by Lemma 4. (d) it follows that $b_k \geq A_k(n+3) - A_k(n+2) = A_k(n+3-k)$. Since the input $B = B_0 = (b_1, \dots, b_k)$ causes the algorithm to perform n steps, it follows that $B_1 := (b_2, b_3, \dots, b_k, b_1 \bmod b_2)$ causes the algorithm to perform $n-1$ division steps. Furthermore $b_1 \bmod b_2 = A_k(n+2) \bmod A_k(n+1) = A_k(n+2-k)$ by Lemma 4.(d), so $b_k \geq b_1 \bmod b_2$. Hence, by Lemma 6. (b) and Lemma 8., we have $b_3 \geq A_k(n)$. But $B \leq A$, so $b_3 = A_k(n)$.

Now we proceed by induction. Suppose $k \geq 4$, and for $j \geq 1$ let B_j denote the k-tuple derived by applying the algorithm j times to $(b_1, b_2, \dots, b_k)$. We will show that

(a) $b_i = A_k(n+3-i)$ for $1 \leq i \leq j+2$;
(b) $b_i \bmod b_{i+1} \leq b_k$ for $1 \leq i \leq j$
(c) The two largest elements of B_i are b_{i+1} and b_{i+2}.

The base case, already verified above, is $j = 1$.

Now assume the result true for j; we prove it for $j+1$. Since $(b_1, b_2, \dots, b_k)$ causes the algorithm to perform n division steps,

$$B_{j+1} = (b_{j+2}, b_{j+3}, \dots, b_k, b_1 \bmod b_2, b_2 \bmod b_3, \dots, b_{j+1} \bmod b_{j+2})$$

causes the algorithm to perform $n-j-1$ division steps. By induction we know $b_{j+1} = A_k(n+2-j)$ and $b_{j+2} = A_k(n+1-j)$. Hence $b_{j+1} \bmod b_{j+2} = A_k(n+2-j) \bmod A_k(n+1-j) \leq b_k$, so b_{j+3} is really the second largest element of B_{j+1}. It then follows from Lemma 6. (b) and Lemma 8. that $b_{j+3} \geq A_k(n-j)$. But since $B \leq A$, $b_{j+3} = A_k(n-j)$.

Continuing in this fashion, we successively deduce that $b_i = A_k(n+3-i)$ for $1 \leq i \leq k$. Hence $B = A$, and so A is the lexicographically least k-tuple that causes the algorithm to perform n division steps.

5   Concluding remarks

This section discusses some observations concerning the number of multiplications performed by the fixed-base Euclidean exponentiation algorithm described in Section 2.

Let $E = \{e_0, e_1, \dots, e_{k-1}\}$ be a k-tuple of positive integers for which GCD performs n division steps on input E. The fixed-base Euclidean method for exponentiation requires at least one multiplication for each division step. Eventually the modified exponent set contains only 0's and 1's and at least one element $a \geq 2$. Hence the fixed-base Euclidean method requires at least $n+1$ exponentiations. Applying the result in Section 4, if E is the lexicographically least k-tuple causing GCD to perform n division steps, then E upon input to the fixed-base exponentiation method will cause it to perform precisely $n+1$ multiplications (more precisely, n multiplications and one squaring). This follows immediately from the recurrence $A_k(n) = A_k(n-1) + A_k(n-k)$ and Theorem 9..

A complete analysis of the expected number of multiplication performed by the fixed-base algorithm is yet to be done.

References

1. E. Bach and J. Shallit. *Algorithmic Number Theory*. The MIT Press, 1996.
2. T. Cormen, C. Leiserson, and R. Rivest. *Introduction to Algorithms*. MIT Press, 1989.
3. P. de Rooij. Efficient exponentiation using precomputation and vector addition chains. In A. de Santis, editor, *Advances in Cryptology—EUROCRYPT '94 Proceedings*, Vol. 950 of *Lecture Notes in Computer Science*, pages 389–399. Springer-Verlag, 1995.
4. P.J.E. Finck. *Traité Élémentaire d'Arithmétique à l'Usage des Candidats aux Écoles Spéciales*. Derivaux, Strasbourg, 1841.
5. G. Lamé. Note sur la limite du nombre des divisions dans la recherche du plus grand commun diviseur entre deux nombres entiers. *C. R. Acad. Sci. Paris* **19** (1844), 867–870.
6. A.J. Menezes, P.C. van Oorschot, and S.A. Vanstone. *Handbook of Applied Cryptography*. CRC Press, 1997.

Zeta Functions of Curves over Finite Fields with Many Rational Points

Kristin Lauter

University of Michigan, Ann Arbor, Mi 48109-1109, USA
klauter@math.lsa.umich.edu

Abstract. Currently, the best known bounds on the number of rational points on an absolutely irreducible, smooth, projective curve defined over a finite field in most cases come from the optimization of the explicit formulae method, which is detailed in [7] and [9]. In [4], we gave an example of a case where the best known bound could not be met. The purpose of this paper is to give another new example of this phenomenon, and to explain the method used to obtain these results. This method gives rise to lists of all possible zeta functions, even in cases where we cannot conclude that no curve exists to meet the bound.

1 Introduction

Curves over finite fields with many rational points and their zeta functions have received widespread attention recently, largely for their applications in areas of modern technology such as coding theory and cryptography. The questions that such applications have generated are interesting in themselves and have led to much progress in the area. The main outstanding problem in the area is to determine the maximum number of points $N_q(g)$ possible on a curve of genus g over a finite field with q elements, for each pair (q, g). (By "curve", we will always mean an absolutely irreducible, smooth, projective curve.)

This is an extremely delicate and subtle question, as demonstrated already by the solution to the problem presented by Serre in the case $g = 2$. The genus one case can be handled via the theory developed by Honda and Tate. Serre builds on this theory, using among other tools the gluing of two elliptic curves to provide a complete solution in the genus 2 case ([9]). For $g \geq 3$, we have no general solution and must rely on specific examples of curves constructed or bounds proved.

The most well-known bound is the Weil bound,

$$N_q(g) \leq q + 1 + 2g\sqrt{q},$$

which was improved by Serre to the bound

$$N_q(g) \leq q + 1 + g[2\sqrt{q}]$$

in [8] using the inequality between the arithmetic and geometric mean. Ihara showed in [3] that the Weil bound cannot be met if the genus is

$$g > \frac{\sqrt{q}(\sqrt{q} - 1)}{2}.$$

When the genus is large compared to q, the best bounds at present are calculated from the optimization due to Oesterlé of the explicit formulae method due to Serre (see [7] or [9]). Since most constructions of curves with many rational points so far have not succeeded in actually meeting the explicit formulae bounds (for the state of the art, see [2]), it is important to develop methods to improve these bounds when at all possible, as Serre did in the case $q = 2$, $g = 7$, when he showed that $N_2(7) < 11$.

In Section 2, we present a method which consists of first, systematically listing the zeta functions which could possibly be associated to a curve having a certain number of rational points, then second, determining which of these zeta functions are actually possible. We give an example of a technique, introduced by Serre in [9], for eliminating some possibilities.

In Section 3, we detail an example of how to successfully apply the method. The result in the particular case that we treat is an improvement of the upper bound by 1, which leads to an exact value for $N_q(g)$. It can be stated as follows:

Theorem 1 *An absolutely irreducible, smooth, projective curve of genus 7 over $\mathbb{F}_3$ with 17 rational points cannot exist. Furthermore, $N_3(7) = 16$.*

2   The Method

We aim to make a finite list of possible zeta functions for any curve with the data (q, g, N_1), that is, a curve defined over the finite field $\mathbb{F}_q$, of genus g, with N_1 rational points. If X is a curve defined over $\mathbb{F}_q$, then the zeta function of X encodes the information on the number of $\mathbb{F}_{q^n}$-points on X for all $n \geq 1$. Recall that the zeta function is defined as

$$Z(X,t) = \exp(\sum_{n \geq 1} \#X(\mathbb{F}_{q^n}) \frac{t^n}{n}),$$

but it is known that (see [5], for example) this power series is actually a rational function. We write

$$Z(X,t) = \frac{h(t)}{(1-t)(1-qt)},$$

where

$$h(t) = \prod_{i=1}^{g} (1 - \alpha_i t)(1 - \bar{\alpha}_i t)$$

is the h-polynomial of the curve whose reciprocal roots α_i and $\bar{\alpha}_i$ are the eigenvalues of Frobenius acting on the Tate module of the Jacobian of the curve. Since the α_j have absolute value $\sqrt{q}$, we write

$$\alpha_j = \sqrt{q} e^{i\theta_j}.$$

The number of points N_n on X over any extension field $\mathbb{F}_{q^n}$ can be expressed in terms of the $\{\alpha_i\}$ as follows:

$$N_n = q^n + 1 - \sum_{i=1}^{g} (\alpha_i^n + \bar{\alpha}_i^{\,n}).$$

If we let a_d be the number of places of degree d on the curve, we can also write

$$N_n = \sum_{d|n} d a_d.$$

Thus, the $\{a_d\}$ can be expressed directly in terms of the $\{\alpha_i\}$ and vice-versa. The relationship between these two quantities will be very important in our method.

The main tool of the method is an auxiliary polynomial related to $h(t)$,

$$F(T) = \prod_{i=1}^{g} (T - u_i), \qquad u_i = \alpha_i + \bar{\alpha}_i.$$

In fact, we will actually make a finite list of possibilities for this polynomial $F(T)$, which can then be easily translated into a finite list of possibilities for the h-polynomial of the curve. Since the $\{u_i\}$ are all real numbers of absolute value less than or equal to $2\sqrt{q}$, it follows that $F(T)$ and all of its derivatives have *all* of their roots *real* and in the interval $[-2\sqrt{q}, 2\sqrt{q}]$. Write

$$F(T) = \sum_{i=0}^{g} b_i T^{g-i},$$

so that the $\{b_i\}$ are the elementary symmetric polynomials in the $\{u_i\}$. We wish to express the $\{b_i\}$ in terms of the $\{a_d\}$. Define the power polynomials

$$s_n = \sum_{i=1}^{g} u_i^n.$$

We can use Newton's identities ([1], A IV.65) to express the elementary symmetric polynomials $\{b_i\}$ in terms of the power polynomials $\{s_n\}$ as follows:

$$b_1 = -s_1,$$

$$b_2 = (s_1^2 - s_2),$$

and in general

$$0 = s_n + b_1 s_{n-1} + b_2 s_{n-2} + \ldots + n b_n.$$

Now we also have the following obvious relationship between the $\{s_n\}$ and the $\{N_n\}$:

$$s_1 = (q + 1) - N_1$$
$$s_2 = (q^2 + 1) - N_2 + 2gq$$
$$s_3 = q^3 + 1 - N_3 + 3qs_1$$
$$s_4 = q^4 + 1 - N_4 + 6gq^2 + 4q(q^2 + 1 - N_2)$$
$$s_5 = q^5 + 1 - N_5 + 5q(q^3 + 1 - N_3) + 10q^2 s_1$$
$$s_6 = q^6 + 1 - N_6 + 6q(q^4 + 1 - N_4) + 15q^2(q^2 + 1 - N_2) + 20gq^3$$
$$s_7 = q^7 + 1 - N_7 + 7q(q^5 + 1 - N_5) + 21q^2(q^3 + 1 - N_3) + 35q^3 s_1$$

$$\ldots$$

These relations lead to an expression for $F(T)$ in which the coefficients, b_i, are written as polynomials in the $\{a_d\}$. From that expression it is immediate that we have a finite number of possibilites for $F(T)$, since we have bounds on a_d for each d. The bounds are obtained from different choices of trigonometric polynomials via the explicit formulae method as follows. For any

$$f(\theta) = 1 + 2 \sum_{n=1}^{\infty} c_n \cos(n\theta),$$

with $\{c_n\}$ non-negative real numbers not all zero and such that $f(\theta) \geq 0$ for all θ, we form the series

$$\Psi_d(t) = \sum_{n=1}^{\infty} c_{nd} t^{nd}.$$

Then for each choice of f, we obtain a bound on the $\{a_d\}$:

$$\sum_{d \geq 2} d a_d \Psi_d(q^{-1/2}) \leq g + \Psi_1(q^{1/2}) - (N_1 - 1)\Psi_1(q^{-1/2}).$$

Theoretically, we could simply check, for each possible choice $[a_2, \ldots, a_g]$, whether the polynomial $F(T)$ had all its roots real and in the correct interval. In fact, there is a function **polsturm** installed in the computational package *pari* which can check this for any square-free polynomial, and polynomials with squares can be checked individually. However, this is computationally inefficient and can be infeasible if the bounds on the a_d are very large. Instead, we proceed by checking successive derivatives for the condition of having all their roots real and in the correct interval. This process progressively narrows the choices for the a_i as i goes from 2 to g. This completes the first stage of the method.

The second stage consists of employing additional techniques to eliminate some of the possibilities from the list for the $F(T)$. The following lemma uses a powerful argument due to Serre ([9],[10]) to accomplish this task.

Lemma 1 *Let $F(T) = \prod_{i=1}^{g}(T - u_i)$ be the polynomial with $u_i = \alpha_i + \bar{\alpha}_i$, $\{\alpha_i\}$ the set of eigenvalues of Frobenius of an absolutely irreducible, smooth, projective curve. Then $F(T)$ cannot be factored as $F(T) = r(T)s(T)$, with r and s non-constant polynomials in $\mathbb{Z}[T]$ such that $resultant(r, s) = \pm 1$.*

Proof: The proof of this lemma is also given in [4], but is repeated here for the reader's convenience. Suppose to the contrary that the polynomial $F(T)$ associated to such a curve can be written as $F(T) = r(T)s(T)$, with r and s non-constant polynomials in $\mathbb{Z}[T]$ with resultant$(r, s) = \pm 1$. This is equivalent by ([1], A IV.73) to the existence of two polynomials, $a, b \in \mathbb{Z}[T]$ such that $ar + bs = 1$.

Denote by *Frob* the Frobenius endomorphism of the Jacobian of the curve, and by V the unique endomorphism satisfying the relation

$$Frob \circ V = V \circ Frob = q.$$

Then the endomorphism $C = Frob + V$ has as characteristic polynomial F^2, and in fact $F(C) = 0$. Consider the two endomorphisms $p = a(C)r(C)$ and $p' = b(C)s(C)$. Then p and p' are idempotents since $pp' = p'p = 0$ and $p+p' = 1$. These two idempotents decompose the Jacobian into a direct sum

$$J = B \oplus B', \text{ where } B = ker\ p,\ B' = ker\ p'.$$

The fact that this decomposition is compatible with the principal polarization of the Jacobian variety follows from the fact that p and p' are Hermitian. Such a decomposition is impossible for a Jacobian variety since it contradicts the irreducibility of the theta divisor. $\square$

3   Example: $q = 3$, $g = 7$, $N_1 = 17$

This section is devoted to treating the case $q = 3$, $g = 7$, $N_1 = 17$, and thereby proving the theorem stated in the introduction. There are two possible zeta functions which must be eliminated by applying Lemma 1. The details are as follows.

Oesterlé's optimization of the explicit formulae method shows that a curve of genus 7 over $\mathbb{F}_3$ cannot have more than 17 rational points. In fact, we can find a specific choice for f which yields this bound. The trigonometric polynomial,

$$f(\theta) = \cos^2 \theta (1 + \frac{2}{\sqrt{3}} \cos \theta)^2,$$

found by Serre to show that the Ree curves are optimal, has coefficients:

$$c_1 = \frac{\sqrt{3}}{2}, \quad c_2 = \frac{7}{12}, \quad c_3 = \frac{\sqrt{3}}{6}, \quad c_4 = \frac{1}{12}.$$

In this case it leads to the bounds

$$a_2 \leq 0.86364, \quad a_3 \leq 2.11112, \quad a_4 \leq 9.5$$

and

$$N_1 \leq 17.463415.$$

Other choices for f lead to bounds on a_5, a_6, and a_7.

The polynomial $F(T)$ can be written as

$$F(T) = T^7 + 13T^6 + (a_2 + 67)T^5 + (13a_2 + a_3 + 174)T^4 + ((1/2)a_2^2 + (147/2)a_2 + 13a_3 + a_4 + 234)T^3 + ((13/2)a_2^2 + (a_3 + (517/2))a_2 + 76a_3 + 13a_4 + a_5 + 33)T^2 + ((1/6)a_2^3 + 40a_2^2 + (13a_3 + a_4 + 4325/6)a_2 + (1/2)a_3^2 + (583/2)a_3 + 79a_4 + 13a_5 + a_6 - 1329)T + (13/6)a_2^3 + ((1/2)a_3 + 343/2)a_2^2 + ((165/2)a_3 + 13a_4 + a_5 + 5494/3)a_2 + (13/2)a_3^2 + (a_4 + 1849/2)a_3 + 330a_4 + 82a_5 + 13a_6 + a_7 - 8862.$$

Now, since $a_2 = 0$, this reduces to the simpler expression

$$F(T) = T^7 + 13T^6 + 67T^5 + (a_3 + 174)T^4 + (13a_3 + a_4 + 234)T^3 + (76a_3 + 13a_4 + a_5 + 33)T^2 + ((1/2)a_3^2 + (583/2)a_3 + 79a_4 + 13a_5 + a_6 - 1329)T + (13/2)a_3^2 + (a_4 + 1849/2)a_3 + 330a_4 + 82a_5 + 13a_6 + a_7 - 8862.$$

Here we begin the process of examining successive derivatives of F. All the roots must be real and in the interval $[-3.4641\ldots, 3.4641\ldots]$. The fourth derivative, $840T^3 + 4680T^2 + 8040T + (24a_3 + 4176)$, has roots in this interval for each of the possible values for a_3, $a_3 = 0, 1, 2$. The recursive step is to evaluate the *third* derivative at the roots of the fourth derivative in order to determine conditions on a_4 under which the third derivative could have all its roots be real and in the interval. For example, we determine that $a_3 = 0$ implies $a_4 \leq 17$, $a_3 = 1$ implies $a_4 \leq 8$, and $a_3 = 2$ leaves no choice for a_4.

Starting with the case $a_3 = 1$, we apply the recursive step again, evaluating the second derivative at the roots of the third derivative for each possible value of a_4. We find that the only possibilities for a_4 are $0, 1, 2, 3$, and

$$a_3 = 1,\ a_4 = 0 \Rightarrow a_5 = 73, \ldots, 82$$
$$a_3 = 1,\ a_4 = 1 \Rightarrow a_5 = 64, \ldots, 71$$
$$a_3 = 1,\ a_4 = 2 \Rightarrow a_5 = 55, \ldots, 60$$
$$a_3 = 1,\ a_4 = 3 \Rightarrow a_5 = 50.$$

For $a_3 = 0$, we apply the recursive step, but combine the restrictions on a_5 with those obtained from another choice of trigonometric polynomial. Together, the result is:

$$a_3 = 0,\ a_4 = 2 \Rightarrow a_5 = 88$$
$$a_3 = 0,\ a_4 = 3 \Rightarrow a_5 = 80, \ldots, 83$$
$$a_3 = 0,\ a_4 = 4 \Rightarrow a_5 = 72, \ldots, 77$$
$$a_3 = 0,\ a_4 = 5 \Rightarrow a_5 = 62, \ldots, 71$$
$$a_3 = 0,\ a_4 = 6 \Rightarrow a_5 = 54, \ldots, 66$$
$$a_3 = 0,\ a_4 = 7 \Rightarrow a_5 = 47, \ldots, 57$$
$$a_3 = 0,\ a_4 = 8 \Rightarrow a_5 = 43, \ldots, 46.$$

The remaining steps of the computation are summarized in the following tables. A star $*$ at any point in a column indicates that this possibility is eliminated, either because not all the roots are in the correct interval or because there is no possible choice for the corresponding a_i.

This computation was also checked using the `polsturm` function in the computational package *pari*.

Of the three possibilities for $[a_2, a_3, a_4, a_5, a_6, a_7]$, the first one, $[0, 0, 5, 67, 108, 314]$, corresponds to

$$F(T) = T^7 + 13T^6 + 67T^5 + 174T^4 + 239T^3 + 165T^2 + 45T,$$

which does not have all 7 roots in the interval. The second one, $[0, 0, 6, 60, 131, 265]$ corresponds to

$$F(T) = T^7 + 13T^6 + 67T^5 + 174T^4 + 240T^3 + 171T^2 + 56T + 6,$$

which factors as

$$F(T) = (T^3 + 5T^2 + 6T + 1)(T^2 + 4T + 2)(T + 1)(T + 3).$$

This polynomial does not correspond to the zeta function of a curve by Lemma 1, since

Table 1. $a_3 = 1$

a_4	0				1	2	3
a_5	73-82				64-71	55-60	50
a_5	77	78	79	70	*	*	
a_6	103-106	92-94	81-82	124			
a_7	*	*	*	*			

Table 2. $a_3 = 0$

a_4	0	1	2	3				4				
a_5	*	*	88	80-83				72-77				
a_5			*	80	81	82	83	73	74	75	76	77
a_6			74	64	53-54	43-45	97	86-88	76-78	65-69	57-60	
a_7			*	*	*	*	*	*	*	*	*	

Table 3. $a_3 = 0$, continued

a_4	5					6		7	8
a_5	62-71					54-66		47-57	43-46
a_5	66	67	68	69	70	60	61	53	
a_6	119	108-112	97-102	88-93	80-81	129-132	119-122	150-151	*
a_6		108				131	122		
a_7	*	314	*	*	*	265	302	*	

$$\text{resultant}((T^3 + 5T^2 + 6T + 1)(T^2 + 4T + 2), (T + 1)(T + 3)) = 1.$$

The third possibility, $[0, 0, 6, 61, 122, 302]$, corresponds to

$$F(T) = T^7 + 13T^6 + 67T^5 + 174T^4 + 240T^3 + 172T^2 + 60T + 8,$$

and factors as

$$F(T) = (T^2 + 3T + 1)(T + 2)(T^2 + 4T + 2)^2.$$

Again, this does not correspond to a curve by Lemma 1, since

$$\text{resultant}((T^2 + 3T + 1), (T + 2)(T^2 + 4T + 2)^2) = 1.$$

This concludes the proof that an absolutely irreducible, smooth, projective curve of genus 7 over $\mathbb{F}_3$ with 17 rational points does not exist. Since a curve of genus 7 over $\mathbb{F}_3$ with 16 rational points was constructed in ([6], p.65), we conclude that indeed $N_3(7) = 16$.

References

1. N. Bourbaki, *Algèbre*, chap. IV, Hermann, Paris, 1958.
2. G. van der Geer, M. van der Vlugt, Tables of Curves with Many Points. Regularly updated tables at: http://www.wins.uva.nl/g̃eer

3. Y. Ihara, *Some remarks on the number of rational points of algebraic curves over finite fields*, J. Fac. Sci. Tokyo **28** (1981), p. 721-724.

4. K. Lauter, *Non-existence of a Curve over* $\mathbb{F}_3$ *of Genus 5 with 14 Rational Points*, Max Planck Institut Preprint **98-51**, submitted for publication March, 1998.

5. J. Milne, *Etale Cohomology*. Princeton University Press: Princeton, NJ, 1980.

6. H. Niederreiter, C.P. Xing, Cyclotomic function fields, Hilbert class fields and global function fields with many rational places. *Acta Arithm.* **79** (1997), p.59-76.

7. R. Schoof, *Algebraic curves and coding theory*, UTM **336**, Univ. of Trento, 1990.

8. J.-P. Serre, *Sur le nombre des points rationnels d'une courbe algébrique sur un corps fini*, C.R. Acad. Sc. Paris Sér. I Math. **296** (1983), p.397-402.

9. J.-P. Serre, *Rational Points on curves over finite fields*. Notes by F. Gouvea of lectures at Harvard University, 1985.

10. J.-P. Serre, Letter to K. Lauter, December 3, 1997.

Codes on Drinfeld Modular Curves

Bartolomé López[1],[*] and Ignacio Luengo[2]

[1] Dpto. de Matemáticas. Universidad de Cádiz. 11510 Puerto Real (Cádiz), Spain
[2] Dpto. de Álgebra. Universidad Complutense de Madrid. 28040 Madrid, Spain

Abstract. In this work, we improve the estimate given by Manin and Vladut (cf. [7]) of the complexity of constructing codes on Drinfeld modular curves (Proposition 15). This improvement is mainly a consequence of Propositions 8, 10 and 11.

Introduction

In 1985, in their work "Linear Codes and Modular Curves" (cf. [7]), Manin and Vladut presented a polynomial algorithm of constructing algebraic-geometric codes on the reduction $\mathrm{mod}T$ of a family of Drinfeld modular curves, denoted by $X_0(I)$, where I is an ideal in the polynomial ring $\mathbf{F}_q[T]$. Those codes have very good asymptotic parameters; this is based on the fact that the reduction of a family of curves $X_0(I)$ attains the Drinfeld-Vladut bound (cf. [9], p. 181).

The construction of these codes given in [7] is not made on a model of the reduction $\mathrm{mod}T$ of $X_0(I)$, but the authors use a plane model of another curve: the reduction $\mathrm{mod}T$ of the Drinfeld modular curve $X_1(I)$; we denote by $C_1(I)$ such a plane model (it is defined by the polynomial given in equation (2)). The estimate of the complexity given in [7] is $O(n^{32})$, where n is the length of the codes. Even when we use the same basic algorithm as in [7], we improve the estimate given there.

There is another approach for the construction of codes on Drinfeld modular curves. The idea is to use the reduction $\mathrm{mod}(T + \zeta)$, for any $\zeta \in \mathbf{F}_q^*$, of the Drinfeld modular polynomial, which defines a plane model of $X_0(I)$ (cf. [1] and [8]). In that case, we have obtained the estimate $O(n^{17})$ (cf. [6], Chapter 4), which is slightly higher than those we obtain in the present work.

Hoholdt, Voss and Haché (cf. [4] and [3]) have also studied the construction of codes with good asymptotic parameters. They use the explicit family of curves considered by García and Stichtenoth in [2].

In Sect. 1, we present some results on Drinfeld modular curves. These results are in [7] and [6]. In Sect. 2, we briefly present Brill-Noether algorithm and we give a bound of the conductor of $C_1(I)$ (cf. Proposition 10). In Sect. 3, we estimate the complexity of constructing codes on Drinfeld modular curves.

We would like to thank Ruud Pellikaan for proposing us the study of modular codes and Sergei Vladut for useful conversations on Drinfeld modular codes.

---

[*] Both authors thank DGICYT (projects PB94-0291 and PB97-0284-C02-C01) and the Universidad de Cádiz (Plan Propio) for financial support.

1   Preliminaries

The main object that we consider in this work is the plane curve $C_1(I)$ (as we mentioned in the introduction, $C_1(I)$ is a model of the reduction $\bmod T$ of the Drinfeld modular curve $X_1(I)$). This model can be easily defined from the polynomial $P(x,y)$ (cf. equation (2)), but proving some properties of this model is a difficult task; it requires the use of some results on Drinfeld modular curves.

We will introduce the plane model $C_1(I)$ from the polynomial $P(x,y)$ and then, we will borrow some results on this model from [7] and [6]. We refer to [7] and [6] for the definitions of $X_0(I)$ and $X_1(I)$, and the reduction $\bmod T$ of these curves.

Let $\mathbf{F}_q[T]$ be the polynomial ring over the finite field $\mathbf{F}_q$. Consider a *prime* ideal I in $\mathbf{F}_q[T]$ generated by a monic polynomial f:

$$f = T^m + a_{m-1}T^{m-1} + \ldots + a_0 .$$

In order to simplify the calculations, we also assume that f is chosen in such a way that the image of T in $\mathbf{F}_q[T]/(f)$ is a generator of the multiplicative group $(\mathbf{F}_q[T]/(f))^*$.

We now define some polynomials $P_i(x,y)$ recursively,

$$P_0 = 1, \quad P_{i+1} = xP_i^q + yP_i^{q^2} , \tag{1}$$

and the polynomial

$$P(x,y) = P_m + a_{m-1}P_{m-1} + \ldots + a_0 , \tag{2}$$

where the $a_i \in \mathbf{F}_q$ are the coefficients of the polynomial f above (note that, with respect to [7], p. 2627, we interchange the roles of x and y).

$P(x,y)$ defines a plane curve that we denote by $C_1(I)$. The curve $C_1(I)$ is absolutely irreducible, is smooth in $\mathbf{A}^2$ and has two singular points on the line at infinity, $(1:0:0)$ and $(1:-1:0)$ (cf. [7], p. 2627).

There are exactly $\frac{q^m-1}{q-1}$ places of $C_1(I)$ on the line at infinity. In the sequel, U will denote this subset of $\frac{q^m-1}{q-1}$ places of $C_1(I)$. Consider now the projection

$$C_1(I) \xrightarrow{\lambda} \mathbf{P}^1$$
$$(x,y) \longmapsto x .$$

For each place $Q \in U$, the symbols e_Q and v_Q will denote the ramification index of λ at Q and the valuation associated to Q, respectively. We have the following results.

Lemma 1.1. *The ramification index of λ is q^l at exactly q^l places of U, for $l = 0, 1, \ldots, m-1$.*

Proof. (cf. [6], Lemma 3.2.1)   □

Lemma 1.2. *Let $Q \in U$. If $e_Q = q^l$, then $v_Q(y) = q^m - q^l(q+1)$.*

Proof. (cf. [6], Lemma 3.2.2)   □

There are also estimates of the valuation of the polynomials $P_i(x, y)$ defined in equation (1).

Lemma 1.3. *Let $Q \in U$. We have that,*
if $e_Q = q^l$, $l \leq m-2$, then $v_Q(P_i) \geq -(q^l + q^{l+1} + \ldots + q^{m-2})$, for $i = 0, 1, \ldots, m$.
If $e_Q = q^{m-1}$, then $v_Q(P_i) \geq 0$, for $i = 0, 1, \ldots, m$.

Proof. (cf. [6], Lemma 3.2.3)   □

The map $(x, y) \mapsto (x(x + y)^{q-1}, y(x + y)^{q^2-1})$ induces an automorphism θ of $C_1(I)$ (cf. [7], p. 2628, Lemma 3; note that $T + (f)$ generates the group $(\mathbf{F}_q[T]/(f))^*$). The action of θ on $C_1(I)$ extends to an action on the normalization $Z_1(I) \to C_1(I)$ of the curve $C_1(I)$. The automorphism θ generates a group Θ of order $\frac{q^m-1}{q-1}$. The quotient of $Z_1(I)$ by the action of Θ is denoted by $Z_0(I)$ (this curve is the reduction $\bmod T$ of the Drinfeld modular curve $X_0(I)$).

The curve $Z_1(I)$ can be identified with the set of places of $C_1(I)$. This allows us to interchange the roles of $Z_1(I)$ and $C_1(I)$. We have the following result.

Proposition 1.4. *(a) The points on $C_1(I) \cap \{x = 0\}$ lie above $\mathbf{F}_{q^2}$-rational points on $Z_0(I)$.*
(b) The places of $C_1(I)$ on the line at infinity lie above a $\mathbf{F}_q$-rational point ("cusp") w on $Z_0(I)$.

Proof. (a) (cf. [7], Proposition 19 and its corollary, p. 2625, and Lemma 22 and its corollary, p. 2633.)
(b) (cf. [7], Proposition 20 and its corollary, p. 2626, and Lemma 3, p. 2628.)
                                                                                       □

Remark 1.5. The projection $\lambda : C_1(I) \to \mathbf{P}^1$, defined by $\lambda(x, y) = x$, is unramified above $\mathbf{A}^1$; this follows from the fact that $P = a_0 + xP'_x + yP'_y$ and $P'_y = (P'_x)^q$. Hence, there are $\frac{q^{2m}-1}{q^2-1} = \deg_y P$ points on

$$C_1(I) \cap \{x = 0\} = \{(0, \alpha) : \alpha \text{ root of } P(y) = P(0, y)\}.$$

The action of Θ on this set is given by $\theta(0, \alpha) = (0, \alpha^{q^2})$. Thus, the set of factors of $P(y)$ irreducible over $\mathbf{F}_{q^2}$ is in bijection with the set of orbits of the action of Θ on the set $C_1(I) \cap \{x = 0\}$.

2   Brill-Noether algorithm and a bound of the conductor of $C_1(I)$

We briefly recall Brill-Noether algorithm and we refer for example to [5] for a more detailed exposition.

Let $P(X, Y, Z)$ be a homogeneous polynomial in $\mathbf{F}_q[X, Y, Z]$ of degree d. We assume that P is absolutely irreducible; let C be the plane curve defined by P. As usual, $\overline{\mathbf{F}}_q(C)$ will denote the function field of C.

The places of C can be represented by irreducible parametrizations of the form $\alpha(t) = (X(t), Y(t), Z(t))$ of $P(X, Y, Z)$, where $X(t)$, $Y(t)$ and $Z(t)$ are formal power series with coefficients in $\overline{\mathbf{F}}_q$ and

$$\min\left(\operatorname{ord} X(t), \operatorname{ord} Y(t), \operatorname{ord} Z(t)\right) = 0 .$$

We will also use parametrizations of the form $(x(t), y(t))$ of $P(x, y) = P(x, y, 1)$, where $x(t)$ and $y(t)$ are formal Laurent series with coefficients in $\overline{\mathbf{F}}_q$, to represent the places of C (cf. [10], chapter 4).

A *divisor* D on C is a formal sum $D = \sum_{Q \in C} n_Q\, Q$, where the Q are places of C, $n_Q \in \mathbf{Z}$, and only finitely many n_Q are different from zero.

Let us now consider the projection

$$C \xrightarrow{\ \lambda\ } \mathbf{P}^1$$
$$(x, y) \longmapsto x .$$

The conductor of C is defined as it follows.

Definition 2.1. *The conductor of C is the divisor*

$$E = (P'_y) + (d - 3)(Z) + 2(x)_\infty - B_x ,$$

where (P'_y) is the divisor of the function P'_y on C, (Z) is the intersection divisor of Z and P, $(x)_\infty$ is the divisor of poles of the function x and B_x is the ramification divisor of the projection λ defined above.

The following result is the base of Brill-Noether algorithm. It is a consequence of Noether theorem.

Proposition 2.2. *Let $D = \sum_Q n_Q\, Q$ be an effective divisor on C. Let G_0 be a homogeneous polynomial of degree l such that P does not divide G_0 and $(G_0) = E + D + D'$ for some effective divisor D'. Then $L(D) = \{\overline{G}/\overline{G_0} : \deg G = l, (G) \geq E + D'\}$, where $\overline{G}$ denotes the image of G in $\overline{\mathbf{F}}_q[X, Y, Z]/(P)$.*

Proof. (cf. [5], p. 245.)   □

In some cases, it is possible to consider a polynomial G_0 of relatively low degree which satisfies the conditions of Proposition 7. This idea can be found in Tsfasman and Vladut's work on codes on classical modular curves (cf. [9], pp. 475-477).

Proposition 2.3. *Let D be an effective divisor such that $\deg D \leq b_0 d$, where $d = \deg P$ and $b_0 \in \mathbf{N}$. There exists a polynomial G_0 satisfying the conditions of Proposition 7 and such that $\deg G_0 \leq (1 + b_0)d$.*

Proof. Since $\deg D \leq b_0 d$, we can consider a product H of linear forms such that $\deg H \leq b_0 d$ and $H \geq D$. Furthermore, we have that

$$(P'_Y) = 2(Z) + E - 2(x)_\infty + B_x .$$

Hence, $G_0 = H P'_Y$ satisfies the required condition.   □

In the sequel, we will assume that the places of C that we will consider are defined over $\mathbf{F}_q$, i.e., they can be represented by parametrizations whose coefficients lie in $\mathbf{F}_q$.

Let us now consider the vector space

$$M = \{G \in \overline{\mathbf{F}}_q[X,Y,Z] : \deg G = l, (G) \geq E + D'\} , \tag{3}$$

where E and D' are as in Proposition 7. A basis for M can be easily obtained as it follows. Let $E + D' = \sum_Q c_Q\, Q$ and let U be the set of places Q such that $c_Q \neq 0$. For each place $Q \in U$, we consider a representative $\alpha_Q(t) = (X_Q(t), Y_Q(t), Z_Q(t))$. Then, a homogeneous polynomial G satisfies that $(G) \geq E + D'$ if and only if

$$\operatorname{ord} G(X_Q(t), Y_Q(t), Z_Q(t)) \geq c_Q , \quad \text{for all } Q \in U .$$

Let Γ be the set of monomials H of degree l with coefficient 1. For each $Q \in U$ and for each $H \in \Gamma$, we consider the formal power series $H(\alpha_Q(t)) = \sum_{i \geq 0} a_{(H,Q,i)}\, t^i$. From a basis for the set of solutions of the system of linear equations

$$\left\{ \sum_{H \in \Gamma} a_{(H,Q,i)}\, u_H = 0 : Q \in U,\, 0 \leq i \leq c_Q - 1 \right\} , \tag{4}$$

where the u_H are indeterminates, we get a basis for M.

Remark 2.4. (a) We assume that the places $Q \in U$ are defined over $\mathbf{F}_q$. Thus, we can manage the coefficients $a_{(H,Q,i)}$ to be in $\mathbf{F}_q$. Hence, we can construct a basis for M whose elements are polynomials with coefficients in $\mathbf{F}_q$.

(b) If $\mathcal{B}$ is a basis for M, then $\{\overline{G}/G_0 : G \in \mathcal{B}\}$ is a set of generators of $L(D)$.

The last result of this section is a consequence of the Lemmas of Sect. 1; it gives some information about the conductor of the curve $C_1(I)$. Note that, due to the fact that $C_1(I)$ is smooth in $\mathbf{A}^2$, the non-zero coefficients of the conductor E of $C_1(I)$ correspond to the places of $C_1(I)$ on the line at infinity. Recall that e_Q denotes the ramification index of the projection $\lambda(x,y) = x$ at Q.

Proposition 2.5. *Let $E = \sum_Q d_Q\, Q$ be the conductor of $C_1(I)$. If $e_Q = q^s$, then*

$$d_Q \leq q^s \frac{q^{2m} - q^2}{q^2 - 1} + q^{2m} .$$

Proof. Let $E = (P'_y) + (d-3)(Z) + 2(x)_\infty - B_x$. For each place Q of $C_1(I)$ at infinity, we give a bound of each non-negative coefficient of Q in the divisors (P'_y), (Z) and $(x)_\infty$.

The points on $C_1(I) \cap Z = 0$ are $(1:0:0)$ and $(1:-1:0)$. Hence, $(Z) = (x)_\infty$. Consider now a place such that $e_Q = q^s$ (cf. Lemma 1). Then, the coefficient of Q in the divisor $(d-3)(Z) + 2(x)_\infty = (d-1)(x)_\infty$ is $q^s \dfrac{q^{2m} - q^2}{q^2 - 1}$. On the other hand, we have that

$$P'_y = P^{q^2}_{m-1} + b_{m-2} P^{q^2}_{m-2} + \ldots + b_0 ,$$

where $b_i \in \mathbf{F}_q$. Hence, by Lemma 3, the number of poles of P'_y is bounded by q^{2m}. Thus, the coefficient of Q in the divisor $(P'_y) + (d-3)(Z) + 2(x)_\infty$ is bounded by $q^s \dfrac{q^{2m} - q^2}{q^2 - 1} + q^{2m}$. Since B_x is an effective divisor, this number is also a bound of the coefficient of Q in the divisor E.   □

3   Complexity

We first recall Manin and Vladut's construction of codes on Drinfeld modular curves.

Let us consider the natural projection $\pi : C_1(I) \to Z_0(I)$ (cf. Sect. 1). The points on $C_1(I) \cap \{x = 0\}$ lie above $\mathbf{F}_{q^2}$-rational points on $Z_0(I)$ (cf. Proposition 4). In what follows, $\mathcal{P}$ will denote the set of these $\mathbf{F}_{q^2}$-rational points on $Z_0(I)$. There is a bijection between $\mathcal{P}$ and the set of factors of the polynomial $P(y) = P(0, y)$ irreducible over $\mathbf{F}_{q^2}$ (cf. Remark 5).

Let now w be the $\mathbf{F}_q$-rational point on $Z_0(I)$ considered in Proposition 4, part (b). For each integer b with $1 \le b \le q^m$, we consider the divisor $D = bw$ on $Z_0(I)$. To construct a generator matrix of the code associated to $\mathcal{P}$ and D, we must obtain the value $h(p)$ for each $p \in \mathcal{P}$ and each function h of a basis for $L(D)$. Manin and Vladut determine these values working with the plane curve $C_1(I)$ as it follows.

Let θ be the automorphism of $C_1(I)$ defined in Sect. 1. We have that

$$\pi^* L(D) = (1 + \theta + \ldots + \theta^{\epsilon-1})(L(\pi^* D)),$$

where $\epsilon = \frac{q^m - 1}{q - 1}$ (cf. [7], Lemma 7, p. 2629). Thus, if $h \in L(D)$, then there exists $H \in L(\pi^* D)$ such that $(1 + \theta + \ldots + \theta^{\epsilon-1})(H) = \pi^* h$. Let $p \in \mathcal{P}$; consider a point $(0, \alpha) \in C_1(I) \cap \{x = 0\}$ such that $\pi(0, \alpha) = p$. Then,

$$h(p) = h(\pi(0, \alpha)) = \pi^* h(0, \alpha) =$$

$$(1 + \theta + \ldots + \theta^{\epsilon-1})(H)(0, \alpha) = \sum_{i=0}^{\epsilon-1} H(\theta^i(0, \alpha)). \tag{5}$$

Thus, we must explicitly determine the vector space $L(\pi^* D)$ and obtain the roots of the polynomial $P(y)$.

Next, we describe the steps of the algorithm of constructing the code associated to $\mathcal{P}$ and D.

Step 1

In this step we present a simple method to obtain the parametrizations which represent the places of $C_1(I)$ on the line at infinity. Note that, since $\text{supp}(\pi^* D) \cup \text{supp}(E) \subset U$, where E is the conductor of $C_1(I)$ and U is the set of places of $C_1(I)$ on the line at infinity, in order to obtain a set of generators of $L(\pi^* D)$ we only have to consider the set U.

We now consider the projection $\lambda : C_1(I) \to \mathbf{P}^1$, where $\lambda(x,y) = x$. By Lemma 1, there is a place $Q \in U$ such that $e_Q = 1$. By Lemma 2, the center of Q is $(1:0:0)$. So, the Newton polygon of the polynomial $R(x_1, y) = x_1^\delta P(1/x_1, y)$, where $\delta = \deg_x P$, has an edge whose vertices are $(a,1)$ and $(b,0)$, with $b > a$; this edge corresponds to the place Q. Now, taking $x_1 = t$ as a local parameter at Q, we get a parametrization $(x(t) = 1/t, y(t))$ of P which represents Q. From this parametrization, we can recursively get parametrizations corresponding to the remaining places of U, as it follows. The places of U are in the same orbit under the action of Θ. Hence, each parametrization of the form

$$(x_i(t), y_i(t)) = \theta^i(x(t), y(t)), \quad \text{for } i = 0, 1, \ldots, \epsilon - 1,$$

represents a place Q_i of U. For practical purposes, we also consider the parametrizations of the form

$$(X_i(t), Y_i(t), Z_i(t)) = t^{-\delta_i}(x_i(t), y_i(t)), \quad \text{for } i = 0, 1, \ldots, \epsilon - 1,$$

where $\delta_i = \operatorname{ord} x_i(t)$. Note that $(X_i(t), Y_i(t), Z_i(t))$ also represents Q_i.

As it follows from the description of Brill-Noether algorithm, for each place $Q_i \in U$, we only have to compute a segment of certain length of the parametrization $(X_i(t), Y_i(t), Z_i(t))$. In Lemma 12, we estimate the complexity of this computation for $(X_0(t), Y_0(t), Z_0(t))$ and then, we estimate the complexity for the remaining parametrizations.

Note that, even when the procedure of Lemma 12 is essentially the same as procedure 6 in [7], p. 2641, we get a better estimate of the complexity. This improvement is based on the following observation (we will also use it in Step 2).

Proposition 3.1. *Let $R(x,y)$ be a polynomial in $\mathbf{F}_q[x,y]$ of degree d. Let $x(t)$ and $y(t)$ be polynomials in $\mathbf{F}_q[t]$ of degree N. There exists an algorithm with complexity $O(d^2N^2)$ which computes the first N coefficients of the polynomial $R(x(t), y(t))$.*

Proof. For the polynomials $x(t)^i$, we obtain the first N coefficients in a recursive way. We first compute the polynomial $x(t)^2$ and define $x_1(t)$ to be the polynomial formed by the terms of $x(t)^2$ up to degree N. Then, inductively, we compute the polynomials $x_i(t)x(t)$ and define $x_{i+1}(t)$ to be the polynomial formed by the terms of $x_i(t)x(t)$ up to degree N. So, the number of multiplications of elements in $\mathbf{F}_q$ in order to get the first N coefficients of $x(t)^i$, for $i = 1, \ldots, d$, is at most dN^2. The same follows for the polynomials $y(t)^j$. Now, to compute the products $x_i(t)y_j(t)$, N^2 multiplications are needed. Since $R(x,y)$ has at most $\frac{d(d-1)}{2}$ non-zero monomials, the complexity of the algorithm is $O(d^2N^2)$. $\qquad\square$

Now, reasoning similarly as in the proof of Proposition 11, we get the following lemma.

Lemma 3.2. *There exits an algorithm with complexity $O(q^{2m}N^3)$ which computes a segment of length N of the parametrization $(X_0(t), Y_0(t), Z_0(t))$.*

Once a segment of length $N + q^{2m+2}$ of $(X_0(t), Y_0(t), Z_0(t))$ has been calculated, it is possible to compute segments of the length N for all the remaining parametrizations, using the action of Θ. This algorithm has complexity $O(q^m(N + q^{2m+2})^2)$.

Step 2

Here, we estimate the complexity of computing a set of generators of $L(\pi^*D)$ using the algorithms of Step 1.

Lemma 3.3. *There exists an algorithm (Brill-Noether) with complexity $O(q^{12m})$ which constructs a set of generators of $L(\pi^*D)$.*

Proof. We first have to find a polynomial G_0 such that $(G_0) = E + \pi^*D + D'$, where D' is an effective divisor. By Proposition 8, we can choose $G_0 = Z^b P_Y'$; thus, $(G_0) = \pi^*D + E + B_x$. Let us now consider the vector space

$$M = \{G \in \overline{\mathbf{F}}_q[X, Y, Z] : \deg G = \deg G_0, (G) \geq E + D'\}.$$

Note that $\deg G_0 = b + \deg P - 1$ and $E + D' = E + B_x$. Let $E + D' = \sum_{Q \in U} c_Q Q$; by the proof of Proposition 10, we have that $c_Q \leq q^{3m}$ (we recall that $E + B_x = (P_y') + (d - 3)(Z) + 2(x)_\infty$).

To construct a basis for M, we first have to calculate the coefficients of the system given in equation (4). So, we have to compute segments of length q^{3m} for each parametrization $(X_i(t), Y_i(t), Z_i(t))$ and then, we apply the algorithm of Proposition 11. We get complexity $O(q^{11m})$ in this case.

Finally, we have to get the solutions of that system of linear equations. The number of equations is at most q^{4m} and the number of indeterminates is at most q^{4m+2}. Hence, applying the algorithm of procedure 4 in [7], pp. 2640-2641, we get complexity $O(q^{12m})$.   □

Step 3

In this step, we consider the remaining computations in order to construct a generator matrix of the code $\mathcal{C}$ associated to $\mathcal{P}$ and D.

Lemma 3.4. *There exists an algorithm with complexity $O(q^{12m})$ which constructs a generator matrix of the code $\mathcal{C}$.*

Proof. We use equation (5) to construct a generator matrix of $\mathcal{C}$. For each factor $R_i(y)$, $i = 1, \ldots, s$, of $P(y) = P(0, y)$ irreducible over $\mathbf{F}_{q^2}$, we consider one of its roots, α_i. Then, for each $G \in M$, and for each root α_i, $i = 1, \ldots, s$, we have to compute $G(\theta^k(0, \alpha_i))$, for $k = 0, 1, \ldots, \epsilon - 1$. The factors $R_i(y)$ can be computed using an algorithm with complexity $O(q^{8m})$ (cf. [7], procedure 5, p. 2641). The complexity of computing the expressions $G(\theta^k(0, \alpha_i))$ is $O(q^{6m})$.

Now, we consider the matrix A whose rows are the vectors

$$\sum_{k=0}^{\epsilon-1}(G/G_0)(\theta^k(0, \alpha_i))_{i=1,\ldots,s}.$$

The set of rows of A is a set of generators of the code $\mathcal{C}$. By Gauss method, we can easily derive from A a generator matrix of $\mathcal{C}$. The complexity of this procedure is $O(q^{6m})$.   □

In summary, one can construct a family of codes $\{\mathcal{C}_{m,b} : m \in \mathbf{N}, 1 \leq b \leq q^m\}$ over $\mathbf{F}_{q^2}$ which attains the Tsfasman-Vladut-Zink bound (cf. [7], proposition 6, p. 2620, or [4]). The complexity of this construction is $O(n^{12})$, where $n := \frac{q^m+1}{q+1}$ is the length of the codes.

Proposition 3.5. *A family of codes over $\mathbf{F}_{q^2}$ attaining the Tsfasman-Vladut-Zink bound is constructed with complexity $O(n^{12})$.*

References

1. Bae, S.: On the Modular Equation for Drinfeld Modules of Rank 2, *J. Number Theory* **42** (1992), 123-133.
2. García, A., Stichtenoth, H.: A tower of Artin-Schreier extensions of function fields attaining the Drinfeld-Vladut bound, *Invent. Math.* **121** (1995), 211-222.
3. Haché, G.: Construction Effective des Codes Géométriques, *Thése*, Univ. Paris VI, 1996.
4. Hoholdt, T., Voss, C.: An explicit construction of a sequence of codes attaining the Tsfasman-Vladut-Zink bound. The first steps. *Rapport technique 1995-10*, Tech. Univ. of Denmark, 1995.
5. Le Brigand, D., Risler, J.: Algorithme de Brill-Noether et Codes de Goppa, *Bull. Soc. Math. France* **116** (1988), 231-253.
6. López, B.: Plane Models of Drinfeld Modular Curves, *Tesis*, Univ. Complutense de Madrid, 1996.
7. Manin, Y., Vladut, S.: Linear Codes and Modular Curves, *J. Soviet Math.* **30** (1985), 2611-2643.
8. Schweizer, A.: On the Drinfeld Modular Polynomial $\Phi_T(X, Y)$, *J. Number Theory* **52** (1995), 53-68.
9. Tsfasman, M., Vladut, S.: Algebraic-Geometric Codes, *in* "Math. and its Applications," Vol. 58, Kluwer Acad. Publ., Dordrecht, 1991.
10. Walker, R.: Algebraic Curves, Springer, New York, 1978.

Elliptic Curves, Pythagorean Triples
and Applications

J. Miret[1], J. Tena[2], and M. Valls[1]

[1] Dept. Matemàtica. Universitat de Lleida, Spain,
{miret,magda}@eup.udl.es
[2] Dept. de Álgebra, Geometría y Topología.
Universidad de Valladolid, Spain,
tena@wamba.cpd.uva.es

Abstract. In this paper we study the properties of a particular family of elliptic curves defined over $\mathbb{Z}_p$. These curves have the property that their order is always a multiple of eight. Moreover, a sub-family of them is also introduced: the curves which are associated with pythagorean triples, for which there exists a rational distinguished point. The number of such triples in $\mathbb{Z}_p$ is given. Finally, a refinement on the Goldwasser-Kilian primality test is suggested.

1 Introduction

In some number theoretic algorithms, such as primality testing [1, 2], it is needed to find the order of a given elliptic curve. In general, it is still quite a difficult problem, although some algorithms have been given [4–6]. This problem could be partially overcome if one could find a family of elliptic curves whose order were easy to determine or, at least, if some information about this order could be given.

In this paper, we study a family of elliptic curves whose order is always a multiple of eight. We show that the number of curves in this family is high enough and the number of isomorphism classes is also given. But moreover, an important sub-family of them can be found: the elliptic curves which are associated with pythagorean triples, which have a rational distinguished point. In many applications it is needed to find a rational point of an elliptic curve, which involves an additional cost. In these curves, such a point can be automatically given. In addition, in order to show that the number of these curves is also high, we need to study the number of pythagorean triples in $\mathbb{Z}_p$ and, as a result, the number of isomorphism classes can be also given. Finally, we suggest a possible application of these curves to primality proving: the Goldwasser-Kilian primality test is slightly changed in order to use these curves.

We have studied this family of curves defined over the prime field $\mathbb{Z}_p$ because this field is the one used in the primality tests and is the most commonly used in cryptography, although the results given could be easily generalized over a finite field $\mathbb{F}_q$, where q is a primer power.

2   A Family of Elliptic Curves

Given a prime number p and a pair (a, b) such that

$$a, b \in \mathbb{Z}_p, \ a^2 \neq b^2, a, b \neq 0$$

we consider the associated cubic curve over $\mathbb{Z}_p$, called $E_{ab}(\mathbb{Z}_p)$, given by the points $(x, y) \in \mathbb{Z}_p^2$ such that verify the equation

$$y^2 = x(x - a^2)(x - b^2) \quad (\text{mod } p)$$

and the point $\mathcal{O}$ at the infinity.

Observe that, in such conditions, $E_{ab}(\mathbb{Z}_p)$ is an elliptic curve. Moreover, it can be easily shown that the Weierstrass form of these curves corresponds to the equation

$$y^2 = x^3 + \frac{1}{3}(a^2b^2 - a^4 - b^4)x + \frac{1}{27}(a^2 + b^2)(5a^2b^2 - 2a^4 - 2b^4), \qquad (1)$$

and that two curves $E_{ab}(\mathbb{Z}_p)$ and $E_{cd}(\mathbb{Z}_p)$ are isomorphic if there exists $u \in \mathbb{Z}_p$ such that $c^2 = u^2a^2$ and $d^2 = u^2b^2$.

2.1   The Group Structure

Let $P = (x_1, y_1)$ and $Q = (x_2, y_2)$ be two points on the elliptic curve $E_{ab}(\mathbb{Z}_p)$, and the point $\mathcal{O}$ be considered the identity. Then, the elliptic sum modulo p

$$(x_1, y_1) \oplus (x_2, y_2) = (x_3, y_3)$$

can be calculated in the following way:

i) If $x_1 \equiv x_2 \ (\text{mod } p)$ and $y_1 \equiv -y_2 \ (\text{mod } p)$, then

$$P \oplus Q = \mathcal{O}.$$

ii) If $x_1 \neq x_2 \ (\text{mod } p)$, consider s the inverse of $x_1 - x_2$ modulo p. Then,

$$x_3 \equiv (y_1 - y_2)^2 s^2 - x_1 - x_2 + c^2 \quad (\text{mod } p)$$
$$y_3 \equiv -((y_1 - y_2)s(x_3 - x_1) + y_1) \quad (\text{mod } p).$$

iii) Finally, if $x_1 \equiv x_2 \ (\text{mod } p)$, $y_1 \equiv y_2 \ (\text{mod } p)$ and $\gcd(2y_1, p) = 1$, consider s the inverse of $2y_1$ modulo p. Then

$$x_3 \equiv (3x_1^2 - 2c^2x_1 + a^2b^2)^2 s^2 - 2x_1 + c^2 \quad (\text{mod } p)$$
$$y_3 \equiv (3x_1^2 - 2c^2x_1 + a^2b^2)s(x_3 - x_1) + y_1 \quad (\text{mod } p)$$

It is already known that with such a law $E_{ab}(\mathbb{Z}_p)$ has the structure of an abelian group [7].

The following theorem and its corollary give us some properties of this family of elliptic curves. It provides the structure of the subgroups which contain the points of orders two and four.

Theorem 1 *The abelian group $E_{ab}(\mathbb{Z}_p)$ verifies the following properties:*

i) *There only exist three points of order two, which generate a subgroup isomorphic to $\mathbb{Z}_2 \times \mathbb{Z}_2$.*

ii) *If $p \equiv 3 \pmod 4$, there are exactly four points of order 4, which generate a subgroup isomorphic to $\mathbb{Z}_2 \times \mathbb{Z}_4$.*

iii) *If $p \equiv 1 \pmod 4$ there are four or twelve points of order 4, according to whether $\left(\frac{a^2-b^2}{p}\right)$ is -1 or 1, respectively, which generate a subgroup isomorphic to $\mathbb{Z}_2 \times \mathbb{Z}_4$ or to $\mathbb{Z}_4 \times \mathbb{Z}_4$.*

Proof. Firstly, notice that the only three points of order two are $P_1 = (0,0), P_2 = (a^2, 0)$ and $P_3 = (b^2, 0)$, which, together with $\mathcal{O}$, generate the subgroup $\mathbb{Z}_2 \times \mathbb{Z}_2$.

In order to find the points of order four, we must look for those P such that the second coordinate of $P \oplus P$ becomes 0. We must study under which conditions the equation

$$(x - ab)(x + ab)(x^2 - 2b^2x + a^2b^2)(x^2 - 2a^2x + a^2b^2) \equiv 0 \pmod p$$

is verified. So the points of order four are those $P = (x, y)$ such that their first coordinate satisfies one of the following equations

$$x \equiv ab \pmod p, \tag{2}$$

$$x \equiv -ab \pmod p, \tag{3}$$

$$x^2 - 2b^2x + a^2b^2 \equiv 0 \pmod p, \tag{4}$$

$$x^2 - 2a^2x + a^2b^2 \equiv 0 \pmod p. \tag{5}$$

Studying them, one can realize that there will exist a point with first coordinate (2) or (3) if, and only if, -1 is a quadratic residue, that is, if $p \equiv 1 \pmod 4$, and there are exactly two points in the curve with each coordinate. The conditions required for the existence of points whose first coordinate satisfies the relations (4) or (5) are, respectively, that $\left(\frac{b^2-a^2}{p}\right) = 1$ and $\left(\frac{a^2-b^2}{p}\right) = 1$, and then there are four points that satisfy each of them.

Finally, it is necessary to study which of these points are in the curve $E_{ab}(\mathbb{Z}_p)$, according to the value of p:

a) On the one hand, if $p \equiv 3 \pmod 4$, then $\left(\frac{a^2-b^2}{p}\right) = -\left(\frac{b^2-a^2}{p}\right)$, so either the third or the fourth equation is 1. Therefore, there only exist four points of order 4. Then, we have a subgroup isomorphic to $\mathbb{Z}_2 \times \mathbb{Z}_4$, which contains these points of order 4, the points of order 2 and $\mathcal{O}$.

b) On the other hand, if $p \equiv 1 \pmod 4$, both the first and the second equation are satisfied, so there are, at least, four points of order 4. But besides, as now $\left(\frac{a^2-b^2}{p}\right) = \left(\frac{b^2-a^2}{p}\right)$, two new cases must be considered: if they are equal to 1, there are eight more points of order 4 and we have a subgroup of order 16 that is isomorphic to $\mathbb{Z}_4 \times \mathbb{Z}_4$. Otherwise, the are only four points of order four, so we have a subgroup isomorphic to $\mathbb{Z}_2 \times \mathbb{Z}_4$.

For a more detailed proof see [3].   □

From this theorem we can easily deduce the following corollary.

Corollary 2 *The number of points of the elliptic curve $E_{ab}(\mathbb{Z}_p)$ is always a multiple of 8. Moreover, the group of points of the curve $E_{ab}(\mathbb{Z}_p)$ is isomorphic to $\mathbb{Z}_{2n} \times \mathbb{Z}_{4m}$, where $2n|p-1$ and $n|2m$.*

2.2   Cardinality and Isomorphism Classes of $E_{ab}(\mathbb{Z}_p)$

Owing to the fact that each curve $E_{ab}(\mathbb{Z}_p)$ is determined by a pair of quadratic residues, we get the following result.

Lemma 3 *The number of different curves in this family is*

$$\binom{\frac{p-1}{2}}{2} = \frac{1}{2}\left(\frac{p-1}{2}\right)\left(\frac{p-3}{2}\right).$$

The following lemmas show the number of curves in this family with $j = 0$ or $j = 1728$.

Lemma 4 *If 3 and -1 are quadratic residues in $\mathbb{Z}_p$, there exist $p-1$ curves with $j = 0$, where*

i) half of them, which are isomorphic, are of the form $y^2 = x(x-a^2)(x-u^2a^2)$,
ii) the rest, which are also isomorphic, are of the form $y^2 = x(x-a^2)(x-v^2a^2)$,

with $u = \frac{\sqrt{3}+\sqrt{-1}}{2}$ and $v = \frac{\sqrt{3}-\sqrt{-1}}{2}$.

Proof. The curves $E_{ab}(\mathbb{Z}_p)$ with $j = 0$ are those for which the linear coefficient in the Weierstrass form (see (1)) equals 0. That is to say

$$a^2b^2 - a^4 - b^4 = 0.$$

If 3 is a quadratic residue, it decomposes into $(a^2+\sqrt{3}ab+b^2)(a^2-\sqrt{3}ab+b^2) = 0$, and there will exist a solution if, and only if, -1 is also a quadratic residue. So then

$$b = \frac{-\sqrt{3} \pm \sqrt{-1}}{2}a \quad \text{or} \quad b = \frac{\sqrt{3} \pm \sqrt{-1}}{2}a,$$

and the result follows.

                                                                          □

Lemma 5

i) *If -1 is a quadratic residue, there exist $\frac{p-1}{4}$ isomorphic curves $E_{ab}(\mathbb{Z}_p)$ with $j = 1728$, with the equation $y^2 = x(x-a^2)(x+a^2)$.*
ii) *If 2 is a quadratic residue, there exist $\frac{p-1}{2}$ isomorphic curves $E_{ab}(\mathbb{Z}_p)$ with $j = 1728$, with the equation $y^2 = x(x-a^2)(x-2a^2)$.*

Proof. The curves $E_{ab}(\mathbb{Z}_p)$ with $j = 1728$ are those for which the independent coefficient in the Weierstrass form (see (1)) equals 0. That is to say

$$a^2 + b^2 = 0 \quad \text{or} \quad 5a^2b^2 - 2a^4 - 2b^4 = 0. \tag{6}$$

So if -1 is a quadratic residue, there will exist $\dfrac{p-1}{4}$ curves with equation $y^2 = x(x - a^2)(x + a^2)$.

The second equation in (6) decomposes into $(a^2 - 2b^2)(b^2 - 2a^2) = 0$. So, if 2 is a quadratic residue then there will exist $\dfrac{p-1}{2}$ curves with the equation $y^2 = x(x - a^2)(x - 2a^2)$.

$\square$

Proposition 6 *The number of isomorphism classes can be found in the table*

(mod 24)	$j = 0$	$j = 1728$	$j \neq 0, 1728$
$p \equiv 1$	2	$1, 1$	$\frac{p-17}{4}$
	$\frac{p-1}{2}$	$\frac{p-1}{2}, \frac{p-1}{4}$	$\frac{p-1}{2}$
$p \equiv 5$		1	$\frac{p-5}{4}$
		$\frac{p-1}{4}$	$\frac{p-1}{2}$
$p \equiv 7, 23$		1	$\frac{p-7}{4}$
		$\frac{p-1}{2}$	$\frac{p-1}{2}$
$p \equiv 11, 19$			$\frac{p-3}{4}$
			$\frac{p-1}{2}$
$p \equiv 13$	2	1	$\frac{p-13}{4}$
	$\frac{p-1}{2}$	$\frac{p-1}{4}$	$\frac{p-1}{2}$
$p \equiv 17$		$1, 1$	$\frac{p-9}{4}$
		$\frac{p-1}{2}, \frac{p-1}{4}$	$\frac{p-1}{2}$

which is classified according to the congruence of p modulo 24, where the first row corresponds to the number of isomorphism classes and the second one to the cardinality of each.

Proof. Taking into account the previous lemmas and that $-1, 2$ and 3 are quadratic residues in $\mathbb{Z}_p$ if, and only if, p is, respectively, congruent to 1 (mod 4), ± 1 (mod 8) and ± 1 (mod 12), it follows that the number of isomorphism classes can be classified modulo 24.

The columns corresponding to the curves with $j = 0$ and $j = 1728$ follow easily from lemmas 4 and 5, respectively. If $j \neq 0, 1728$, each class contains $\frac{p-1}{2}$ isomorphic curves and the number of classes is deduced from the number of curves in this family, given in lemma 3.

$\square$

3   The Sub-family of the Elliptic Curves Associated with Pythagorean Triples

Let p be a prime number. Then $(a, b, c) \in (\mathbb{Z}_p)^3$ is a pythagorean triple in $\mathbb{Z}_p$ if $a^2 + b^2 = c^2$.

In this section, we will study the properties of the sub-family of elliptic curves associated with pythagorean triples:

$$E_{abc}(\mathbb{Z}_p) = \{(x, y) \in \mathbb{Z}_p^2 \mid y^2 = x(x - a^2)(x - b^2) \pmod{p}\}$$

with $a^2 \neq b^2, a, b, c \neq 0$.

For such a purpose, firstly we need to compute the number of pythagorean triples in $\mathbb{Z}_p$.

3.1   Pythagorean Triples over $\mathbb{Z}_p$

In this subsection we want to determine the number of expressions like

$$a^2 + b^2 = c^2, \quad a, b, c \in \mathbb{Z}_p.$$

Such a computation can be more generally carried out over the finite field $\mathbb{F}_q$ with $q = p^r$, where p is a prime number and $p \neq 2$. This last condition is considered because, if the characteristic of this field were 2, counting them would be of no interest, owing to the fact that then every element in $\mathbb{F}_q$ would be a quadratic residue.

Moreover, we are interested in counting those expressions of this kind that are non-trivial, that is, $a, b, c \neq 0$, and those for which $a^2 \neq b^2$ (this second condition is required in the following applications). We will denote them by *strict pythagorean triples*.

For such a purpose we will use the following lemma:

Lemma 7 *There exist $q - 1$ solutions for the equation $x^2 - y^2 = 1$ in $\mathbb{F}_q$.*

Proof. This lemma is a particular case of the well-known result of Weil [8] which states that the number N of solutions of the equation

$$a_1 x_1^{d_1} + \ldots + a_n x_n^{d_n} = b, \quad a_i, b \in \mathbb{F}_q^*, \quad d_i | q - 1$$

is given by the expression

$$N = q^{n-1} + \sum_{\substack{i_1, \ldots, i_n \\ 1 \leq i_j \leq d_j - 1}} \overline{\chi}_1^{i_1}(b^{-1}a_1) \cdots \overline{\chi}_n^{i_n}(b^{-1}a_n) \pi(\chi_1^{i_1}, \ldots, \chi_n^{i_n}),$$

which is the extended sum for $i = (i_1, \ldots, i_n)$, $1 \leq i_j \leq d_j - 1$, and where χ_j is the multiplicative character of order d_i in $\mathbb{F}_q$, $\overline{\chi}_j(x) = \overline{\chi_j(x)}$ (complex conjugation), and π denotes the Jacobi sum

$$\pi(\chi_1, \ldots, \chi_n) = \sum_x \chi_1(x_1) \cdots \chi_n(x_n),$$

which is the extended sum to those $x = (x_1, \ldots, x_n)$ such that $x_1 + \cdots + x_n = 1$.

In the case we are interested in, we can take $\chi = \chi_1 = \chi_2$, which is the only character of order 2 in $\mathbb{F}_q$ (if q is prime, it is the Legendre's symbol). Then, we get

$$N = q + \chi(-1)\pi(\chi, \chi).$$

Finally, as $\chi = \overline{\chi} = \chi^{-1}$, we have

$$\pi(\chi, \chi) = \sum_{x \neq 0, 1} \chi(x)\chi(1 - x) = \sum_{x \neq 0, 1} \chi(x)\chi^{-1}(1 - x) =$$

$$= \sum_{x \neq 0, 1} \chi\left(\frac{x}{1 - x}\right) = \sum_{z \neq 0, 1} \chi(z) = -\chi(-1),$$

and the desired result is obtained.

$\square$

Notice that, geometrically, the $q - 1$ solutions of this expression correspond to those points at finite distance from the curve $x^2 - y^2 = 1$ (with two points at infinity) that can be obtained by rationally parameterizing such a conic.

As a result, we can state the following proposition:

Proposition 8 *The number of strict pythagorean triples over $\mathbb{F}_q$ can be given by*

$$N = \frac{(q - 1)(q - k)}{16},$$

where

$$k = \begin{cases} 5 & \text{if } -1 \text{ is a quadratic residue but 2 is not,} \\ 7 & \text{if } 2 \text{ is a quadratic residue but } -1 \text{ is not,} \\ 9 & \text{if both } -1 \text{ and 2 are quadratic residues,} \\ 3 & \text{otherwise.} \end{cases}$$

Proof. Firstly we will consider the number N' of strict pythagorean triples of the kind $1 + b^2 = c^2$. The previous lemma shows that there exist exactly $q - 1$ solutions (b, c). But notice that the following solutions must be removed:

i) The solutions $(0, \pm 1)$.
ii) The solutions $(\pm w, 0)$, with $w^2 = -1$, if $q \equiv 1 \pmod 4$.
iii) The solutions $(\pm 1, \pm w)$, $w^2 = 2$, if r is even or if $p \equiv \pm 1 \pmod 8$ (when $r = 1$).

Then, taking account of the possible signs of b and c, we get $N' = \frac{q-k}{4}$. Finally, the number N of pythagorean triples can be obtained by multiplying it by $\frac{q-1}{2}$ (the number of quadratic residues in $\mathbb{F}_q^*$) and then dividing it by 2, because the expression $a^2 + b^2 = c^2$ is considered equal to $b^2 + a^2 = c^2$.

$\square$

As a particularization, we get the following result.

Corollary 9 *The number of strict pythagorean triples in $\mathbb{Z}_p$ is $N = \frac{(q-1)(q-k)}{16}$, where*

$$
k = \begin{cases} 5 & \text{if } p \equiv 5(\bmod\ 8) \\ 7 & \text{if } p \equiv 7(\bmod\ 8) \\ 9 & \text{if } p \equiv 1(\bmod\ 8) \\ 3 & \text{if } p \equiv 3(\bmod\ 8) \end{cases}
$$

3.2 Cardinality and Isomorphism Classes of $E_{abc}(\mathbb{Z}_p)$

In this section we will determine the number of isomorphism classes (and their cardinality) of the curves associated with pythagorean triples. In order to take benefit from the results given in section 2.2, we obtain the relationship between the number of curves $E_{ab}(\mathbb{Z}_p)$ which are associated with pythagorean triples and those which are not. The following results show such a relation.

Lemma 10 *Let $x, x' \in \mathbb{Z}_p^*$ such that $\left(\frac{x}{p}\right) = \left(\frac{x'}{p}\right)$. Then, the number of different decompositions of x and x' in sums of two quadratic residues is exactly the same.*

Proof. If x and x' have the same quadratic character, there exists $d \in \mathbb{Z}_p$ such that $x' = xd^2$. Hence, for each decomposition of x into a sum of two quadratic residues, $x = a^2 + b^2$, one can obtain a decomposition of x' given by $x' = a^2d^2 + b^2d^2$. Notice that it works equally the other way round.

Owing to the fact that two different decompositions of x lead us to two different decompositions of x', the result follows immediately.

$\square$

Proposition 11 *Let $x, y \in \mathbb{Z}_p^*$ such that $\left(\frac{x}{p}\right) = 1$ and $\left(\frac{y}{p}\right) = -1$. Let N and M be the number of different decompositions of x and y, respectively. Then $M = N$, if 2 is not a quadratic residue in $\mathbb{Z}_p$, and $M = N - 1$, otherwise.*

Proof. If 2 is not a quadratic residue, then y can be written as $y = 2x'$ with $\left(\frac{x}{p}\right) = \left(\frac{x'}{p}\right)$. Then, each decomposition of x' leads to a decomposition of y in the following way

$$
x' = a^2 + b^2 \quad \rightsquigarrow \quad y = 2x' = (a - b)^2 + (a + b)^2,
$$

so, in this case, $N = M$.

Else, if 3 is not a quadratic residue, $y = 3x'$ with $\left(\frac{x}{p}\right) = \left(\frac{x'}{p}\right)$. Then,

$$
x' = a^2 + b^2 \quad \rightsquigarrow \quad y = 3x' = \begin{cases} (a - mb)^2 + (ma + b)^2 \\ (a + mb)^2 + (ma - b)^2 \end{cases}
$$

where $2 = m^2$. These two decompositions are equal if, and only if, $ab = 0$ or $a^2 = b^2$. But there are always two decompositions of x'

$$
x' = a^2 + b^2
$$
$$
x' = \left(\frac{(1 - m^2)a - 2mb}{1 + m^2} \right)^2 + \left(\frac{(m^2 - 1)b - 2ma}{1 + m^2} \right)^2 \Big\}
$$

which lead to the same decomposition of $y = (a - mb)^2 + (ma + b)^2$. So, in this case, $M = \frac{2N-2}{2} = N - 1$.

In the general case, let s be the first odd non-quadratic residue. Then $y = sx'$ and

$$x' = a^2 + b^2 \quad \rightsquigarrow \quad y = sx' = \begin{cases} (na - mb)^2 + (ma + nb)^2 \\ (na + mb)^2 + (ma - nb)^2 \end{cases}$$

with $n^2 = 2$ and $m^2 = s - 2$. The recount works in the same way as before. Finally, notice that such an s will exist. $\qquad\square$

Proposition 12 *The number of isomorphism classes can be found in the table*

(mod 24)	$j = 1728$	$j \neq 1728$
$p \equiv 1$	1	$\frac{p-17}{8}$
	$\frac{p-1}{2}$	$\frac{p-1}{2}$
$p \equiv 5, 13$		$\frac{p-5}{8}$
		$\frac{p-1}{2}$
$p \equiv 7$		$\frac{p-7}{8}$
		$\frac{p-1}{2}$
$p \equiv 11, 19$		$\frac{p-3}{8}$
		$\frac{p-1}{2}$
$p \equiv 17$		$\frac{p-9}{8}$
		$\frac{p-1}{2}$
$p \equiv 23$	1	$\frac{p-15}{8}$
	$\frac{p-1}{2}$	$\frac{p-1}{2}$

which is classified according to the congruence of p modulo 24, where the first row corresponds to the number of isomorphism classes and the second one to the cardinality of each.

Proof. Notice that the curves of the form $y^2 = x(x - a^2)(x + a^2)$ (see lemma 5) do not belong to the family of curves associated with pythagorean triples. Nevertheless, the curves of the form $y^2 = x(x - a^2)(x - 2a^2)$ will belong to this sub-family if, and only if, 3 is a quadratic residue, that is, if $p \equiv \pm 1$ (mod 12). In this manner, the isomorphism classes of the curves with $j = 1728$ are completely determined. Then taking account of corollary 9 or the lemma and proposition 10 and 11, the rest of isomorphism classes are given.

But, in order to determine which curves given in lemma 4 (corresponding to $j = 0$) belong to this sub-family, we must study when $1 + u^2$ or $1 + v^2$ are quadratic residues. Notice that $(1 + u^2)(1 + v^2)$ equals 3, which is a quadratic residue (because of the condition required in the lemma), so both of them are quadratic residues or none of them is. But we cannot exactly determine when each situation will happen. For instance, for $p = 37$ or $p = 61$ they are both quadratic residues, but when $p = 73$ or $p = 97$ they are not. So, we must take into account that when $p \equiv 1$ or $p \equiv 13$ modulo 24, perhaps two isomorphism classes could correspond to curves with $j = 0$. $\qquad\square$

4   Application: Refinements on the Goldwasser-Kilian Primality Test

In the Goldwasser-Kilian primilaty test [2] it is required to use Schoof's algorithm in order to obtain the cardinality of some curves. But as such an algorithm is not efficient in practice, one wonders about the existence of a family of curves for which their order is known or, at least, one has some information. That was what motivated the search for this family of curves.

When studying them, we found out that their order is always multiple of eight and that there exists a sub-family for which there is a distinguished point (so in the algorithm it is not needed to find out one). Moreover, it has already been shown that the number of curves in such a sub-family is high enough.

The algorithm can be slightly modified, and we suggest using Schoof's algorithm just partially, in order to get some more information about the order of the curve.

Outline of the algorithm   Let N be the probable prime and $E_{abc}(\mathbb{Z}_N)$ an elliptic curve, then

1. Take some odd primes and consider $p_1 p_2 \cdots p_s = l$. Compute $\tau \equiv t \pmod{l}$ by using Schoof's algorithm partially.
2. Compute τ' taking into account that $N \equiv 0 \pmod{8}$.
3. Take $P = (c^2, abc)$ in $E_{abc}(\mathbb{Z}_N)$.
4. Consider the successive integers $m \equiv N + 1 - \tau' \pmod{8l}$.
5. Find m such that $mP = \mathcal{O}$.
6. Partially factorize $m = fm'$, $m' \geq (\sqrt[4]{N} + 1)^2$ and $fP \neq \mathcal{O}$. If m' were prime, N would be prime. To determine its primality, a downrun process is started.

This algorithm could be improved if one could find more information about this family of curves. It would be interesting to study the properties of the order of the distinguished point or of the cardinality of these curves. In fact, although empirical results seem to show it, it would be important to prove that for every integer in Hasse's interval which is a multiple of eight, there exists a curve in this family with such an order.

References

1. A.O.L. Atkin and F. Morain. Elliptic Curves and Primality Proving. *Math. Comp.*, 61(203): 29–68, 1993.
2. S. Goldwasser and J. Kilian. Almost All Primes Can Be Quickly Certified. *Proc. 18th STOC*, pages 316–329, 1986.
3. J. Miret, M. Valls, R. Moreno, J. Gimbert, and J. Conde. Un test de primalidad sobre curvas elípticas generadas por ternas pitagóricas. *Actas V Reunión Española de Criptografía*, pages 135–141, 1998.
4. V. Müller and J. Buchmann. Computing the number of points of elliptic curves over finite fields. *Proceedings of ISSAC'91*, pages 179–182, 1991.

5. R. Schoof. Elliptic curves over finite fields and the computation of square roots mod p. *Mathematics of Computation*, 44: 483–494, 1985.

6. D. Shanks. Class number, a theory of factorization and genera. *Proc. Symposium of Pure Math. 20, Amer. Math. Soc.*, pages 415–440, 1970.

7. J.H. Silverman. The arithmetic of elliptic curves. Springer-Verlag, New York, 1986.

8. A. Weil. Number of solutions of equations in finite fields. *Bull. Amer. Math. Soc.*, 55: 497–508, 1949.

Exponential Sums and Stationary Phase (I)

Carlos Julio Moreno

CUNY, N.Salem NY 10560, USA
carlos@kepler.baruch.cuny.edu

Abstract. With a view toward applications to number theory, coding theory, and combinatorics, we consider in this article Gauss sums defined over finite rings, especialy those arising from p-adic fields. The principal theme is the application of the classical method of stationary phase in a number theoretic context. After surveying the well known results of Kummer, Lamprecht, and Dwork, we indicate some generalizations which include and extend certain results of Hua in the one variable case, and Igusa in the case of several variables. In general, Gauss sums over finite rings possess asymptotic expansions whose leading term can be expressed in terms of quadratic Gauss sums. The principal tool turns out to be an arithmetic analogue of Morse's Lemma.

1 Introduction

The method of stationary phase has a long and distinguished history in connection with its applications in number theory. It was discussed extensively in the nineteenth century by the physisists, especially those working in electromagnetism. It was also discovered independently by Riemann in his profound work on the Riemann zeta function where it was used as a basis for his third proof of the functional equation. The clarification of this aspect of Riemann's work was carried out by Siegel, who in the twenties greatly extended these ideas to derive the functional equations for Dirichlet's L-functions and for theta functions. An outcome of this line of work was an elucidation of the important role played by Gauss sums in the derivation of different types of functional relations. A salient point in these results is the role played by certain Gauss sums appearing in the leading term of these asymptotic expansions. In this note we exploit certain analogies between the classical method of stationary phase and the calculation of exponential sums over finite rings.

The classical method of stationary phase is used in the study of the asymtotic behavior of integrals of the type $\int_M a(y)e^{ikf(y)}dy$, where M is a Riemannian manifold with volume element dy and $a(y)$, $f(y)$ are suitable functions on M. If $C_f := \{s \in M : (df)_s = 0\}$ is the critical set, i.e., the points in M where the gradient of f vanishes, and if $H(s) = (d^2f)_s$ denotes the matrix associated to the Hessian of f, i.e., the matrix of the quadratic form defined by the second order terms in the Taylor expansion of f about the point s, then under suitable assumptions on the data, e.g., f is a Morse function, so that C_f is a finite set of isolated points, the principle of stationary phase can be formulated in terms of the following beautiful formula

(1)
$$\int_M a(y)e^{2\pi i k f(y)}dy = \left\{\sum_{s\in C_f} a(s)e^{2\pi i\, sgnH(s)/8}\frac{e^{2\pi i k f(s)}}{\sqrt{|\det H(s)|}}\right\} k^{-n} + O(k^{-n-1}).$$

(See Hormander [5], p. 220 or Sternberg and Guillemin [4], p. 6) A central theme of our discussion is that the method of stationary phase is at the heart of the explicit evaluation of exponential sums over finite rings which are not finite fields.

One example of the of type of results that can be established by analytic means is the evaluation of the quadratic Gauss sum

(2)
$$\sum_{s\in \mathbf{Z}/n\mathbf{Z}} e^{\frac{2\pi i s^2}{n}} = \frac{1 + (-i)^n}{1-i} \cdot \sqrt{n},$$

which follows by an application of the distributional identity (Poisson sumation formula): $\delta_0 = \sum_{s\in\mathbf{Z}} e^{2\pi i s}$ applied to the function $f_s(\xi) = e^{2\pi i(s+\xi)^2/n}$, where δ_0 is Dirac's measure concentrated at 0. From the classical work on quadratic Gauss sums we should point out that Mertens's well known proof of (2), as presented in Landau's elementary number theory book [1], brings out very clearly the necessity of some sort of "primary decomposition" for a finite ring R which replaces the use of the Chinese Remainder Theorem, as well as the need for a graded filtration for the set of ideals in R which mimics the parametrization "$p^{\ell-1}z + t$". This latter requirement on the ring R will be facilitated by a systematic use of the Jordan-Holder composition series on finite rings which possess a non-trivial theory of Gauss sums.

In this paper we shall review some of the classical results about Gauss sums over finite rings, especially those of Kummer, Lamprecht and Dwork. We shall also discuss the Stationary phase method for Carlitz-Uchiyama sums as well as Kloosterman sums. In a final section, I have included a brief description of the recent work of Igusa in the context of finite p-adic rings $\mathbf{Z}_p/p^k\mathbf{Z}_p$.

There is a relation between the problem of counting the number N of rational points on an algebraic curve $\mathcal{C} : f(x,y) = 0$ over the field $\mathbf{F}_p = \mathbf{Z}/p\mathbf{Z}$ (the degree one places of the function field) and the number N_ℓ of such points over the finite ring $\mathbf{Z}/p^\ell\mathbf{Z}$ for $\ell = 1, 2, 3, \ldots$. The precise statement of such a relation is found in Igusa's study of the Zeta function $Z(\mathcal{C}, t) = \sum_{\ell=1}^{\infty} N_\ell t^\ell/\ell$ (see Igusa [6], Denef [7]), which is based on a suitable modification of the mechanism which underlies the generalized Hensel Lemma for polynomials in several variables. We shall not pursue this direction here.

2 Gauss sums for the ring $R = \mathbf{Z}/p^\ell\mathbf{Z}$

In this section we describe the results of Kummer concerning Gauss sums over the ring $R = \mathbf{Z}/p^\ell\mathbf{Z}$ with p an odd prime. Analogous results can be derived for

$p = 2$. The multiplicative group $R^\times$ is cyclic of order $\phi(p^\ell)$ and we fix a generator (primitive root) g. We fix a generic additive character $\psi : R^+ \to \mathbf{C}^\times$ so that the Pontrjagin dual of R^+, viewed as an R-module satisfies the reflexive property

$$Hom(R^+, \mathbf{C}^\times) \simeq R \cdot \psi.$$

We let $\chi : R^\times \to \mathbf{C}^\times$ be a multiplicative character. We choose an element $r \in R$ that satisfies the relation $\chi(g)^{p-1} = \psi(r(p-1))^p$. Let ρ be an integer that satisfies $r(p-1) + \frac{g^p}{p}\log_p(g^{p-1}) \equiv 0 \mod. p^{\ell'}$, where $\ell' = [\frac{\ell}{2}]$ and log_P is the usual p-adic logarithmic function. The Gauss sum is defined by

$$G(\chi, \psi) := \sum_{s \in R^\times} \chi(s)\psi(s).$$

The following result was proved by Kummer in a slightly different notation. (See Kummer [10]).

THEOREM. *With notations and assumptions as above, we have*

-   $G(\chi, \psi) = 0$ *if* $r \equiv 0 \mod. p$,

Suppose $r \not\equiv 0 \mod. p$. *Then*

-   $|G(\chi, \psi)|^2 = p^\ell$, *and*

-   $G(\chi, \psi) = \begin{cases} p^{\ell/2}\chi(g^\rho)\psi(g^\rho) & \text{if } \ell \text{ is even} \\ G(\psi)p^{(\ell-1)/2}\chi(g^\rho)\psi(g^\rho) & \text{if } \ell \text{ is odd} , \end{cases}$

    where

$$G(\psi) = \sum_{s \in \mathbf{Z}/p\mathbf{Z}} \psi'(s^2), \qquad \psi'(s) = \psi(rp^{\ell-1}s/2).$$

The importance of this result lies in that it reveals how the general case of exponential sums over finite rings should behave. This remark will be amplified in the next section where we discuss the generalization to arbitrary finite rings. The extension of the above result to $p = 2$ necessitates a more carefull definition of the character ψ' of order two with a corresponding change in the definition of $G(\psi)$.

3 Gauss Sums over Finite Rings

The theory of finite rings used in this section is due to Krull and appears explicitly in the work of Lamprecht [9] in connection with the study of root numbers in the theory of Artin L-functions. The same fundamental ideas were discovered independently by Dwork and appear in his thesis [8].

Let R be a finite commutative ring with identity. If S is a subset of R, we denote by $N(S)$ its cardinality. Let $R^\times$ denote the multiplicative group of invertible elements in R. Given an additive character $\psi : R^+ \to \mathbf{C}^\times$ and a multiplicative character $\chi : R^\times \to \mathbf{C}^\times$, we define the Gauss sum by

$$G(\chi, \psi) := \sum_{s \in R^\times} \chi(s)\psi(s),$$

and note that its value is an algebraic integer. The study of this sum will be based on the structure theory of finite rings, a theory whose rudiments we now review.

We recall that a ring R is called *primary* when its zero divisors are all nilpotent in which case there is a unique prime ideal. The Noether-Lasker theorem gives for finite rings a direct product decomposition

$$R = R_1 \oplus \cdots \oplus R_n,$$

where each component R_i is primary. The identity element in R has a decomposition $1_R = 1_1 \oplus \cdots \oplus 1_n$ into idenpotents. The multiplicative and additive character groups possess analogous decompositions. In particular, we can write

$$\psi(s) = \prod_{i=1}^n \psi_i(s_i), \qquad \chi(s) = \prod_{i=1}^n \chi_i(s_i),$$

where $\psi_i(s_i) = \psi(1_i s)$ and $\chi_i(s_i) = \chi(s_i^*)$ with $s_i^* = s_i + \sum_{k \neq i} 1_k$. In analogy with Merten's use of the Chinese Remainder Theorem to evaluate the quadratic Gauss sum (see Landau [1], p. 213) we have the Noether-Lasker decomposition which provides a product factorization for the Gauss sums (see Lamprecht [9], Theorem 1.):

$$G(\chi, \psi) = \prod_{i=1}^n G_i(\chi_i, \psi_i).$$

Here each of the factors $G_i(\chi_i, \psi_i)$ is a Gauss sum associated to R_i. This reduces the study of Gauss sums over finite rings to the case of finite primary rings.

We suppose in the following that R is a finite primary commutative ring and let $\wp$ denote the its unique prime ideal with residue field $\mathbf{F}_\wp = R/\wp$ of cardinality $q = p^f$. The theory works well if we impose the following condition on R:

Condition G: *The character group* $\mathrm{Hom}(R^+, \mathbf{C}^\times)$ *is a rank one R-module.*

This simply means that if $\mathrm{Hom}(R^+, \mathbf{C}^\times) \simeq R \cdot \psi$, then for any other character ψ' of R^+ we have $\psi'(s) = \psi(r \cdot s)$ for some $r \in R$. We will call such characters ψ *generic* ("echt" in Lamprecht's terminology). If $\mathsf{q} \subseteq \wp$, then $s \equiv 1$ mod. q implies $s \equiv 1$ mod. $\wp$ and $s \in R^\times$. We say that a multiplicative character χ is defined modulo q if $\chi(s) = 1$ for all $s \equiv 1$ mod. q. The conductor $\mathcal{F}_\chi$ of a character χ is defined as the largest common divisor of all ideals q with respect to which χ is defined modulo q. When the conductor $\mathcal{F}_\chi = (0)$, we say that χ is proper. The first basic result of the theory is that proper characters exist only in the case where the condition **(G)** is satisfied. Also, a simple change of variables in the definition of $G(\chi, \psi)$ shows that if the condition **G** is not satisfied and $\psi : R^+ \to \mathbf{C}^\times$ is an additive character with kernel $\mathsf{q}_\psi := \{s \in R : \psi(sx) = 1$ for all $x \in R\}$, then the existence of a single element $r \equiv 1 \bmod. \mathsf{q}_\psi$ with $\chi(r) \neq 1$ implies $G(\chi, \psi) = 0$. Also, if $\chi(r) = 1$ for all elements $r \equiv \bmod. \mathsf{q}_\psi$, then one obtains that

$$G(\chi, \psi) = \frac{N((0))}{N(\mathsf{q}_\psi)} \sum_{s \in R^+, s \bmod. \mathsf{q}_\psi} \chi(s)\psi(s).$$

(See Lamprecht [9], Regel 2, p. 155). This reduces the problem of calculating Gauss sums over finite rings to the same problem over primary rings that possess the property (**G**).

DEFINITION. *A finite primary ring R is called reducible if its zero ideal can be expressed as the intersection of proper primary ideals; otherwise it is called irreducible.*

It is not difficult to show that the irreducible rings are identical to those that satisfy the condition (**G**). If q is an ideal in R, we denote its annihilator by $ann_R(\mathsf{q}) := \{s \in R : s\mathsf{q} = (0)\}$. In an irreducible ring R, the operation $\mathsf{q} \mapsto ann_R(\mathsf{q})$ defines an involution on the family of all ideals. If $\wp$ is the prime ideal of an irreducible primary ring R, the integer $n = n(R)$ such that $\wp^n = (0)$ but $\wp^{n-1} \neq (0)$ is an important invariant of the ring called the *exponent* of the ring R. It is not difficult to see that $ann_R(\wp) = \wp^{n-1}$ and $ann_R(\wp^{n-1}) = \wp$. The length $m = m(R)$ of a Jordan-Holder composition series (c.s.) for a finite irreducible ring R

$$(0) = \mathsf{q}_0 \subset \mathsf{q}_1 \subset \mathsf{q}_2 \subset \cdots \subset \mathsf{q}_{m-1} = \wp \subset \mathsf{q}_m = R,$$

with

$$\mathsf{q}_{i+1}/\mathsf{q}_i \simeq \mathbf{F}_\wp, \quad 0 \leq i \leq m-1$$

is called the length of the composition series. Note that $\mathrm{Card}(R) = q^{m(R)}$.

DEFINITION. *If the two ideals $\mathsf{r} := ann_R(\mathsf{q}) \subset \mathsf{q}$ appearing in a (c.s.) for a finite irreducible ring R have the property that no element $s \in R$ satisfies $s \equiv 0 \bmod. \mathsf{q}$, $s \not\equiv 0 \bmod. \mathsf{r}$ and $s^2 = 0$, then they are called a pair of characteristic ideals of the ring.*

With these definitions in place, we can now state the main structure theorem for irreducible primary rings. It is the essential ingredient in the explicit evaluation of Gauss sums. It is explicitly stated in Lamprecht's work (see [9], Theorem 5, p. 160) and is certainly proved by Dwork (see Thesis, [8] Chap. III) in the case of p-adic rings.

THEOREM. *Let R be a finite irreducible primary ring with prime ideal $\wp$ and residue field $\mathbf{F}_\wp = R/\wp$ with q elements. Then R has a pair of characteristic* *ideals* q *and* r = ann_R(q); *such a pair satisfies* $\wp \cdot q \subseteq r$, $r^2 = (0)$ *and the structure of the ring is described by one of the following cases:*

- *If* $m(R) = 2\mu + 1$, *then* $N(q) = q^\mu$, $q/r \simeq F_\wp$ *as modules and a given representation* $q = r + (\alpha)$ *satisfies* $q^2 = \wp^{n-1} = (\alpha^2) \neq (0)$.
- *If* $m(R) = 2\mu$, *then* $N(q) = q^\mu$ *and hence* $q = r$, $q^2 = (0)$ *(This case can occur only in characteristic 2).*
- *If* $m(R) = 2\mu$ *and* char$(F_\wp) \neq 2$, *then* $N(q) = q^{\mu-1}$, $q^2 = \wp^{n-1}$, $q/r \simeq F_\wp \oplus F_\wp$ *as modules and a given representation* $q = r + (\alpha) + (\beta)$ *satisfies* $\alpha\beta = 0$, $\alpha^2 \neq 0$, $\beta^2 \neq 0$; *there is a relation* $\alpha^2 + \gamma\beta^2 = 0$ *with* γ *not a square modulo* $\wp$.

DEFINITION. *A Jordan-Holder composition series* $(0) = q_0 \subset q_0 \subset q_1 \subset \cdots \subset q_m = R$ *in a finite irreducible ring* R *with the property that each ideal* q_i *and its corresponding annihilator* $ann_R(q_i) = q_{m-i}$ *is said to be inversion invariant (i.i.c.s.).*

In practice, many rings R that are of interest in applications have the (i.i.c.s.) property. This is certainly the case if the residue field $F_\wp$ is of characteristic 2. The following is the main result for the evaluation of Gauss sums. In a sense it states that under very natural conditions, the results of Kummer stated earlier also hold for Gauss sums over finite rings.

Let R be a finite primary ring. Let q and r = ann_R(q) be a characteristic pair with $r^2 = (0)$. Suppose $\chi(1 + s) = \psi(-\varepsilon_\chi s)$ for all $s \in r$ for some $\varepsilon_\chi \in R^\times$ (uniquely determined modulo q).

THEOREM. *With notations as above we have*

- *If* χ *is not proper,* $G(\chi, \psi) = 0$.

Suppose χ *is proper and* q, r = ann_R(q) *is a characteristic pair in an (i.i.c.s.). Then*

- $|G(\chi, \psi)|^2 = N(R)$.
- $G(\chi, \psi) = \begin{cases} N(q)\chi(\varepsilon_\chi)\psi(\varepsilon_\chi) & \text{if } m(R) \text{ is even} \\ G_q(\chi, \psi)N(q)\chi(\varepsilon_\chi)\psi(\varepsilon_\chi) & \text{if } m(R) \text{ is odd,} \end{cases}$

  where
$$G_q = \sum_{s \in q/r} \chi(1 - s)\psi(-\varepsilon_\chi s).$$

A noteworthy contribution by Lamprecht and Dwork is the analysis they are able to make of the so-called theta series

$$G(\Delta) = \sum_{s \in q/r} \Delta(s),$$

constructed with a quadratic character $\Delta : R \to \mathbf{C}^\times$ that satisfies the functional equation

$$\Delta(s)\Delta(s') = \Delta(s + s')\psi(-\varepsilon_\chi w^2 s s')$$

for some suitable element $w \neq 0$. To some extent the value of this sum can be found by the same method as in Kummer's theorem. It should also be noted that the last item in the above theorem is a manifestation of the method of stationary phase applied to finite rings. The appearance of the Gauss sum $G_q(\chi, \psi)$ is analogous to that in the leading term of the classical method of stationary phase. It is important to mention, as was first noted by Chevalley in his review of Lamprech's work (see *Math. Reviews in Number Theory*, T25-6, vol. V), that the evaluation of the Gauss sum in the case when $R/\wp$ is of characteristic 2 requires the use of a certain function Q on $K = R/\wp$ with values in the integers modulo 4 which satisfies the functional equation

$$Q(x + x') = Q(x) + Q(x') + 2Tr(xx'),$$

where Tr is the absolute trace in K and the elements of the ground subfield of K are considered as integers modulo 2. The existence of such a function can be established by making use of basis for K over its prime subfield.

It is possible to construct Q more directly using the properties of (length 2) Witt vector addition together with Teichmuller representatives for the elements of a finite field of characteristic 2.

4   Carlitz-Uchiyama Sums over Finite Rings

We give in this section a brief introduction to the method of stationary phase as it applies to exponential sums of Carlitz-Uchiyama type. These are the exponential sums which have many applications to error correcting codes and to signal processing (m-sequences, etc.). Our presentation is a modification of a well known method in the additive theory of numbers first introduced by L.K.Hua. The main result that we describe, the so-called **Fundamental Lemma**, is in all its essentials a primitive version of the one-variable version of the method of stationary phase which will be discussed in a later section.

Let $R_\ell = \mathbf{Z}/p^\ell \mathbf{Z}$, and define an additive character on R_ℓ by $\psi_\ell(s) = e^{2\pi i s / p^\ell}$. We shall be considering polynomials of degree k with integer coefficients

$$f(x) = b_0 + b_1 x + b_2 + \cdots + b_k x^k.$$

Note that we may assume that the leading coefficient $b_0 = 0$ without loss of generality. The exponential sums of Carlitz-Uchiyama type that we are interested in are defined by

$$G(p^\ell, f) := \sum_{s \in R_\ell} e^{2\pi i f(s)/p^\ell}.$$

The special case of $\ell = 1$, which to some extent is the more difficult case, is dealt with adequately by Weil's estimates of the roots of the L-functions of exponential sums over finite fields (see Moreno [11], Chap. iv). We therefore make the assumption $\ell \neq 1$. We now make a definition which makes precise the notion of multiplicity for the roots of arithmetic polynomials.

DEFINITION. *A polynomial congruence*

$$f(x) \equiv 0 \ mod. \ p$$

has a root of multiplicity m at μ when $f(x) = (x - \mu)^m f_1(x) + p f_2(x)$ with f_1, f_2 polynomials with the property that $p \nmid f_1(\mu)$.

To define the critical set for the polynomial $f(x)$ we assume that $f(x)$ is not trivial modulo p, that is to say, $p \nmid (b_1, b_2, \dots, b_k)$. Let a be the exact power of p which divides the derivative of f, namely $p^a \| (b_1, 2b_2, \dots, kb_k)$. We define the critical set for the polynomial f by

$$C_f(p) := \{s : p^{-a} f'(s) \equiv 0 \bmod. \ p\}.$$

If m_μ denotes the multiplicity of an element $\mu \in C_f(p)$, then by Lagrange's theorem, the sum of the m_μ's is bounded by $k - 1$. Given an element μ in the critical set $C_f(p)$ we consider the polynomial $f(px + \mu) - f(\mu) = c_1 x + \dots + c_k x^k$ in which the coefficients are obtained from those of f by the recursive relation

$$c_i = p^i \sum_{h=i}^{k} b_h \binom{h}{i} \mu^{h-1}, \qquad 1 \leq i \leq k.$$

We denote by $\rho(\mu)$ the largerst power of p which divides all the coefficients c_i: $p^{\rho(\mu)} \| (c_1, \dots, c_k)$. Clearly $\rho(\mu) \geq 1$. With the point μ and the integer $\rho(\mu)$, we define a new polynomial of degree k:

$$f_\mu(x) := p^{-\rho(\mu)}(f(px + \mu) - f(\mu)).$$

There is another important exponent associated to the points in the critical set which is defined by the requirement that $p^{\sigma(\mu)}$ is the highest power of p that devides the derivative of $f'_\mu(x)$; it satisfies $1 \leq \sigma(\mu) \leq k$. The basic principle behind the method of stationary phase for sums of the Carlitz-Uchiyama type is contained in the following result which is an elaboration of Hua's proof.

FUNDAMENTAL LEMMA. *With notations and assumptions as above, we have*

$$(3) \qquad \sum_{s \in R_\ell} \psi_\ell(f(s)) = \sum_{\mu \in C_f(p)} \psi_\ell(f(\mu)) p^{\rho(\mu)-1} \sum_{r \in R_{\ell-\rho(\mu)}} \psi_{\ell-\rho(\mu)}(f_\mu(r)).$$

Furthermore, if a critical point $\mu \in C_f(p)$ has multiplicity m_μ, then the number of solutions of the congruence $p^{-\sigma(\mu)} f'_\mu(x) \equiv 0(\ mod. \ p)$, counting multiple solutions with their multiplicities, is at most m_μ; that is, the critical set for the polynomial $f_\mu(x)$ satisfies $Card \, C_{f_\mu}(p) \leq m_\mu$.

The proof of this result is an easy modification of the argument given by Hua in ([2], §4, Lemma 1.4 and §5, Lemma 1.6). A very clear and lucid presentation of Hua's result can also be found in Vaughan ([3], §7.2). One point that merits mention in the proof of the above formula which is not stressed in Hua's original argument is the appearence of an **"Arithmetic Morse Lemma"** in which the iteration

$$f \mapsto f_\mu$$

corresponds to a change of coordinates near the critical point μ so that the end result is a quadratic function in the new variables. (For a review of the classical Morse Lemma, see Guillemin and Sternberg [4], Appendix I, p.16).

Particular applications of the Fundamental Lemma have appeared frequently in the literature of exponential sums as can be easily verified by browsing over the appropriate section of the *"Number Theory Reviews"* of the American Mathematical Society. We only want to point out here that for an arbitrary polynomial in one variable one obtains a representation for the exponential sum in the form

$$\sum_{s \in R_\ell} e^{2\pi i f(s)/p^\ell} = \sum_j a_j (\sqrt{p})^{w(j)},$$

where the complex numbers a_j have absolute value 1 and the "weight exponents" $w(j)$ are integers which in the "Morse Case" lie in the set $\{0, 1, \dots, \ell\}$. The p-adic expansions of Igusa for polynomials in several variable are a generalization of the above representation. We shall briefly discuss these ideas in the last section.

5   Kloosterman Sums over Finite Rings

We will discuss Kloosterman sums only briefly, as these have been treated fully in the literature, particularly in the work of Mordell, Sally, and Davenport. As in the case of Gauss sums, there is a multiplicative relation between Kloosterman sums defined modulo a positive integer n and modulo a prime power p^ℓ, which results from an easy application of the Chinese Remainder Theorem (the Noether-Lasker decomposition theorem in the case of arbitrary finite rings). Therefore we shall describe only the case of the finite ring $R_\ell = \mathbf{Z}/p^\ell\mathbf{Z}$. We shall also assume that p is an odd prime. The case of $p = 2$ is quite interesting in its own right and requires a slightly more delicate consideration which we do not give here.

Let a, b be two non-zero elements in the ring R_ℓ. We do not assume that they are units in $R_\ell^\times$. The **Kloosterman Sum** we consider is defined by

$$K_\ell(a, b) := \sum_{x \in R_\ell^\times} e^{2\pi i (ax + \frac{b}{x})/p^\ell},$$

where the sum is over the units in $R_\ell^\times$.

The group $R_\ell^\times$ has a very simple decomposition given by $R_\ell^\times = R_1 \oplus R_{\ell-1}^\times$ which reflects the obvious parametrization

$$x = t + p^{\ell-1}z, \qquad z = 0, 1, \ldots, p-1, \qquad t = 1, \ldots, p^{\ell-1}, (t, p) = 1.$$

For x, t, and z in their appropriate ranges, we obtain by a formal application of the Taylor expansion that for the function $f(x) = ax + \frac{b}{x}$, we have

$$f(x) \equiv f(t) + f'(t)zp^{\ell-1} + \frac{1}{2}f''(t)z^2p^{2(\ell-1)} \text{ mod. } p^\ell.$$

(Note that the second term is significant only when $p = 2$). This provides the following decomposition of the sum:

$$K_\ell(a, b) := \sum_{z \in R_1} \sum_{t \in R_{\ell-1}^\times} e^{2\pi i(f(t)+f'(t)zp^{\ell-1}+\frac{1}{2}f''(t)z^2p^{2(\ell-1)})/p^\ell}.$$

We now note that unless $f'(t) = 0$, the sum over z vanishes. Hence we are lead to introduce the critical set

$$C_f := \{t \in R_{\ell-1}^\times : f'(t) = 0\}.$$

In other words, these are the unit solutions to the congruence $at^2 - b \equiv 0$ (mod. $p^{\ell-1}$) As we are assuming p odd, $\ell \geq 2$, and $p^{2(\ell-1)}/p^\ell = p^{\ell-2}$ is an integer, the term with f'' does not contribute anything to the sum. We thus have the following result.

THEOREM. *With notation and assumptions as above, we have for $\ell \geq 2$*

$$(4) \qquad K_\ell(a, b) = p \sum_{t \in C_f} e^{2\pi i f(t)/p^\ell}.$$

REMARK. We note that for integers a, b and fixed ℓ, the values of the Kloosterman sums $K_\ell(a, b)$ do not lie in a fixed algebraic number field. They in fact lie in cyclotomic fields of increasing degree. This makes their study particularly difficult from a global perspective. We also note that the structure of the critical set C_f is relatively simple and that exactly the same argument as indicated above works for other rational functions f. We leave the formulation of the appropriate analogues to the interested reader. The case of $\ell = 1$ is covered by the Riemann Hypothesis for the hyperelliptic function field $y^p - y = f(x)$ and is beyond the elementary methods that enter into the method of stationary phase (See Moreno [11], chap. iv). Another fact that is worth mentioning here is that, as in the finite field case ($\ell = 1$), there is a relationship between Gauss sums and Kloosterman sums which is dictated by the harmonic analysis associated to the ring R_ℓ and which allows one to view one sum as the Fourier transform of the square of the second one.

6   P-adic Stationary Phase

In this section we review the p-adic methods introduced by Igusa in his development of p-adic asymptotic expansions for local Mellin transforms and which are capable of yielding fairly precise results on the asymptotics of exponential sums for polynomials in several variable.

We work over the ring $\mathbf{Z}_p$ and take for local uniformizing parameter $\pi = p$ so that the residual rings are given by $\mathbf{Z}_p/p^\ell \mathbf{Z}_p = \mathbf{Z}/p^\ell \mathbf{Z} = R_\ell$. We denote the canonical reduction map by

$$\mathbf{Z}_p \mapsto \mathbf{F}_p, \quad x \mapsto \bar{x}.$$

Let f be a polynomial in $\mathbf{Z}_p[x]$ and make the assumption that not all its coefficients lie in $p\mathbf{Z}_p$ so that its reduction $\bar{f}(x)$ mod.p is not the zero polynomial. For any $\mathbf{F}_p$-algebra A, we denote by $V_f(A)$, the set of A-valued points, and by $C_f(A)$ the A-valued critical set, that is to say, the A-valued singular points. We let $T \subset \mathbf{Z}_p$ be a fixed set of representatives of $\mathbf{F}_p$ in $\mathbf{Z}_p$ (a Teichmuller system).

We now consider a polynomial f in $\mathbf{Z}_p[x_1, x_2, \ldots, x_n]$ and a point $\mu_1 = (y_1, \ldots, y_n)$ in affine n-space $\mathbf{A}^n(\mathbf{Z}_p) = (\mathbf{Z}_p)^n$. In analogy with the definition of the polynomial $f_\mu(x)$, we introduce the polynomial

$$f_{\mu_1}(x) = p^{-e_{\mu_1}} f(px + \mu_1),$$

called the dilatation of f at μ_1, and where e_{μ_1} is the minimum order of p in expansion of the polynomial $f(px + \mu_1)$. We call the $\mathbf{Q}_p$-hypersurface $V_{f_{\mu_1}}$ the dilatation of V_f at μ_1, and the set $C(f_{\mu_1})$, the liftings of $C_{f_{\mu_1}}(\mathbf{F}_p)$, the first generation of descendants of μ_1. Given a sequence $(\mu_i)_{i\geq 1}$, we define inductively $e_{\mu_1 \cdots \mu_k}$ and $f_{\mu_1 \cdots \mu_k}(x)$, by the following procedure:

$$f_{\mu_1 \cdots \mu_k}(x) = \begin{cases} f(x) & \text{if } k = 0 \\ p^{-e_{\mu_1 \cdots \mu_k}} f_{\mu_1 \cdots \mu_{k-1}}(px + \mu_k) & \text{if } k \geq 1. \end{cases}$$

The union of the sets $C(f_{\mu_1 \cdots \mu_k})$, as the μ_k runs through the descendants of the (k-1)-th generation of μ_1, is called the k-th generation of descendants of μ_1.

Let $\bar{X}$ be a subset of affine n-space $\mathbf{A}^n(\mathbf{F}_p)$ and X its pre-image under the canonical map $\mathbf{Z}_p \to \mathbf{F}_p$. Denote by $C(f, X)$ the subset of T^n (T the Teichmuller representatives) which maps bijectively to the set $C_f(\mathbf{F}_p) \cap \bar{X}$. We define two integers associated to the reduced polynomial $\bar{f} \equiv f$ mod. p:

$$\nu(\bar{f}) := \text{Card } \{\bar{\mu} \in \bar{X} : \bar{\mu} \notin V_f(\mathbf{F}_p)\}$$

and

$$\sigma(\bar{f}) := \text{Card } \{\bar{\mu} \in \bar{X} : \bar{\mu} \in C_{\bar{f}}(\mathbf{F}_p)\}.$$

The following is Igusa's version of stationary phase for p-adic integrals.

STATIONARY PHASE FORMULA. (Igusa [6], p. 177) *With notations as above we have*

$$\int_X |f(x)|_p^s dx = \{\nu(\bar{f}) + \sigma(\bar{f})\frac{(1 - p^{-1})p^{-s}}{(1 - p^{-1}p^{-s})} + \sum_{\mu \in C(f, X)} q^{-e_\mu s} \int_{\mathbf{Z}_p^n} |f_\mu(x)|_p^s dx\} q^{-n},$$

where s is a complex number with $Re(s) > 0$.

The essential idea of the proof is not difficult to describe. Suppose the set $\bar{X} = \{\bar{\mu}_1, \ldots, \bar{\mu}_N\}$ and let μ_i be the lifting of $\bar{\mu}_i$. Then the set X is the disjoint union $\cup_\mu X_\mu$, where $\bar{\mu} = (\bar{y}_1, \ldots, \bar{y}_n) \in \bar{X}$ and X_μ is defined as

$$X_\mu = \{(x_1, \ldots, x_n) \in X : x_i = y_i + pz_i, z_i \in \mathbf{Z}_p, i = 1, \ldots, n\}.$$

With this notation we can now write

$$\int_X |f(x)|_p^s dx = \sum_{\mu \in X_\mu} \int_{X_\mu} |f(x)|_p^s dx = \sum_{\mu \in X_\mu} \int_{\mathbf{Z}_p^n} |f(px + \mu)|_p^s dx$$

$$= \sum_{\mu \in X_\mu} p^{-e_\mu s} \int_{\mathbf{Z}_p^n} |f_\mu(x)|_p^s dx.$$

The determination of the integrals arising from the points μ for which $\bar{\mu} \notin V_f(\mathbf{F}_p)$ are easily computable. The integrals corresponding to the $\mu's$ for which $\bar{\mu}$ is a non-singular point of $V_f(\mathbf{F}_p)$ are calculated by using an appropriate version of Hensel's Lemma together with the implicit function theorem (see [6], p. 177).

The p-adic integrals which arise in the applications to exponential sums over finite p-adic rings are of a slightly more general shape. The integrand contains an additional term of the form $\psi(f(x))$; since these are locally constant functions, suitable partitions of the domain of integration will lead to integrals similar to those in Igusa's formula. Let V be a smooth variety defined by polynomial equations over $\mathbf{Z}_p$ of relative dimension n and f a function on V. We suppose that f is a Morse function, i.e., the critical set

$$C_f := \{s \in V : \text{grad}_s f = 0\}$$

is finite and etale[1] set over $\mathbf{Z}_p$. If ψ denotes the canonical additive character on $\mathbf{Q}_p^+$ and Φ denotes a Schwartz function on the $\mathbf{Q}_p$-points of V, then we define a function F_Φ^* on $\mathbf{Q}_p$ by

$$F_\Phi^*(i^*) = \int_{V(\mathbf{Q}_p)} \Phi(x)\psi(i^* f(x))dx, \quad (i^* \in \mathbf{Q}_p).$$

It is easy to see that if $V = \mathbf{A}^n$ is affine n-space and Φ is the characteristic function of the set $\mathbf{A}^n(\mathbf{Z}_p)$, and f is a polynomial in $\mathbf{Q}_p[x_1, \ldots, x_n]$, then

$$F_\Phi^*(i^*) = p^{-nm} \sum_{i \in \mathbf{A}^n(\mathbf{Z}_p/p^m \mathbf{Z}_p)} \psi(i^* f(i)), \quad i^* = u \cdot p^{-m},$$

---

[1] This is a condition on the relative smoothness of the Galois action of $Gal(\bar{\mathbf{Q}}_p/\mathbf{Q}_p)$.

where $u \in \mathbf{Z}_p^{\times}$.

Several important results of Igusa revolve around the asymptotic expansions of the integrals $F_{\Phi}^*(i^*)$. Two particularly interesting cases arise as follows. When the critical set $C_f(\mathbf{Z}_p)$ is not empty and $i^* = up^{-m}$ with $m \geq 2$, the integral $F_{\Phi}^*(i^*)$ can be expressed as a sum of local terms indexed by the finite number of elements in $C_f(\mathbf{Z}_p)$ analogous to those that appear in the classical stationary phase. For example if m is even, we actually get an exact formula

$$(5) \qquad \int_{V(\mathbf{Z}_p)} \psi(i^* f(x)) dx = \{ \sum_{s \in C_f(\mathbf{Z}_p)} \psi(i^* f(s)) \} |i^*|_p^{-n/2}.$$

For the case $i^* = up^m$, $m \geq 3$ odd, we have

$$(6) \qquad \int_{V(\mathbf{Z}_p)} \psi(i^* f(x)) dx = \{ \sum_{s \in C_f(\mathbf{Z}_p)} \zeta_8^{sgn\, H(s)} \psi(i^* f(s)) \} |i^*|_p^{-n/2},$$

where $H(s)$ is the Hessian matrix $(d^2 f)_s$ of f at the critical point $s \in C_f(\mathbf{Z}_p)$, and $\zeta_8^{sgn\, H(s)}$ is just a symbolic way of writing the fourth root of unity whose value is given by the multidimensional Gauss sum

$$\zeta_8^{sgn\, H(s)} = p^{-n/2} \sum_{z \in (\mathbf{Z}_p/p\mathbf{Z}_p)^n} \psi(p^{m-1} i^* H(z)).$$

The Work of Igusa has been carried quite far by Denef and his collaborators, particularly with respect to the determination of Igusa's zeta function and the monodromy properties of the set $C_f(\mathbf{Z}_p)$. See Denef's Bourbaki Report [7].

In a future publication we will pursue the implications arising from Igusa's stationary phase formula for the explicit determination of exponential sums in several variables over finite rings of the type $\mathbf{Z}_p/p^{\ell}\mathbf{Z}_p$.

References

1. Landau. E.: *Elementary Number Theory*, Chelsea Publishing Co., New York, 1966.
2. Hua, L.K.: *Additive Theory of Prime Numbers*, American Mathematical Society, R. Island, 1965.
3. Vaughan, R.C.: *The Hardy-Litlewood Method*, Cambridge University Press, Cambridge, England, 1981.
4. Guillemin, V. and Sternberg, S.: *Geometric Asymptotics*, American Mathematical Society, Providence, R. Island, 1977.
5. Hormander, L.: *The Analysis of Linear Partial Differential Operators I*, Springer-Verlag, New York, 1983.
6. Igusa, J.-I.: *A stationary phase formula for p-adic integrals and its applications*, Algebraic Geometry and its Applications, Springer-Verlag, 1994, pp. 175-194.
7. Denef, J.: *Report on Igusa's Zeta Function*, Seminaire Bourbaki, n° 741, Asterisque 201-203 (1991) pp. 359-386.

8. Dwork, B.: *On the Root Number in the Functional Equation of the Artin-Weil L-function*, (Thesis, Columbia University, July 19, 1954).

9. Lamprecht, E.: *Allgemeine Theorie der Gaussschen Summen in endlichen kommutativen Ringen*, Math. Nachr., vol. 9 (1953), pp. 149-196.

10. Kummer, E.: *Note sur une expression analogue a la resolvante de Lagrange pour la equation $z^p = 1$*, Atti dell'Accademia Pontifica de Nuovi Lincei VI, (1852-1853), pp. 237-241.

11. Moreno, Carlos Julio: *Algebraic Curves Over Finite Fields*, Cambridge Tract in Mathematics, No. 97, Cambridge, England 1991.

Exponential Sums in Several Variables
Over Finite Fields

Oscar Moreno[1], Francis N. Castro[1], and Alberto Cáceres[2]

[1] Univ. of Puerto Rico, Rio Piedras
[2] Univ. of Puerto Rico, Humacao

Abstract. [1] In this paper, we explore the relationship between exponential sums over finite fields of characteristic 2 and exponential sums over the binary field following the method of Moreno and Moreno (see [3] and [6]). We begin combining the reduction to the ground field method and the Newton polyhedra method of Adolphson and Sperber to improve the partial improvement given by Moreno and Moreno of the Ax-Katz's theorem (see theorem 4). We give a new and better lower bound for the 2-divisibility of exponential sums over the binary field (see theorem 5). As an application we compute the 2-divisibility for infinite families of polynomials over the binary field. We compare the lower bounds given by Adolphson an d Sperber (see [1]) and Moreno and Moreno (see [3] and [6]) for exponential sums over the binary field. We also give an elementary new proof of Moreno-Moreno's partial improvement to the Ax-Katz's theorem (see [5]) for finite fields of characteristic 2.

1 Combination of the Reduction to the Ground Field Method and Newton Polyhedra Method

In this section, we will combine the reduction to the ground field method and the Newton Polyhedra method of Adolphson and Sperber to give an improvement to the Moreno-Moreno's result (see [4] and [5]).

Let $\mathbf{F}_q$ be a finite field with $q = p^f$ elements. We denote by $N(F_1, \ldots, F_t)$ the number of simultaneous solutions of the polynomials $F_1, \ldots, F_t$ over $\mathbf{F}_q$.

Ax's theorem gives information about the p-divisibility of the number of solutions of a system of polynomials. Katz improved the Ax's theorem in [2]. The Ax-Katz's theorem states the following:

Theorem 1. *Let $F_1, \ldots, F_t$ be polynomials in n variables with coefficients in F_q of total degrees $d_1, \ldots, d_t$. Let μ be the least integer that satisfies*

$$\mu \geq \frac{n - \sum_{i=1}^{t} d_i}{\max_i d_i}.$$

[1] This work was supported by the grants: 1. "Multi-Media Transmission in Fiber-Optic Networks Using Optical CDMA", DEPSCoR Office of Naval Research, Grant No. N00014-96-1-1192, 2. Infrastructure for Computer Science Research in Puerto Rico: NSF CISE, 1994-99, Grant No. CDA-9417362.

Then q^μ divides $N(F_1, \ldots, F_t)$.

The proof of the Ax-Katz's theorem can be found in [2].

Moreno and Moreno proved in [4] a partial improvement to the Ax-Katz's theorem. Before we state the Moreno and Moreno result, we need to give a definition.

Definition 1. *For each integer n with p-expansion*

$$n = a_0 + a_1 p \cdots + a_s p^s \text{ where } 0 \le a_i < p,$$

we denote its p-weight by $\sigma_p(n) = \sum_{i=0}^s a_i$. The p-weight degree of a monomial $x^d = x_1^{d_1} \cdots x_n^{d_n}$ is $w_p(x^d) = \sigma_p(d_1) + \cdots + \sigma_p(d_n)$. The p-weight degree of a polynomial $F(x_1, \ldots, x_n) = \sum_d a_d x^d$ is $w_p(F) = \max_{x^d, a_d \ne 0} w_p(x^d)$.

The Moreno-Moreno's result is the following:

Theorem 2. *Let $F_1, \ldots, F_t$ be polynomials in n variables with coefficients in F_q, a finite field with $q = p^f$ elements. Let $w_p(F_i)$ be the p-weight degree of F_i and let μ be the smallest integer such that*

$$\mu \ge f\left(\frac{n - \sum_{i=1}^t w_p(F_i)}{\max_i w_p(F_i)}\right).$$

Then p^μ divides $N(F_1, \ldots, F_t)$.

In the proof of theorem 2, Moreno and Moreno used the reduction to ground field method. It consists in transforming a polynomial in n variables over $\mathbf{F}_{p^f}$ to a polynomial in fn variables over $\mathbf{F}_p$ by applying the trace map. We are going to illustrate the reduction to the ground field method for a polynomial over a finite field of characteristic 2. Given a monomial $x_1^{d_1} \cdots x_n^{d_n}$ over $\mathbf{F}_{2^f}$, we can choose a $\mathbf{F}_2$-basis $\alpha_1, \ldots, \alpha_f$ of $\mathbf{F}_{2^f}$ such that

$$x_j = \sum_{i=1}^f x_{ji}\alpha_i \text{ for } j = 1, \ldots, n$$

If we substitute the above in $x_1^{d_1} \cdots x_n^{d_n}$, we get

$$x_1^{d_1} \cdots x_n^{d_n} = \left(\sum_{i=1}^f x_{1i}\alpha_i\right)^{d_1} \cdots \left(\sum_{i=1}^f x_{ni}\alpha_i\right)^{d_n}$$

$$= \left(\sum_{i=1}^f x_{1i}\alpha_i\right)^{2^{j_{11}}} \left(\sum_{i=1}^f x_{1i}\alpha_i\right)^{2^{j_{12}}} \cdots$$

$$= \left(\sum_{i=1}^f x_{1i}\alpha_i^{2^{j_{11}}}\right)\left(\sum_{i=1}^f x_{ji}\alpha_i^{2^{j_{21}}}\right) \cdots \quad (3)$$

where $d_s = 2^{j_{s1}} + 2^{j_{s2}} + \cdots$. Applying to (3) the trace map from $\mathbf{F}_{2^f}$ to $\mathbf{F}_2$, we get a polynomial in nf variables over $\mathbf{F}_2$. We can transform any polynomial over $\mathbf{F}_{p^f}$ by using the ground field method to a polynomial over $\mathbf{F}_p$ since any polynomial is a sum of monomials and the trace map is linear.

Let $F(x) = F(x_1, \dots, x_n)$ be a polynomial in n variables over $\mathbf{F}_q$. We define the exponential sum associated to F as follows:

$$S(F) = \sum_{(x_1,\dots,x_n)\in\mathbf{F}_q^n} e^{2\pi i Tr(F(x_1,\dots,x_n))/p},$$

where Tr is the trace map from $\mathbf{F}_q$ to $\mathbf{F}_p$.

Let F' be the polynomial corresponding to F after we apply the reduction to ground field method to F. Therefore

$$\begin{aligned} S(F) &= \sum_{(x_1,\dots,x_n)\in\mathbf{F}_{p^f}^n} e^{2\pi i Tr(F(x_1,\dots,x_n))/p} \\ &= \sum_{(x_1,\dots,x_{nf})\in\mathbf{F}_p^{nf}} e^{2\pi i F'(x_1,\dots,x_{nf})/p} = S(F'). \end{aligned}$$

The above implies that any estimate of $S(F')$ gives an estimate of $S(F)$.

Let p be a prime number. For any integer number a, we denote as $\mathrm{ord}_p(a)$ the highest power of p which divides a, i.e.,

$$\mathrm{ord}_p(a) = \max\{\, k \mid p^k \text{ divides } a\,\}.$$

Adolphson and Sperber in [1] gave a lower bound for the p-divisibility of exponential sums. They used the theory of Newton Polyhedra in the proof of their theorem.

Let $F(x) = F(x_1, \dots, x_n)$ be a polynomial over $\mathbf{F}_q$ and let D be the set of all the integral vectors d corresponding to the monomials $a(d)x^d = a_{d_1,\dots,d_n} x^{d_1} \cdots x_n^{d_n}$ of F. The Newton polyhedron $\Delta(F)$ is defined to be convex hull in $\mathbf{R}^n$ of the set $D\bigcup\{(0,\dots,0)\}$. Let $\omega(F)$ be the smallest positive rational number such that $\omega(F)\Delta(F)$ contains at least one point with positive integral coordinates. Now, we state without proof (see [1]) the Adolphson and Sperber's theorem:

Theorem 3. *With the above notation and assumptions, we have*

$$\mathrm{ord}_q(S(F)) \geq \omega(F).$$

Now, we state our result:

Theorem 4. *Let $F_1, \dots, F_t$ be polynomials in n variables with coefficients in F_q, a finite field with $q = p^f$ elements. Let $w_p(F_i)$ be the p-weight degree of F_i. Let $(y_i F_i)'$ be the polynomial coresponding to $y_i F_i$ after we apply the reduction to the ground field to $y_i F_i$. Then*

$$\omega(\sum_{i=1}^t (y_i F_i)') - t \geq f\left(\frac{n - \sum_{i=1}^t w_p(F_i)}{\max_i w_p(F_i)}\right).$$

Remark. $\omega(\sum_i^t(y_iF_i)')$ is the smallest positive rational number such that $\omega(\sum_i^t(y_iF_i)')\ \Delta(\sum_i^t(y_iF_i)')\ (\Delta(\sum_i^t(y_iF_i)')$ is the Newton polyhedron associated to $\sum_i^t(y_iF_i)'$, where $\sum_i^t(y_iF_i)'$ is the polynomial over $\mathbf{F}_p$ corresponding to $\sum_i^t(y_iF_i)$. contains at least one point with positive integral coordinates.

Proof. Adolphson and Sperber proved in [1] that theorem 3 is an improvement to the Ax-Katz's theorem. The Ax-Katz's theorem and Moreno-Moreno's theorems coincide when $q = p$. This completes the proof.

Examples 1.

a. Let $F(x,y) = \alpha x^{23}y^{23} + \beta x^7y^7 + \epsilon xy$ over $\mathbf{F}_{2^{3f}}$. Using Moreno-Moreno's result, we get that $\mathrm{ord}_2(S(F)) \geq \frac{3f}{4}$. Using Adolphson-Sperber result, we get that $\mathrm{ord}_2(S(F)) \geq \frac{9f}{31}$ since $(\frac{31}{3}, \frac{31}{3}) \in \Delta(F)$.

Example 1.a proved that the Adolphson-Sperber's result is not better than the Moreno- Moreno's result in general. To get an improvement of the Moreno-Moreno's result we need to combine the Adolphson–Sperber's result and the reduction to the ground field method.

b. Let $F(x,y) = \alpha x^3y + \beta xy$ over $\mathbf{F}_{2^f}$. Using Moreno-Moreno's result, we get that $\mathrm{ord}_2(S(F)) \geq \frac{2f}{3}$. Using Adolphson-Sperber result, we get that $\mathrm{ord}_2(S(F)) \geq f$ since $\omega(F) = 1$.

Example 1.b proved that the Moreno-Moreno's result is not better than the Adolphson- Sperber's result in general.

2   Exponentials Sums over the Binary Field

Let $\mathbf{F}_2$ be the binary field and $\mathbf{F}_2^n = \{(x_1,\dots,x_n)\,|\,x_i \in \mathbf{F}_2,\ i = 1,\dots,n\}$. Let $F(x) = F(x_1,\dots,x_n)$ be a polynomial in n-variables over $\mathbf{F}_2$. The exponential sum associated to F is the following:

$$S(F) = \sum_{(x_1,\dots,x_n)\in\mathbf{F}_2^n} (-1)^{F(x_1,\dots,x_n)}$$

Throughout the rest of the paper, we will say that a polynomial $F(x) = F(x_1,\dots,x_n)$ is not a polynomial in some proper subset of the variables $x_1,\dots,x_n$ if any expression of F contains all the variables. If that were not the case, we can consider F in a polynomial ring with fewer variables and not lose of generality.

We will write $F(x) = F(x_1,\dots,x_n)$ as sum of monomials of the form $a(d)x^d$ where $d = (d_1,\dots,d_n)$ is a vector of nonnegative integers such that $x^d = x_1^{d_1}\cdots x_n^{d_n}$. Let D be the set of all the vectors d corresponding to the monomials of F. Let $\mathcal{C}$ be a minimal set of the monomials of F covering all variables, that is, every variable x_i is in at least one monomial in $\mathcal{C}$ and $\mathcal{C}$ is minimal with that property. We call this set $\mathcal{C}$ a *minimal covering of F* and we assume that its cardinality is r.

Throughout the rest of the section, we will use the identities $(-1)^{a(d)x^d} = 1 - 2a(d)x^d$ and $x_i^l = x_i$ for $l > 0$, which hold only on binary variables. Therefore, if $F(x_1, \ldots, x_n) = \sum_{d \in D} a(d)x^d$ then

$$S(F) = \sum_{x_1, \ldots, x_n \in \mathbf{F}_2^n} \prod_{d \in D} (1 - 2a(d)x^d).$$

If we expand this equation, we get

$$S(F) = \sum_{x_1, \ldots, x_n} (1 + \sum_{\lambda} 2^{n(\lambda)} g_\lambda(x_1, \ldots, x_n))$$

$$= 2^n + \sum_{\lambda} 2^{n(\lambda)} \sum_{(x_1, \ldots, x_n) \in \mathbf{F}_2^n} g_\lambda(x_1, \ldots, x_n) \qquad (1)$$

where the g_λ's are monomials.

The following lemma (see [3] and [6]) establishes divisibility properties of an exponential sum which will determine the divisibility of the number of zeros of a polynomial equation. For the sake of completeness, we include their proof.

Lemma 1. Let $F(x) = F(x_1, \ldots, x_n)$ be a polynomial in n variables over $\mathbf{F}_2$. Let C be a minimal set of monomials of $F(x)$ covering all the variables and r be the cardinality of C. Then 2^r divides $S(F)$.

Proof. We claim that the minimum power of 2 that divides

$$2^{n(\lambda)} \sum_{(x_1, \ldots, x_n) \in \mathbf{F}_2^n} g_\lambda(x_1, \ldots, x_n),$$

for all the λ in (1), happens when $n(\lambda) = r$ and $g_\lambda(x_1, \ldots, x_n) = \prod_{x^d \in C} x^d$. Recall that if $g_\lambda(x_1, \ldots, x_n)$ lacks l variables, then

$$2^{n(\lambda)} \sum_{(x_1, \ldots, x_n) \in \mathbf{F}_2^n} g_\lambda(x_1, \ldots, x_n) = 2^{n(\lambda)+l}.$$

We assume that there is a term $2^{n(\lambda)} \sum_{(x_1, \ldots, x_n) \in \mathbf{F}_2^n} g_\lambda(x_1, \ldots, x_n) = 2^a$ where $a < r$. Hence $a = n(\lambda) + l$, where l is the number of variables that does not appear in g_λ. This is a contradiction to the minimality of the set C. That ends the proof.

Remark. Lemma 1 has applications to the number of zeros of a system of polynomials . Let $F_1, \ldots, F_t$ be a system of polynomials over $\mathbf{F}_2$ and let r be minimal number of monomials of $F_1, \ldots, F_t$ that covers $x_1, \ldots, x_n$. Then 2^{r-t} divides $N(F_1, \ldots, F_t)$, where $N(F_1, \ldots, F_t)$ is number of simultaneous solutions of $F_1, \ldots, F_t$.

Example 2. Let $F_1(x_1, \ldots, x_n) = x_1 x_2 + x_2 x_3 + \cdots + x_{n-1} x_n + x_1 x_n$ and $F_2(x_1, \ldots, x_n) = x_1 x_2 x_3 + x_1 + \cdots + x_n$ over $\mathbf{F}_2$. We assume that n is an odd number. In this case $r = \frac{n-1}{2}$, hence $2^{(n-5)/2}$ divides $N(F_1, F_2)$. The Moreno-Moreno's and Ax-Katz's theorems (see theorem 2 and theorem 1) imply that $\mathrm{ord}_2(N(F_1, F_2)) \geq \frac{n-5}{3}$. In particular, we obtain the following:

n	$N(F_1, F_2)$
15	$2^5 \times 7 \times 37$
17	$2^6 \times 3^3 \times 19$
19	$2^7 \times 1021$
21	$2^8 \times 23 \times 89$
23	$2^9 \times 4099$

Note, that the 2-divisibility given for $N(F_1, F_2)$ in the example 2 is tight for the cases calculated in the above table.

Since r is the cardinality of a minimal set of monomials covering all the variables $x_1, \ldots, x_n$, we will call a *generalized minimal covering of F to any monomial $g_\lambda(x_1, \ldots, x_n)$* in (1) such that

$$2^{n(\lambda)} \sum_{(x_1, \ldots, x_n) \in \mathbf{F}_2^n} g_\lambda(x_1, \ldots, x_n) = 2^r.$$

We assume that the number of generalized minimal coverings of F is m. The following theorem is an improvement of Lemma 1 and expresses:

Theorem 5. *Let $F(x) = F(x_1, \ldots, x_n)$ be a polynomial in n-variables over $\mathbf{F}_2$. We assume that F is not a polynomial in some proper subset of variables $x_1, \ldots, x_n$. We assume further that the number of generalized minimal coverings of F is m and the cardinality of C is r. Then, $\mathrm{ord}_2(S(F)) = r$ when m is odd and $\mathrm{ord}_2(S(F)) > r$ when m is even.*

Proof. Suppose that the following terms

$$g_{\lambda_1}(x_1, \ldots, x_n), \ldots, g_{\lambda_m}(x_1, \ldots, x_n) \qquad (2)$$

are the generalized mimimal coverings of F in (1), i.e.,

$$2^{n(\lambda_i)} \sum_{(x_1, \ldots, x_n) \in \mathbf{F}_2^n} g_{\lambda_i}(x_1, \ldots, x_n) = 2^r$$

for $i = 1, \ldots, m$. Let m_1 the number of generalized minimal coverings of F with positive sign in (1) and m_2 be generalized minimal coverings of F with negative sign in (1). We assume that the first m_1 terms in (2) are the generalized coverings of F with positive sign in (1) and the other m_2 in (2) are the generalized coverings of F with negative sign in (1). Then

$$\overbrace{2^{n(\lambda_1)} \sum_{x \in \mathbf{F}_2^n} g_{\lambda_1}(x_1, \dots, x_n) + \dots + 2^{n(\lambda_{m_1})} \sum_{x \in \mathbf{F}_2^n} g_{\lambda_m}(x_1, \dots, x_n)}^{m_1-terms} -$$

$$\overbrace{\left(2^{n(\lambda_{m_1}+1)} \sum_{x \in \mathbf{F}_2^n} g_{\lambda_1}(x_1, \dots, x_n) + \dots + 2^{n(\lambda_{m_1})} \sum_{x \in \mathbf{F}_2^n} g_{\lambda_m}(x_1, \dots, x_n) \right)}^{m_2-terms}$$

$$= 2^r(m_1 - m_2).$$

If m is odd, then only one of m_1 or m_2 is odd, therefore $\mathrm{ord}_2(S(f)) = r$. If m is even, then both m_1 and m_2 are simultaneously even or odd and in any case $\mathrm{ord}_2(S(f)) > r$.

Corollary 1. *Let $F(x_1, \dots, x_n)$ be a polynomial in n-variables over $\mathbf{F}_2$ that is not a polynomial in some proper subset of variables $x_1, \dots, x_n$. We assume that the number of generalized minimal coverings of F is m and the cardinality of C is r. Let $N(F)$ be the number of zeros of F over $\mathbf{F}_2$. Then, $\mathrm{ord}_2(N(F)) = r - 1$ when m is odd and $\mathrm{ord}_2(N(F)) > r - 1$ when m is even.*

Proof. Just apply theorem 5 to the formula $S(F) = -2^n + 2N(F)$.

Now we will show some applications of theorem 5.

Corollary 2. *Let $F(x_1, \dots, x_n)$ be a polynomial over $\mathbf{F}_2$ containing the monomial $x_{i_1} \cdots x_{i_{n-1}}$ and assume that the other monomials of F have degree $< \frac{n}{2}$. Let m be the number of monomials of F that do contain the missing variable of the monomial $x_{i_1} \cdots x_{i_{n-1}}$. Then, if m is odd, $\mathrm{ord}_2(S(f)) > 2$, and if m is even, $\mathrm{ord}_2(S(f)) = 2$.*

Proof. Without loss of generality, we assume that x_n is the variable not appearing in $x_{i_1} \cdots x_{i_{n-1}}$. Let $g_{\lambda_1}, \dots, g_{\lambda_m}$ be the monomials of F containing x_n. Note that in this case $r = 2$. Then

$$S(F) = 2^n + \sum_{\lambda} 2^{n(\lambda)} \sum_{(x_1, \dots, x_n) \in \mathbf{F}_2^n} g_{\lambda}(x_1, \dots, x_n)$$

$$= -2 \sum_{(x_1, \dots, x_n)} x_1 \cdots x_{n-1} + 2^2 \sum_{(x_1, \dots, x_{n-1})} x_1 \cdots x_{n-1} g_{\lambda_1} + \dots +$$

$$2^2 \sum_{(x_1, \dots, x_{n-1})} x_1 \cdots x_{n-1} g_{\lambda_m} + 2^s(\pm 1 + \cdots \text{ where } s > 2$$

$$= \overbrace{-2^2 + 2^2 + \dots + 2^2}^{m-\text{times}} + 2^s(\pm 1 + \cdots \qquad \text{where } s > 2.$$

We simplify the last expression and we get $\overbrace{2^2 + \dots + 2^2}^{(m-1)\text{times}}$. If m is odd, $m-1$ is even, hence $\mathrm{ord}_2(S(F)) > 2$. If m is even, $m-1$ is odd and hence $\mathrm{ord}_2(S(F)) = 2$.

Example 3. Let

$$F(x_1, \ldots, x_n) = x_1 \cdots x_{n-1} + x_1 x_2 + x_2 x_3 + \cdots + x_{n-1} x_n + x_1 x_n.$$

Here $\mathrm{ord}_2(S(F)) > 2$ since the variable x_n appears three times, hence 2^2 divides $N(F)$. Ax-Katz's and Moreno-Moreno's results implies that 2 divides $N(F)$. The Adolphson-Sperber (see theorem 3) implies that 2 divides $N(F)$ since $(\frac{1}{2}, \ldots, \frac{1}{2}, 1, 1) \in \Delta(F)$.

Numerical Example 4.
Let $F(x_1, \ldots, x_6) = x_1 x_2 + x_3 x_4 + x_5 x_6 + x_1 x_2 x_3 x_4 x_5$ over $\mathbf{F}_2$ then by corollary 2 we have $\mathrm{ord}_2(S(F)) > 2$. It happens that the exact value of $S(F)$ is 2^3.

Corollary 3. Let $F(x_1, \ldots, x_n)$ be a polynomial over $\mathbf{F}_2$. $F(x_1, \ldots, x_n)$ is a polynomial of degree n if and only if $\mathrm{ord}_2(S(F)) = 1$.

Proof. If F has degree n, then there is one generalized minimal covering of F. Therefore $\mathrm{ord}_2(S(F)) = 1$. By Lemma 1, if $\mathrm{ord}_2(S(F)) = 1$, then F has degree n. This completes the proof.

Corollary 3 implies that 2 does not divide the number of solutions of F.

Corollary 4. Let $F(x_1, \ldots, x_n)$ be a polynomial over $\mathbf{F}_2$ of degree $n - 1$ containing k monomials in which all of them are of degree $n - 1$. If $k + \binom{k}{2}$ is odd then $\mathrm{ord}_2(S(F)) = 2$, otherwise $\mathrm{ord}_2(S(F)) > 2$.

Proof. In (1) there are k terms with $n(\lambda) = 1$ and missing one variable. Also, there are $\binom{k}{2}$ with $n(\lambda) = 2$ and no variables missing. We apply theorem 5 to the $k + \binom{k}{2}$ generalized minimal coverings of F. This completes the proof.

Corollary 4 implies that if the number of monomials of F is odd then 2^2 divides the number of solutions of F. Ax-Katz's and Moreno-Moreno's theorems imply that 2 divides the number of solutions of F (see theorems 1 and 2). Adolphson-Sperber's estimate implies that at most 2 divides the number of zeros of F since $\omega(F)$ (see theorem 3) ≤ 2.

Numerical Example 5. Let

$$F(x_1, \ldots, x_8) = x_1 x_2 x_3 x_4 + x_2 x_3 x_4 x_5 + x_1 x_3 x_4 x_5,$$

over $\mathbf{F}_2$. Corollary 4 implies that 4 divides $N(F)$, since $3 + \binom{3}{2}$ is even. The actual value of $N(F) = 2^2 \times 5$.

Corollary 5. Let l be a proper divisor of n. Let $g_1(x_1, \ldots, x_n), \ldots, g_l(x_1, \ldots, x_n)$ be monomials in $\mathbf{F}_2[x_1, \ldots, x_n]$. We assume that $g_i(x_1, \ldots, x_n)$ and $g_j(x_1, \ldots, x_n)$ do not have any variable in common for $i \neq j$. Then if

$$F(x_1, \ldots, x_n) = g_1(x_1, \ldots, x_n) + \cdots + g_l(x_1, \ldots, x_n) + x_1 + \cdots + x_n, \text{ then,}$$

$$\mathrm{ord}_2(S(F)) = \frac{n}{l}.$$

Proof. In this case there is just one generalized minimal covering of F. This ends the proof of Corollary 5

3   Adolphson-Sperber's Theorem over the Binary Field

Let $\mathbf{F}_2$ be the binary field and $F(x) = F(x_1, \ldots, x_n)$ be a polynomial in n variables over $\mathbf{F}_2$. We keep the notation of the sections 1 and 2. Let D be the set of all the integral vectors d corresponding to the monomials of F. The Newton polyhedron $\Delta(F)$ is defined to be convex hull in $\mathbf{R}^n$ of the set $D \bigcup \{(0, \ldots, 0)\}$. Let $\omega(F)$ be the smallest positive rational number such that $\omega(F)\Delta(F)$ contains at least one point with positive integral coordinates. The Adolphson and Sperber's theorem for the binary field is the following: $\mathrm{ord}_2(S(F)) \geq \omega(F)$.

Let C be a minimal set of monomials of F covering all the variables, that is, every variable x_i appears in at least one monomial in C. Let r be the cardinality of C.

Theorem 6. *With the above notation and assumptions, we have*

$$r \geq \omega(F).$$

Proof. We have $r \geq \omega(F)$, since

$$r\left(\sum_{x^d \in C} \frac{1}{r} d\right)$$

has nonnegative integral coordinates.

The following is an example where $r > \omega(F)$.

Example 6. Let

$$F(x_1, x_2, \ldots, x_n) = x_1 x_2 + x_2 x_3 + x_3 x_4 + \cdots + x_{n-2} x_{n-1} + x_{n-1} x_n + x_1 x_n$$

over $\mathbf{F}_2$, where n is odd. Therefore

$$D = \{d_1 = (1, 1, 0, 0, \ldots, 0, 0), \ldots, d_n = (1, 0, 0, \ldots, 0, 1)\}.$$

We take $t_1 = t_2 = t_3 = \cdots = t_n = \frac{1}{n}$, therefore

$$t_1 d_1 + t_2 d_2 + \cdots + \cdots + t_n d_n = (\tfrac{2}{n}, \tfrac{2}{n}, \ldots, \tfrac{2}{n}, \tfrac{2}{n}) \in \Delta(F)$$

We can conclude that $\frac{n}{2} \geq \omega(F)$ since $\frac{n}{2}(\frac{2}{n}, \frac{2}{n}, \ldots, \frac{2}{n}) = (1, \ldots, 1)$. Also $\omega(F) \leq \frac{n}{2}$, implies $\omega(F) = \frac{n}{2}$. Note that $r = \frac{n+1}{2}$.

In all the cases that we have computed the least integer $\geq \omega(F)$ is equal to r (the cardinality of minimal set of monomials of F coverings all the variables).

4 An Elementary Proof of the Moreno-Moreno Partial Improvement to the Ax-Katz's theorem over Finite Fields of Characteristic 2

In this section, we give an elementary new proof of the Moreno-Moreno's result for finite fields of characteristic 2. Their proof depends on p-adic analysis.

In our proof, we use the ground field method (see section 1) and Lemma 1. Now we are ready to prove the main theorem of this section.

Theorem 7. *Let* $F_1, \ldots, F_t$ *be polynomials in* n *variables with coefficients in* $\mathbf{F}_{2^f}$ *whose 2-weight degrees* $l_1, \ldots, l_t$. *If* $\mathrm{ord}_2(N(F_1, \ldots, F_t)) = \mu$ *then*

$$\mu \geq \frac{f(n - \sum_{i=1}^{t} l_i)}{\max_i l_i}.$$

Proof. Assume that $l_1 \geq l_2 \geq \cdots \geq l_t$ and apply the reduction to the ground field method to $y_i F_i$ for $i = 1, \ldots t$. Let $\alpha_1, \ldots, \alpha_f$ be a $\mathbf{F}_2$-basis of $\mathbf{F}_{2^f}$. Let G_i be the polynomial over $\mathbf{F}_2$ corresponding to $y_i F_i$ for $i = 1, \ldots, t$. Consider the exponential sum:

$$S(F_1, \ldots, F_t) = \sum_{\substack{x_1 \ldots, x_n \in \mathbf{F}_{2^f} \\ y_1, \ldots, y_t \in \mathbf{F}_{2^f}}} (-1)^{Tr(\sum_{i=1}^{t} y_i F_i)} = \sum_{\substack{x_{11} \ldots, x_{nf} \in \mathbf{F}_2 \\ y_{11}, \ldots, y_{tf} \in \mathbf{F}_2}} (-1)^{\sum_{i=1}^{t} G_i},$$

where $x_j = \sum_{i=1}^{f} x_{ji} \alpha_i$ for $j = 1, \ldots, n$ and $y_k = \sum_{i=1}^{f} y_{ki} \alpha_i$ for $k = 1, \ldots, t$.
Recall that
$$\mathrm{ord}_2(N(F_1, \ldots, F_t)) = \mathrm{ord}_2(S(F)) - tf.$$

We will give a lower bound for the minimum number of monomials necessary to cover all the variables of $\sum_{i=}^{t} G_i$. Note that the minimum number of monomials of $\sum_{i=1}^{t} G_i$ that covers $x_{11}, \ldots, x_{nf}, y_{11}, \ldots, y_{tf} \geq$ the minimum number of monomials of $\sum_{i=1}^{t} G_i$ covering the variables $x_{11}, \ldots, x_{nf}$, where we choose at least f monomials of each G_i. The last sentence is true since in G_i does not appear the variables $y_{j1}, \ldots, y_{jf}$ for $j \neq i$ (*). Now, we estimate the number of monomials necessary to cover $x_{11}, \ldots, x_{nf}$. Recall that $Tr(\sum_{i=1}^{t} y_i F_i) = \sum_{i=1}^{t} G_i$. Given a minimal set $\mathcal{C}$ of monomials of $\sum_{i=1}^{t} G_i$ covering all the variables x_{ij}'s, let m_i be the number of monomials of G_i that are present in set $\mathcal{C}$ ($m_1 + \cdots + m_t$ is equal to the cardinality of $\mathcal{C}$). Then

$$m_1 l_1 + \cdots l_t m_t \geq nf \qquad (4).$$

By (*), we have

$$m_1 \geq f$$

$$\vdots \qquad \vdots$$

$$m_t \geq f$$

These inequalities imply that

$$(l_1 - l_2)m_2 \geq (l_1 - l_2)f$$

$$\vdots \qquad\qquad \vdots$$

$$(l_1 - l_t)m_t \geq (l_1 - l_t)f$$

We sum the above to (4) and we get the following:

$$l_1(m_1 + m_2 + \cdots + m_t) \geq (n + (t-1)l_1 - \sum_{i=2}^{t} l_i)f,$$

We have that the number of monomials of $\sum_{i=1}^{t} G_i$ coverings all the variables is

$$m_1 + m_2 + \cdots + m_t \geq \frac{(n + (t-1)l_1 - \sum_{i=2}^{t} l_i)f}{l_1}$$

and Lemma 1 implies that

$$\operatorname{ord}_2(S(\sum_{i=1}^{t} y_i F_i)) \geq \frac{f(n + (t-1)l_1 - \sum_{i=2}^{t} l_i)}{l_1}.$$

Therefore

$$\operatorname{ord}_2(N(F_1,\dots,F_t)) \geq \frac{fn + (t-1)l_1 - \sum_{i=2}^{t} l_i}{l_1} - tf.$$

This completes the proof since

$$\frac{f(n + (t-1)l_1 - \sum_{i=2}^{t} l_i)}{l_1} - tf = \frac{f(n - \sum_{i=1}^{t} l_i)}{l_1}.$$

5   Conclusion

In section 1, for arbitrary characteristic p, we have shown how combining methods of Moreno-Moreno and Adolphson-Sperber we can improve upon previous improvements to the Ax-Katz's result. In section 2 and 3, for characteristic 2 we proved how the method of coverings of [3] and [6] should be used instead to the methods of section 1 since it give better or at least as good divisibility and it is simpler and easier to compute. In other words in characteristic 2 once the reduction to the ground field method, explained in section 1 has been applied, it is better to apply coverings than to use the Newton polyhedra method. We point out that the divisibility of the exponential sums can be used to give the best Moreno-Moreno type of bound for the absolute value of the exponential sums (see [7]).

References

1. Adolphson and Sperber, p-adic Estimates for Exponential Sums and the of Chevalley-Warning, *Ann. Sci. Ec. Norm. Super.*, 4e série, vol **20**, pp. 545-556, 1987.
2. N. M. Katz, On a Theorem of Ax, *Amer. J. Math.*, **93** (1971), 485-499.
3. O. Moreno, C. Cáceres and M. Alonso, An Improved and Simplified Binary Ax Theorem, *Proceedings 1994 IEEE International Symposium on Information Theory*, Trondheim, Norway June-July 1994.
4. O. Moreno and C.J. Moreno, Improvement of the Chevalley-Warning and the Ax-Katz theorems, *Amer. J. Math.* **117:1** (1995), 241-244.
5. O. Moreno and C.J. Moreno, An elementary proof of a partial improvement to the Ax-Katz theorem, *Proc. of Applied Algebra, Algebraic Algorithms an Error Correcting Codes*-AAECC-10, (*Lectures Notes In Comput. Sci.* **673**), 25 7-268, Springer-Verlag, Berlin,1993.
6. O. Moreno and C. J. Moreno, The MacWilliams-Sloane conjecture on the tightness of the Carlitz-Uchiyama bound and the weights of dual of BCH codes, *IEE Trans. Inform. Theory* **40:6** (1994), 1894).
7. O. Moreno and C.J. Moreno,*A p-adic Serre Bound*, Finite Fields and their Applications, vol. **40**, pp. 201-217, 1998.

Decoding Reed-Solomon Codes Beyond Half the Minimum Distance

R. Refslund Nielsen and T. Høholdt

Technical University of Denmark
Department of Mathematics, Bldg 303
DK-2800 Lyngby
Denmark

Abstract. We describe an efficient implementation of M. Sudan's algorithm for decoding Reed-Solomon codes beyond half the minimum distance. Furthermore, we calculate an upper bound of the probability of getting more than one codeword as output.

1 Introduction

In a recent paper M. Sudan [1] presented an algorithm for correcting more than $\frac{d_{min}-1}{2}$ errors in a Reed-Solomon code with low rate and in [2] he extended his algorithm to higher rates. The algorithm produces a list of codewords closest to the received word. In this paper we present an efficient implementation of Sudan's extended algorithm by speeding up the two crucial steps, namely interpolation and factorization. Based on the weight distribution of MDS codes we also calculate an upper bound on the probability that the list contains more than one codeword. The paper is organized as follows: Section 2 contains some basic definitions and in Section 3 we present Sudan's extended algorithm and prove that it works. Section 4 calculates the asymptotic error-correcting capability of the algorithm and Section 5 gives an upper bound on the number of candidates. Section 6 contains the efficient method to get the interpolation polynomial, this is partly based on R. Kötter [3], and Section 7 is devoted to the factorization. Finally we give some examples in Section 8 and Section 9 contains the conclusion and some remarks.

2 Basic definitions

Let $\mathbb{F}_q$ denote a finite field with q elements. If $a, b \in \mathbb{F}_q^n$ then $d(a, b)$ will denote the Hamming-distance between the words a and b. The Hamming-weight of a word, a, will be denoted by $w(a)$.

Definition 1 (Reed-Solomon codes) *Let* $P = \{P_1, \ldots, P_n\} \subseteq \mathbb{F}_q$ *with* $|P| = n$. *For* $k \leq n$ *the set*

$$\mathrm{RS}(P, k) = \{f(P) \mid f \in \mathbb{F}_q[x] \wedge \deg(f) < k\}$$

where $f(P) = (f(P_1), \ldots, f(P_n))$ *is called a **Reed-Solomon code**.*

It is a well-known fact that $\text{RS}(P,k)$ has length n, dimension k, and minimum distance $d = n - k + 1$. Furthermore, the weight distribution of Reed-Solomon codes is well-known (in fact, for given n, k, and q all MDS-codes have the same weight distribution). The following proposition can be found in various textbooks, for example [7], Theorem 14.1.2.

Proposition 2 (Weight distribution of Reed-Solomon codes) *If*

$$A_u = |\{c \in \text{RS}(P,k) \mid \text{w}(c) = u\}|$$

then $A_0 = 1$, $A_u = 0$ for $1 \le u < d$ and

$$A_u = \binom{n}{u} \sum_{i=0}^{u-d} (-1)^i \binom{u}{i} (q^{u-d+1-i} - 1)$$

for $d \le u \le n$.

Let $\text{B}(w,r)$ denote the **ball** with radius r and center w. That is

$$\text{B}(w,r) = \{u \in \mathbb{F}_q^n \mid \text{d}(w,u) \le r\}$$

The **sphere** with radius r and center w will be denoted by

$$\text{S}(w,r) = \{u \in \mathbb{F}_q^n \mid \text{d}(w,u) = r\}$$

Suppose that w is a received word and that $\text{RS}(P,k)$ is the code in use. Then decoding up to τ errors from w can be specified as calculating the following set:

$$\text{dec}_\tau(w) = \text{B}(w,\tau) \cap \text{RS}(P,k)$$

Notice that if $|\text{dec}_\tau(w)| = 0$ then more than τ errors have been detected. If $\tau \le \lfloor \frac{d-1}{2} \rfloor$ then for any w, $|\text{dec}_\tau(w)| \le 1$. If $\tau > \lfloor \frac{d-1}{2} \rfloor$ then decoding τ errors is often referred to as list-decoding, since $\text{dec}_\tau(w)$ may be a list of several codewords, all within distance τ from the received word. In this text, the codewords in $\text{dec}_\tau(w)$ will be called the **candidates** of decoding τ errors from w.

The decoding algorithm uses bivariate polynomials and we need an ordering of these.

Let $M = \{x^\alpha y^\beta \in \mathbb{F}_q[x,y] \mid (\alpha, \beta) \in \mathbb{N}^2\}$ be the set of monomials in $\mathbb{F}_q[x,y]$. A **monomial ordering** is a relation, $\le_m$, on M, which satisfies the following:

- $\le_m$ is total.
- $\forall f, g, h \in M : f \le_m g \Rightarrow fh \le_m gh$
- $\le_m$ is a well-ordering.

One monomial ordering is the **lexicographic order** defined by

$$x^\alpha y^\beta \le_\text{lex} x^a y^b \Leftrightarrow \alpha < a \vee (\alpha = a \wedge \beta \le b)$$

In the following this will be called the lexicographic order with $x < y$. In a similar way we may define a lexicographic order with $y < x$ by exchanging x and y in the expression above.

Let $f(x, y) \in \mathbb{F}_q[x, y] \backslash \{0\}$. Then the **weighted degree** of $f(x, y)$ is given by

$$\deg^{(a,b)}(f) = \max\{\alpha a + \beta b \mid f_{(\alpha,\beta)} \neq 0\}$$

where $a \in \mathbb{N}$ is called the weight of x and $b \in \mathbb{N}$ is called the weight of y. For any choice of a and b we may define $\deg^{(a,b)}(0) = -\infty$.

Given a weighted degree, $\deg^{(a,b)}$, and a lexicographic order, $\leq_{\text{lex}}$, we can define a corresponding **weighted degree lexicographic order** on M by

$$f \leq_{\text{wdeg}} g \Leftrightarrow \deg^{(a,b)}(f) < \deg^{(a,b)}(g) \vee (\deg^{(a,b)}(f) = \deg^{(a,b)}(g) \wedge f \leq_{\text{lex}} g)$$

for all $f, g \in M$.

In the following, if $f \in \mathbb{F}_q[x, y]$, then $f_{(\alpha,\beta)}$ will be defined for any pair, $(\alpha, \beta) \in \mathbb{N}^2$ by

$$f = \sum_{(\alpha,\beta)\in\mathbb{N}^2} f_{(\alpha,\beta)} x^\alpha y^\beta$$

Furthermore, define

$$\text{coef}(f, x^\alpha y^\beta) = f_{(\alpha,\beta)}$$

3   Sudan's algorithm

The following formulation of Sudan's algorithm is inspired by the presentation of Sudan's original algorithm by W. Feng and R. E. Blahut.

Algorithm 3 (Sudan's extended algorithm) *Input: The code* $\mathrm{RS}(P, k)$, *a received word,* $w \in \mathbb{F}_q^n$, *and a parameter,* $s \geq 1$.
Output: $\text{dec}_{\tau_s}(w)$.

– *Calculate* r_s *so that*

$$\binom{r_s}{2} \leq \frac{n\binom{s+1}{2}}{k-1} < \binom{r_s + 1}{2} \tag{1}$$

and let

$$\ell_s = \left\lfloor \frac{n\binom{s+1}{2}}{r_s} + \frac{(r_s - 1)(k - 1)}{2} \right\rfloor$$

Then

$$\tau_s = n - \left\lfloor \frac{\ell_s}{s} \right\rfloor - 1 \tag{2}$$

– *Calculate* $Q_s(x, y) \in \mathbb{F}_q[x, y] \backslash \{0\}$ *so that*
   1. $\text{coef}(Q_s(x + P_i, y + w_i), x^\alpha y^\beta) = 0$ *for all* $\alpha, \beta \in \mathbb{N}$ *with* $\alpha + \beta < s$.

2. $\deg^{(1,k-1)}(Q_s) \le \ell_s$

- *Factorize Q_s into irreducible factors.*
- *If $y - f(x)$ with $\deg(f) < k$ divides Q_s and $\mathrm{d}(f(P), w) \le \tau_s$ then include $f(P)$ in the set of candidates. That is*

$$\mathrm{dec}_{\tau_s}(w) = \{f \in \mathbb{F}_q[x] \mid \deg(f) < k \wedge (y - f(x))|Q_s \wedge \mathrm{d}(f(P), w) \le \tau_s\}$$

If $s = 1$ then this algorithm is identical to Sudan's original algorithm [1].

Any non-zero polynomial satisfying conditions 1 and 2 on the polynomial Q_s will be called a Q_s-polynomial in the following.

Notice that condition 1 on a Q_s-polynomial states that (P_i, w_i) must be a zero of multiplicity s.

In the following it will be proven that Sudan's extended algorithm gives the promised result. This will be done by proving that a Q_s-polynomial exists and that it has the right factors of the form $y - f(x)$. Furthermore, it will be clear that a Q_s has at most $r_s - 1$ such factors, so the number of codewords in the output is upper bounded by $r_s - 1$.

Lemma 4 (Weighted degree of ith monomial) *Consider the polynomial ring $\mathbb{F}_q[x, y]$ with the weigthed degree $\deg^{(1,k-1)}$ and let the monomials be ordered by a corresponding $\le_{\mathrm{wdeg}}$ order. Suppose that*

$$m_0 \le_{\mathrm{wdeg}} m_1 \le_{\mathrm{wdeg}} m_2 \le_{\mathrm{wdeg}} \cdots$$

is an increasing list of all the monomials in $F_q[x, y]$. Then

$$\deg^{(1,k-1)}(m_i) = \left\lfloor \frac{i}{r} + \frac{(r-1)(k-1)}{2} \right\rfloor \tag{3}$$

where r satisfies

$$\binom{r}{2} \le \frac{i}{k-1} < \binom{r+1}{2}$$

Proof: Group the monomials into the disjoint sets, $M_1, M_2, \ldots$, where

$$M_c = \{m_j \mid (c-1)(k-1) \le \deg^{(1,k-1)}(m_j) < c(k-1)\}$$

Then $|M_c| = c(k-1)$ so $|M_1| + |M_2| + \ldots + |M_{c-1}| = \binom{c}{2}(k-1)$. Since $\binom{r}{2}(k-1) \le i < \binom{r+1}{2}(k-1)$, $m_i \in M_r$. The smallest monomial in M_r has weighted degree $(r-1)(k-1)$ and for each a with $(r-1)(k-1) \le a < r(k-1)$ there are exactly r monomials with weighted degree a in M_r. If the monomials of M_r are listed increasingly with respect to $\le_{\mathrm{wdeg}}$ then m_i, is monomial number $i - \binom{r}{2}(k-1)$, so the weighted degree of m_i must be

$$\deg^{(1,k-1)}(m_i) = (r-1)(k-1) + \left\lfloor \frac{i - \binom{r}{2}(k-1)}{r} \right\rfloor = \left\lfloor \frac{i}{r} + \frac{(r-1)(k-1)}{2} \right\rfloor$$

QED

Lemma 5 (Transformation of polynomial) *Let* $f(x, y) \in \mathbb{F}_q[x, y]$ *and* x_0, y_0 $\in \mathbb{F}_q$ *then*

$$\text{coef}(f(x + x_0, y + y_0), x^\alpha y^\beta) = \sum_{a \geq \alpha \wedge b \geq \beta} \binom{a}{\alpha}\binom{b}{\beta} \text{coef}(f(x, y), x^a y^b) x_0^{a-\alpha} y_0^{b-\beta}$$

for all $(\alpha, \beta) \in \mathbb{N}^2$.

Proof: This is seen by direct calculation.

Theorem 6 (Existence of Q_s-polynomial) *A Q_s-polynomial does exist.*

Proof: Notice that by Lemma 5 the first condition on a Q_s-polynomial form $n\binom{s+1}{2}$ linear equations. Therefore, a Q_s-polynomial can be constructed by solving a system of $n\binom{s+1}{2}$ linear equations with $n\binom{s+1}{2} + 1$ unknowns, where the unknowns are the coefficients of Q_s using the basis monomials $m_0, \ldots, m_n$ as defined in Lemma 4.

By Lemma 4 the weighted degree of Q_s will satisfy

$$\deg^{(1,k-1)}(Q_s) \leq \deg^{(1,k-1)}(m_{n\binom{s+1}{2}}) = \left\lfloor \frac{n\binom{s+1}{2}}{r_s} + \frac{(r_s - 1)(k - 1)}{2} \right\rfloor = \ell_s$$

QED

Lemma 7 (Factors of $Q_s(x, f(x))$) *Let $f(x)$ be a polynomial and suppose that $f(P_i) = w_i$. Then*

$$(x - P_i)^s | Q_s(x, f(x))$$

Proof: Let $p(x) = f(x + P_i) - w_i$. Then $p(0) = 0$ so $x|p(x)$. Now consider $h(x) = Q_s(x + P_i, p(x) + w_i)$. By the definition of Q_s, $h(x)$ has no terms of degree less than s (due to the fact that x divides $p(x)$). So $x^s|h(x)$ and $(x - P_i)^s|h(x - P_i)$. This proves the lemma since $h(x - P_i) = Q_s(x, f(x))$. QED

Theorem 8 (Factors of the Q_s-polynomial) *Let $Q_s(x, y)$ be a Q_s-polynomial corresponding to the received word, w. If $f \in \mathbb{F}_q[x]$ with $\deg(f) < k$ satisfies $d(f(P), w) \leq \tau_s$ then $(y - f(x))|Q_s(x, y)$.*

Proof: Let $g(x) = Q_s(x, f(x))$. Then $\deg(g) \leq \ell_s$ and by Lemma 7 $(x - P_i)^s|g(x)$ for each value of i where $f(P_i) = w_i$. This happens at least $n - \tau_s$ times. This means that a polynomial of degree at least

$$s(n - \tau_s) = s(n - (n - \left\lfloor \frac{\ell_s}{s} \right\rfloor - 1)) = s\left(\left\lfloor \frac{\ell_s}{s} \right\rfloor + 1\right) > \ell_s$$

divides $g(x)$. So $g(x)$ must be the 0-polynomial.

Now consider $Q_s(x, y)$ as a polynomial in y where the coefficients are polynomials in x. As shown, $f(x)$ is a zero of this polynomial, so $y - f(x)$ must divide $Q_s(x, y)$. QED

4   Error-correcting capability of Sudan's algorithm

The error-correcting capability, τ_s, of Sudan's algorithm depends on the code rate, $\frac{k}{n}$, and on the parameter, s. Asymptotically, we have the following result:

Theorem 9 (Asymptotic error-correcting capability) *Suppose that n and s are "big" and that $k = \rho n$. Then*

$$\frac{\tau_s}{n} \approx 1 - \sqrt{\rho}$$

Proof: By the definition of τ_s (Equation 2) we see that

$$\frac{\tau_s}{n} \approx 1 - \frac{\ell_s}{ns}$$

The definition of r_s (Equation 1) gives the following approximation:

$$\frac{r_s^2}{s^2} \approx \frac{1}{\rho}$$

Then look at $\frac{\ell_s}{ns}$:

$$\frac{\ell_s}{ns} \approx \frac{s}{2r_s} + \rho\frac{r_s}{2s} \approx \frac{1}{2}\sqrt{\rho} + \frac{1}{2}\sqrt{\rho} = \sqrt{\rho}$$

QED

The asymptotic error-correcting capability obtained by Sudan's extended algorithm can actually be shown to be the best possible for a list-decoder where the size of the list of candidates is bounded independently of the code length [5].

Figure 1 shows the actual error-correcting capability of Sudan's extended algorithm for $n = q = 64$ and $s \in \{1, 4\}$ together with the asymptotical error-correcting capability.

5   The number of candidates

The next theorem shows that the number of candidates does not increase with the code length nor with the size of the alphabet:

Theorem 10 (Upper bound on number of candidates) *For any $w \in \mathbb{F}_q^n$, $|\text{dec}_{\tau_s}(w)| < r_s$.*

Proof: As seen in the proof of Lemma 4, M_{r_s} only contains monomials with degree in y at most $r_s - 1$ so $\deg^{(0,1)}(Q_s(x, y)) < r_s$. Therefore, Q_s can have at most $r_s - 1$ factors of the form $y - f(x)$. This proves the theorem.   QED

The occurrence of various candidates is a problem in applications, where one would like to end up with a unique answer. In the following we will analyse (under certain assumptions) how often the algorithm actually will return more than one candidate. Similar results can be found in [7], Sect. 14.2.

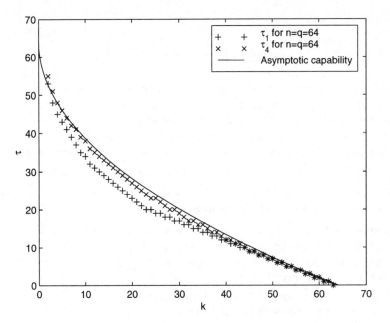

Fig. 1. Error-correcting capability of Sudan's extended algorithm.

Suppose that all error vectors with weight at most τ occur with the same probability regardless of the weight of the vector. Let v be some transmitted codeword and assume that at most τ errors occurred. If w is the received word then $P(|\text{dec}_\tau(w)| > 1)$ will be given by the fraction of words in $B(v, \tau)$ which are also within distance τ from another codeword. Since the code is linear it will be sufficient to analyse the situation in the case $v = 0$. Unfortunately, an exact count of the number of words in $B(0, \tau)$ within distance τ from another codeword is very difficult if not impossible to make, but an upper bound is given by

$$M(\tau) = \sum_{c \in \text{RS}(P,k) \setminus \{0\}} |B(0, \tau) \cap B(c, \tau)|$$

(equality holds when $r_s \leq 3$ in Sudan's algorithm).

In the following an explicit expression for $M(\tau)$ will be found. The first step is taken by this lemma:

Lemma 11 (Intersection of spheres) *Let $c \in \text{RS}(P, k)$ and let $i, a < w(c)$. If $w(c) - a > i$ then*

$$|S(0, a) \cap S(c, i)| = 0$$

If $w(c) - a \leq i$ then

$$|S(0, a) \cap S(c, i)| = \sum_{j=0}^{\ell} \binom{u}{u-a+j} \binom{n-u}{j} (q-1)^j \binom{a-j}{i-(u-a)-2j} (q-2)^{i-(u-a)-2j}$$

where $u = w(c)$ and

$$\ell = \min\left\{\left\lfloor\frac{i-(u-a)}{2}\right\rfloor, n - u\right\}$$

Proof: The problem is to calculate the number of words with weight a which have distance i to c, or put in another way: in how many ways c can be changed on exactly i positions to get a word with weight a.

Suppose that exactly j positions where c was previously 0 are now changed to a non-zero value. Then $j \le n - w(c)$ which is the number of zeroes and $j \le \left\lfloor\frac{i-(w(c)-a)}{2}\right\rfloor$, because to end at the right weight, a, it will be necessary to change $w(c) - a + j$ of the previously non-zero values to zero. The number of ways to choose the $w(c) - a + j$ "non-zero to zero" positions will be $\binom{w(c)}{w(c)-a+j}$. The number of ways to choose the j "zero to non-zero" positions will be $\binom{n-w(c)}{j}$ and each position can be given any of the $q - 1$ non-zero values. What is left is to change $i - (w(c) - a + j) - j$ of the remaining non-zero values to another non-zero value. The positions can be chosen in $\binom{a-j}{i-(w(c)-a)-2j}$ ways and each position can be given any of the $q - 2$ other non-zero values.

This gives the expression in the lemma.   QED

Now $M(\tau)$ can be calculated as

$$M(\tau) = \sum_{u=d}^{\min\{2\tau,n\}} A_u \sum_{a=u-\tau}^{\tau} \sum_{i=u-a}^{\tau} |S(0,a) \cap S(c,i)| \qquad (4)$$

where A_u is given by Proposition 2.

This gives an upper bound for the wanted probability:

Proposition 12 (Probability of multiple candidates) *The probability that Sudan's algorithm returns more than 1 codeword given that at most τ errors have occurred (assuming that all error-patterns occur with the same probability, regardless of the weight) satisfies*

$$P(|dec_\tau(w)| > 1) = \frac{M(\tau)}{|B(0,\tau)|}$$

Proof: This is a direct consequence of the above calculations.

Due to the complexity even of the calculation of the upper bound, $\frac{M(\tau)}{|B(0,\tau)|}$ it is difficult to get an immediate estimate of this probability. However, it turns out that it is usually very small. For example, Fig. 2 shows the upper bound for $n = q = 64$ and $s \in \{1, 4\}$. This figure should be compared to Fig. 1, which shows the corresponding error-correcting capability.

6   Determining a Q_s-polynomial

Algorithm 3 contains two main steps. The first is to determine a Q_s-polynomial and the second is to identify factors of the form $y - f(x)$ with $\deg(f) < k$. This

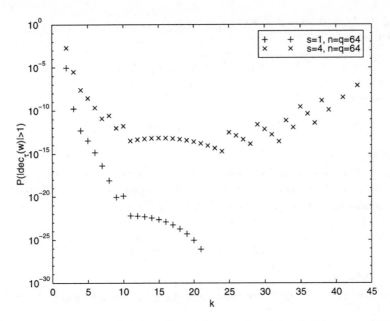

Fig. 2. Upper bound on probability of getting more than one candidate as output from Sudan's extended algorithm when $n = q = 64$ and $s \in \{1, 4\}$. The bound is zero when $\tau_s = \left\lfloor \frac{d_{\min} - 1}{2} \right\rfloor$.

section and the next section describe how to make an efficient implementation of these two steps.

To determine the Q_s-polynomial we must find a polynomial with lowest possible weighted degree which has each point (P_i, w_i) as a zero of multiplicity s. The straight-forward way to do this is to solve a system of $\binom{s+1}{2}n$ equations. However, in the following we present a faster algorithm, which is an application of the Fundamental Iterative Algorithm, along the same lines as the application in [3], Chap. 4. The Fundamental Iterative Algorithm was first presented in [8].

Let the monomials of $\mathbb{F}_q[x, y]$ be $m_0, m_1, \ldots$ as in Lemma 4. Now define $\mathrm{ord} : \mathbb{F}_q[x, y] \to \mathbb{N} \cup \{-\infty\}$ by $\mathrm{ord}(0) = -\infty$, $\mathrm{ord}(\mathrm{span}\{m_0\} \backslash \{0\}) = \{0\}$, and

$$\mathrm{ord}(\mathrm{span}\{m_0, \ldots, m_i\} \backslash \mathrm{span}\{m_0, \ldots, m_{i-1}\}) = \{i\}$$

for $i > 0$. Notice that ord induces both an partitioning of the polynomials in $\mathbb{F}_q[x, y]$ and a total ordering on the set of partitions. As it will be seen later, the polynomials which occur at a given stage in the algorithm all come from different partitions. It will therefore be well-defined to talk about the smallest or the minimal polynomial as the polynomial which belongs to the smallest partition.

Let $g_1, \ldots, g_c \in \mathbb{F}_q[x, y]$ be a set of polynomials so that if

$$A_j = \mathrm{ord}(\{x^{is} g_j \mid i \in \mathbb{N}\}), \quad j = 1, \ldots, c$$

then $A_j \cap A_{j'} = \emptyset$ for $j \neq j'$ and $\bigcup_{j=1}^{c} A_j = \mathbb{N}$.

Furthermore, define

$$G_j = \{g \in \mathbb{F}_q[x, y] \mid \mathrm{ord}(g) \in A_j\}$$

Finally, let

$$M_s = \{x^\alpha y^\beta \mid \alpha + \beta < s\} = \{e_1, \ldots, e_{|M_s|}\}$$

where $|M_s| = \binom{s+1}{2}$.

The following is a mathematical description of the interpolation algorithm. This will be used to show the correctness of the algorithm. A more implementation-oriented description is given in Algorithm 15.

First, let

$$G^{(0)} = \{g_1{}^{(0)}, \ldots, g_c{}^{(0)}\}$$

where $g_j{}^{(0)} = g_j$.

The sets $G^{(1)}, \ldots, G^{(n)}$ are then constructed iteratively using the following method:

Let $G^{(i,0)} = G^{(i-1)}$ and do the following for each $k = 1, \ldots, |M_s|$:

Let $G^\star = \{g \in G^{(i,k-1)} \mid \mathrm{coef}(g(x + x_i, y + y_i), e_k) \neq 0\}$. If $G^\star = \emptyset$ then $G^{(i,k)} = G^{(i,k-1)}$. If not, let $f^{(i,k)}$ be chosen so that

$$\mathrm{ord}(f^{(i,k)}) = \min \mathrm{ord}(G^\star)$$

Now let

$$G^{(i,k)} = \{(x - x_i)^s f^{(i,k)}\} \cup \{\mathrm{coef}(f^{(i,k)}(x + x_i, y + y_i), e_k)g - \\ \mathrm{coef}(g(x + x_i, y + y_i), e_k)f^{(i,k)} \mid g \in G^{(i,k-1)} \setminus \{f^{(i,k)}\}\}$$

where $p_i = (x_i, y_i)$.

Let $G^{(i)} = G^{(i,|M_s|)}$.

The result is now given by the polynomial in $G^{(n)}$ which is smallest with respect to ord.

To prove the correctness of the algorithm, the following lemma is needed:

Lemma 13 (Correctness of one step of the interpolation algorithm) *If* $g_j{}^{(i,k-1)}$ *is a minimal polynomial with respect to* ord *in* G_j *where*

$$\mathrm{coef}(g_j{}^{(i,k-1)}(x + x_{i'}, y + y_{i'}), e_{k'}) = 0$$

for all $i' \leq i$ *and* $k' \leq k - 1$ *then the similar condition is true for* $g_j{}^{(i,k)}$.

Proof: Assume that $g_j{}^{(i,k-1)}$ is a minimal polynomial with respect to ord in G_j where $\mathrm{coef}(g_j{}^{(i,k-1)}(x + x_{i'}, y + y_{i'}), e_{k'}) = 0$ for all $i' \leq i$ and $k' \leq k - 1$.

If $g_j{}^{(i,k-1)} \neq f^{(i,k)}$ then $\mathrm{ord}(g_j{}^{(i,k)}) = \mathrm{ord}(g_j{}^{(i,k-1)})$ and the lemma is true by assumption.

If $g_j{}^{(i,k-1)} = f^{(i,k)}$ then let g be some polynomial with $\mathrm{ord}(g) = \mathrm{ord}(g_j{}^{(i,k-1)})$ and $\mathrm{coef}(g(x + x_{i'}, y + y_{i'}), e_{k'}) = 0$ for $k' \le k$. It may be assumed that both g and $g_j{}^{(i,k-1)}$ are monic. Now consider the polynomial $h = g_j{}^{(i,k-1)} - g$. $\mathrm{ord}(h) < \mathrm{ord}(g_j{}^{(i,k-1)})$ but $\mathrm{coef}(h(x + x_i, y + y_i), e_k) \ne 0$ so $h \in G_j$ since the definition of $f^{(i,k)}$ implies that it cannot be in any $G_{j'}$ for $j' \ne j$. However, this contradicts the assumption that $g_j{}^{(i,k-1)}$ is minimal. So a polynomial like g does not exist. Therefore it is necessary that $\mathrm{ord}(g_j{}^{(i,k)}) > \mathrm{ord}(g_j{}^{(i,k-1)})$ and since $g_j{}^{(i,k)} = (x - x_i)^s g_j{}^{(i,k-1)}$ introduces the smallest possible increase with respect to ord, the theorem is also true for $g_j{}^{(i,k)}$.   QED

The correctness of the algorithm follows from this theorem:

Theorem 14 (Correctness of the interpolation algorithm) *For each $j \in \{1, \ldots, c\}$ and each $i = 0, \ldots, n$, $g_j{}^{(i)}$ is a minimal polynomial with respect to ord in G_j where $\mathrm{coef}(g_j{}^{(i)}(x + x_{i'}, y + y_{i'}), e_{k'}) = 0$ for all $i' \le i$ and $k' \le |M_s|$.*

Proof: The proof is by induction over i, however, most of the work has been done in Lemma 13.

$i' = 0$: The polynomials $g_{i'}{}^{(0)}$ are minimal in the sets $G_{i'}$ by construction so the theorem is true for $i' = 0$.

$i' = i + 1$: Assume that the theorem is true for i. Since $G_j{}^{(i',0)} = G_j{}^{(i,|M_s|)}$ it follows by Lemma 13 and induction over k that the theorem is true for i'.

QED

The following shows how to implement the interpolation algorithm:

Algorithm 15 (Fast interpolation) *Input: Interpolation points, $p = \{(x_1, y_1), \ldots, (x_n, y_n)\}$, a polynomial ordering, ord, and a required root multiplicity, s.*
Output: Interpolation polynomial which is minimal with respect to ord.

```
Result = Interpolate(p, ord, s) :
  d ← max{deg^(0,1) mᵢ | i = 0, ... , n(s+1 choose 2)}
  c ← 0
  for j' = 0, ... , d do
    for k = 1, ... , s do
      c ← c + 1
      g_c ← x^(k-1) y^j
    end
  end
  for i = 1, ... , n do
    for k = 1, ... , |M_s| do
      f ← min_ord {g_j | coef(g_j(x + xᵢ, y + yᵢ), e_k) ≠ 0}
      for j = 1, ... , c do
        if g_j = f then
          g_j ← (x − xᵢ)^s g_j
```

else
$$g_j \leftarrow \mathrm{coef}(f(x+x_i,y+y_i),e_k)g_j - \mathrm{coef}(g_j(x+x_i,y+y_i),e_k)f$$
    end
   end
  end
 end
Result $\leftarrow \min_{\mathrm{ord}}\{g_j\}$

Notice that $\mathrm{coef}(f(x+x_0,y+y_0),e)$ for some polynomial, f, some scalars, x_0 and y_0, and some monomial, e, can be calculated using the result in Lemma 5.

7   Factoring the Q_s-polynomial

Q_s is a bivariate polynomial, but the goal is to identify factors of the form $y - f(x)$ with $\deg(f) < k$. Therefore, let $E = \mathbb{F}_{q^k} = \mathbb{F}_q[x]/\langle e(x)\rangle$ where $e(x)$ is an irreducible polynomial in $\mathbb{F}_q[x]$ of degree k. E is then a finite field with q^k elements.

Consider the map $\phi : \mathbb{F}_q[x,y] \to E[y]$ given by

$$\phi(\sum_i p_i(x)y^i) = \sum_i [p_i(x)]_E \, y^i \tag{5}$$

It is a well-known fact that

Lemma 16 ϕ *is a ring homomorphism.*

Theorem 17 *If $f(x,y)|Q(x,y)$ then $\phi(f)|\phi(Q)$.*

Proof: The theorem follows from Lemma 16:

$$f|Q \Rightarrow Q = fg \Rightarrow \phi(Q) = \phi(fg) = \phi(f)\phi(g) \Rightarrow \phi(f)|\phi(Q)$$

for some $g \in F_q[x,y]$.   QED

Corollary 18 *If $(y-f(x))|Q(x,y)$ then $y-[f]_E$ is an irreducible factor of $\phi(Q)$.*

Proof: If $(y - f(x))|Q(x,y)$ then $y - [f]_E$ is a factor of Q by the theorem. Furthermore, $y - [f]_E$ must be irreducible since it is a polynomial of degree 1.
                                                                         QED

Theorem 17 reduces the problem of factoring the bivariate polynomial Q_s into the problem of factoring the univariate polynomial $\phi(Q_s)$, which is much easier.

The following is a short description of Berlekamp's algorithm for factoring univariate polynomials over a large finite field. This version of the algorithm is from [4], which describes more details and proves the correctness of the algorithm.

First the factoring problem is reduced to factoring monic square-free polynomials by the following algorithm:

Algorithm 19 (Square-free factorization) *Input: A monic polynomial, $a(x)$* $\in \mathbb{F}_{p^m}[x]$.
Output: $a(x)$ factorized on the form $a = \prod_j a_{[j]}^j$ where each $a_{[j]} \in \mathbb{F}_{p^m}[x]$ is square-free.

> Result = SquareFree(a) :
> j ← 1, Result ← 1
> **if** a' = 0 **then**
>> Result ← Result · SquareFree($a^{1/p}$)^p
>
> **else**
>> c ← gcd(a, a')   c holds the repeated factors.
>>
>> w ← a/c   w is a square-free polynomial containing each of the factors except those where p divides the exponent.
>>
>> **while** w ≠ 1 **do**
>>> y ← gcd(w, c)   y becomes a square-free polynomial containing the repeated factors.
>>>
>>> z ← w/y   Now z contains all the non-repeated factors ($a_{[j]} = z$).
>>>
>>> Result ← Result · (z^j), j ← j + 1
>>>
>>> w ← y   Remove the factors which are now contained in the result.
>>>
>>> c ← c/y   Decrease the exponent of each factor by 1 except where p divides the exponent.
>>
>> **if** c ≠ 1 **then**
>>> Result ← Result · SquareFree($c^{1/p}$)^p
>>
>> **end**
>
> **end**

In the while-loop in the algorithm, factors with exponent divisible by p will survive in the polynomial c so that when all the factors of w has been included in the result, c is either 1 or all its factors has an exponent divisible by p.

If $a(x) \in \mathbb{F}_q[x]$ is a monic square-free polynomial then we will define the vector space

$$W(a) = \{v \in \mathbb{F}_q[x]/\langle a(x)\rangle \mid v^q = v\}$$

The following algorithm for factoring a monic square-free polynomial assumes the presence of a function, WBasis, which calculates a basis of $W(a)$.

Algorithm 20 (Berlekamp's factoring algorithm) *Input: A monic square-free polynomial, $a(x) \in \mathbb{F}_{p^m}[x]$.*
Output: The set of irreducible factors of $a(x)$.

> Result = Factor(a) :
> Result ← {a}

$(v_1, \ldots, v_k) \leftarrow \text{WBasis}(a)$ Calculate k polynomials spanning W.

while $|\text{Result}| < k$ **do**

    for $u \in \text{Result}$ **do**

        $v \leftarrow \text{Random}(v_1, \ldots, v_k)$ Get a random polynomial from W.

        $v \leftarrow v + v^2 + \cdots + v^{2^{m-1}}$

        $g \leftarrow \gcd(v, u)$

        if $g \neq 1$ **and** $g \neq u$ **then**

            $\text{Result} \leftarrow \text{Result} \backslash \{u\}$

            $\text{Result} \leftarrow \text{Result} \cup \{\frac{u}{g}, g\}$

            if $|\text{Result}| = k$ **then return** Result

        end

    end

end

This factoring algorithm is probabilistic and requires $O(r^2 kt \log(t) + r^3)$ operations in $\mathbb{F}_{q^k}$. Here, t is the number of irreducible factors, r is the degree of the polynomial, and k is the logarithm of the field size.

Another efficient method for finding factors of the form $y - f(x)$ is given in [6].

8 Examples

This section contains two examples of decoding by Sudan's extended algorithm. The code alphabet is $\mathbb{F}_{16}$ with primitive element, α, satisfying $\alpha^4 + \alpha + 1 = 0$. The Reed-Solomon code is $\text{RS}(\{1, \alpha, \alpha^2, \ldots, \alpha^{14}\}, 7)$ so $(n, k, d) = (15, 7, 9)$. This gives $\tau_4 = 5$ which is one more than half the minimum distance.

In the first example, $c = (0, \ldots, 0)$ is the transmitted codeword, and we generate 5 random errors, so the received word is

$$w = (0, 0, \alpha^{11}, 0, \alpha^{12}, \alpha^{11}, 0, 0, 0, 0, 0, 0, \alpha^3, 0, \alpha^7)$$

The first step of Sudan's extended algorithm with $s = 4$ gives the following interpolation polynomial:

$$
\begin{aligned}
Q_4 = &(\alpha^5 + \alpha^4 x + \alpha^2 x^2 + x^3 + \alpha^7 x^4 + \alpha^3 x^5 + \alpha^6 x^6 + \alpha^3 x^7 + \alpha^7 x^8 + \alpha^6 x^9 + \\
&\alpha^2 x^{10} + \alpha^{13} x^{11} + \alpha^9 x^{12} + \alpha^{10} x^{13} + \alpha^3 x^{14} + \alpha^{11} x^{15} + x^{17} + \alpha^{14} x^{18} + \\
&\alpha^2 x^{19} + \alpha^4 x^{20} + \alpha^{10} x^{21} + \alpha^{12} x^{22} + \alpha^6 x^{24} + \alpha^2 x^{25} + \alpha^2 x^{26} + \alpha x^{27} + \\
&\alpha^3 x^{28} + \alpha^8 x^{29} + \alpha^9 x^{31} + \alpha^{13} x^{32} + x^{33})y + (\alpha^4 x + x^2 + \alpha^{11} x^3 + \alpha^8 x^4 + \\
&x^5 + \alpha^{14} x^6 + \alpha^3 x^7 + \alpha^{14} x^8 + \alpha^5 x^9 + \alpha^7 x^{10} + \alpha^{11} x^{11} + \alpha^6 x^{12} + \alpha x^{13} + \\
&\alpha^9 x^{14} + \alpha^{11} x^{15} + \alpha x^{16} + \alpha^{13} x^{17} + \alpha^{11} x^{18} + \alpha^{11} x^{19} + \alpha^{14} x^{20} + \alpha^{10} x^{21} + \\
&\alpha^5 x^{22} + \alpha^{13} x^{23} + \alpha x^{24} + \alpha^4 x^{25} + \alpha^3 x^{26} + \alpha^9 x^{27})y^2 + (\alpha^9 + \alpha^5 x + \alpha^9 x^2 + \\
&\alpha^3 x^3 + \alpha^{10} x^4 + \alpha^4 x^6 + \alpha^9 x^{15} + \alpha^5 x^{16} + \alpha^9 x^{17} + \alpha^3 x^{18} + \alpha^{10} x^{19} + \\
&\alpha^4 x^{21})y^3 + (\alpha^2 + \alpha^3 x + \alpha^5 x^2 + \alpha^{10} x^3 + \alpha x^4 + \alpha^{10} x^5 + \alpha^{14} x^7 + \alpha^4 x^8 + \\
&\alpha^5 x^9 + \alpha^{12} x^{10} + \alpha^{12} x^{11} + \alpha^{12} x^{12} + x^{13} + x^{14})y^4 + (\alpha^{12} + \alpha^{14} x + \alpha^{13} x^2 + \\
&\alpha^4 x^3 + \alpha^{10} x^4 + \alpha^9 x^5 + \alpha^{11} x^6 + \alpha^4 x^7 + \alpha^8 x^8)y^5 + (\alpha x + \alpha^8 x^2)y^6
\end{aligned}
$$

This is reduced modulo $e(x) = 1+\alpha^{11}x+\alpha^6x^2+\alpha^8x^3+\alpha^{11}x^4+\alpha^{12}x^5+\alpha^{13}x^6+x^7$ which is irreducible over $\mathbb{F}_{16}$:

$$\phi(Q_4) = (\alpha^8 + \alpha^{14}x + \alpha^9x^2 + \alpha^{10}x^3 + x^4 + \alpha^5x^5 + \alpha^{12}x^6)y + (\alpha^8 + \alpha x + \alpha^{11}x^2 + \alpha^4x^3 + \alpha^7x^4 + \alpha^6x^5 + x^6)y^2 + (\alpha^8 + \alpha^{10}x + \alpha^8x^2 + x^3 + \alpha^2x^4 + x^5)y^3 + (\alpha^4 + \alpha^{14}x + \alpha^8x^2 + \alpha^{11}x^3 + \alpha^9x^4 + \alpha^9x^5 + \alpha x^6)y^4 + (\alpha^5 + \alpha^2x + x^2 + \alpha^9x^3 + \alpha x^4 + \alpha^3x^5 + x^6)y^5 + y^6$$

Factoring this into irreducible factors over $\mathbb{F}_{16^7} = \mathbb{F}_{16}/\langle e(x)\rangle$ gives:

$$\phi(Q_4) = y\cdot$$
$$((\alpha^4 + \alpha^{10}x + \alpha^{11}x^2 + \alpha^9x^3 + \alpha^4x^4 + \alpha^9x^5 + x^6) + y)\cdot$$
$$((1 + \alpha^{14}x + \alpha^3x^2 + \alpha^5x^3 + \alpha^7x^4 + \alpha^3x^5 + \alpha x^6) + (\alpha^{13} + \alpha^{10}x + \alpha^7x^2 + \alpha x^3 + \alpha^{10}x^4 + \alpha^7x^5 + \alpha^4x^6)y + (\alpha^8 + \alpha^8x + \alpha^{12}x^2 + \alpha^3x^3 + \alpha^{13}x^4 + \alpha^6x^6)y^2 + (\alpha^8 + \alpha^4x + \alpha^{12}x^2 + x^4 + \alpha x^5)y^3 + y^4)$$

The factor $(\alpha^4 + \alpha^{10}x + \alpha^{11}x^2 + \alpha^9x^3 + \alpha^4x^4 + \alpha^9x^5 + x^6) + y$ corresponds to a word with distance 14 to w. The only other factor of degree 1 is y. This gives

$$\text{dec}_{\tau_4}(w) = \{(0,\dots,0)\}$$

which was the transmitted codeword. So in this case the decoding was not only correct, but also unique. As mentioned in Section 5 this is the normal case.

The next example was suggested by J. Justesen. The code is the same as before and the transmitted codeword is again the 0-word. However, the received word is now

$$v = (1,0,0,1,0,0,1,0,0,1,0,0,1,0,0)$$

Again, we let $s = 4$. The interpolation polynomial is

$$R = (\alpha^{14} + \alpha^{14}x^{10} + \alpha^{14}x^{20})y^2 + (\alpha^{14} + \alpha^{14}x^{10})y^4 + \alpha^{14}y^6$$

which is reduced modulo $e(x)$ to

$$\phi(R) = (\alpha^{11} + \alpha^6x^2 + \alpha^2x^3 + \alpha^{10}x^4 + \alpha^8x^5 + \alpha^2x^6)y^2 + (\alpha^8 + \alpha^6x + \alpha^5x^2 + \alpha^8x^4 + \alpha^{10}x^5 + x^6)y^4 + y^6$$

Factoring this polynomial into irreducibles gives

$$\phi(R) = ((\alpha^5 + \alpha^{10}x^5) + y)^2((\alpha^{10} + \alpha^5x^5) + y)^2y^2$$

Which corresponds to 3 codewords all with distance 5 to v:

$$\text{dec}_{\tau_4}(v) = \{(0,\dots,0),$$
$$(1,\alpha^{10},0,1,\alpha^{10},0,1,\alpha^{10},0,1,\alpha^{10},0,1,\alpha^{10},0),$$
$$(1,0,\alpha^5,1,0,\alpha^5,1,0,\alpha^5,1,0,\alpha^5,1,0,\alpha^5)\}$$

9   Conclusion

We have presented an efficient version of Sudan's algorithm for decoding Reed-Solomon codes. The corresponding algorithm for decoding algebraic geometry codes is the subject of a forthcoming paper.

References

1. M. Sudan: "Decoding of Reed Solomon Codes beyond the Error-correction bound" Journal of complexity 13, pp. 180-193, 1997.
2. V. Guruswami and M. Sudan: "Improved Decoding of Reed Solomon and Algebraic Geometric Codes" MIT, 1998.
3. Ralf Kötter: "On Algebraic Decoding of Algebraic-Geometric and Cyclic Codes" Department of Electrical Engineering, Linköping University, 1996.
4. K. Geddes, S. Czapor, and G. Labahn: "Algorithms for Computer Algebra" Kluwer Academic Publishers, 1992.
5. J. Justesen: "Bounds on list decoding of MDS codes" Manuscript, 1998.
6. M. A. Shokrollahi: "Computing Roots of Polynomials over Function Fields of Curves" Draft, June 1998.
7. R. E. Blahut: "Theory and Practice of Error Control Codes" Addison-Wesley Publishing Company, 1983.
8. G.-L. Feng and K. K. Tzeng: "A Generalization of the Berlekamp-Massey Algorithm for Multisequence Shift-Register Synthesis with Applications to Decoding Cyclic Codes", IEEE Trans. Inform. Theory, vol. 37, pp. 1274-1287, Sept. 1991.

The proof of this lemma is easy and is omitted (cf.[11]).

Let K be a field and let $\mathbb{P}_m(K)$ be the m-projective space over K. If $\underline{x} = (x_0, ..., x_m)$ is an element of $\mathbb{P}_m(K)$ the *Veronese map* is defined as

$$v_n : \mathbb{P}_m(K) \to \mathbb{P}_N(K), \quad v_n(\underline{x}) = (..., M(\underline{x}), ...)$$

where $M(\underline{X})$ runs over all the monomials of degree n in the variables $\underline{X} = (X_0, ..., X_m)$, and $N = \binom{n+m}{n} - 1$. If K is algebraically closed, it is well known that v_n is a smooth embedding with the property that every hypersurface of degree n in $\mathbb{P}_m(K)$ becomes a hyperplane section of the Veronese variety, the image of $\mathbb{P}_m(K)$ under the mapping v_n. For $m = 1$ the image of v_n is the *rational normal curve* of degree n, this case was extensively treated in (cf. [11]) and for $m = 2$ the image is the *Veronese surface*. In the sequel an element of $\mathbb{P}_N(K)$ will be denoted by $y = (y_0, ..., y_N)$, the coordinate ring of $\mathbb{P}_m(K)$ will be denoted by $A_m = K[X_0, ..., X_m] = \oplus_{j \geq 0} A_m(j)$ and the corresponding ring of $\mathbb{P}_N(K)$ by $A_N = K[Y_0, ..., Y_N] = \oplus_{j \geq 0} A_N(j)$, both with the standard graduation.

From now on we restrict the above construction to the case where K is the finite field $GF(q)$ with $q = p^r$ elements (p prime and r a positive integer), and let $S = v_n(\mathbb{P}_m(K)) = \{v_n(P) \in \mathbb{P}_N(K) : P \in \mathbb{P}_m(K)\}$. Observe that since v_n is an embedding then $\#(S) = \pi_m = \# (\mathbb{P}_m(K)) = \frac{q^{m+1}-1}{q-1}$. Let I_S be the graded vanishing ideal of the set S, i.e., $I_S = \{f(\underline{Y}) \in A_N : f(P) = 0 \; \forall \; P \in S\} = \oplus_{j \geq 0} I_S(j)$. In [10] the a-invariant of the vanishing ideal of the affine and projective space was found, and in [11] the a-invariant and a set of generators for the defining ideal of the rational normal curve were determined. In this note the a-invariant, the dimension and minimal distance of the code $C_S(d)$ as well as the vanishing ideal of the set S will be determined.

3   The a-invariant of S

For a positive integer d and for an element $f \in A_N(d)$, let $\varphi_f(\underline{X}) = f(M_0(\underline{X}), \dots, M_N(\underline{X}))$. Then φ_f is homogeneous of degree nd and it defines a mapping $\varphi : A_N(d) \to A_m(nd)$ which induces an isomorphism between the K-vector spaces $A_N(d)/I_S(d)$ and $A_m(nd)/I_{\mathbb{P}_m}(nd)$. In particular it follows that

$$H_S(d) = H_{\mathbb{P}_m}(nd) = \sum_{j=0}^{m} \sum_{i=0}^{j} (-1)^i \binom{j}{i} \binom{j+nd-1-iq}{nd-1-iq} \tag{1}$$

which is the dimension of the Reed-Muller type code $C_S(d)$ (for the last equality see [10], where we assume that $\binom{d}{d} = 1$ for all integers d and that $\binom{k}{t} = 0$ for $k \neq t$ and $t < 0$). We recall that the a-invariant of the projective space $\mathbb{P}_m(K)$ is $a_m = m(q - 1)$ and that $I_{\mathbb{P}_m} = < X_i^q X_j - X_i X_j^q, \; i < j; \; i, j = 0, 1, ..., m >$ (cf.[8], [10]).

Lemma 2. *The a-invariant a_S of the set $S = v_n(\mathbb{P}_m(K))$ is equal to:*

$\frac{a_m + 1 - j}{n}$ *if* $a_m + 1 \equiv j \mod n$ *and* $j > 0$

$\frac{a_m + 1}{n} - 1$ *if* $a_m + 1 \equiv 0 \mod n$

Proof: If $a_m + 1 \equiv j \mod.n$ with $0 < j < n$ let $d = \frac{a_m - 1 - j}{n}$, then $H_S(d+1) = H_{\mathbb{P}_m}(n(d+1))$. Since $n(d+1) = a_m + 1 - j + n$ and $n - j > 0$, it follows that $n(d+1) > a_m$ and consequently $H_S(d+1) = \#(S)$. Furthermore, since $j > 0$, we have that $a_m + 1 - j < a_m + 1$ and hence $H_S(d) = H_{\mathbb{P}_m}(nd) = H_{\mathbb{P}_m}(a_m + 1 - j) < \#(S)$. From the definition of the a-invariant it follows that $d = a_S$. The case $j = 0$ is similar since if $d = \frac{a_m + 1}{n} - 1$, then $H_S(d) = H_{\mathbb{P}_m}(nd) = H_{\mathbb{P}_m}(a_m + 1 - n) < \#(\mathbb{P}_m) = \#(S)$, and $H_S(d+1) = H_{\mathbb{P}_m}(n(d+1)) = H_{\mathbb{P}_m}(a_m + 1) = \#(\mathbb{P}_m) = \#(S)$, showing that $a_S = \frac{a_m + 1}{n} - 1$.

4   The vanishing ideal of S

In this section the vanishing ideal of S, i.e., the image of $\mathbb{P}_m(K)$ under the Veronese mapping is described and the minimal distance of the corresponding code is determined. In order to do this let $\overline{K}$ be the algebraic closure of K and let $\overline{S}$ be the Veronese variety, i.e., the image of $\mathbb{P}_m(\overline{K})$ under the mapping $v_n : \mathbb{P}_m(\overline{K}) \to \mathbb{P}_N(\overline{K})$, $v_n(\underline{x}) = (..., M(\underline{x}), ...)$. If $A = K[Z_0, ..., Z_t]$ let $\overline{A} = \overline{K}[Z_0, ..., Z_t]$ and let $I_{\overline{S}}$ be the vanishing ideal of $\overline{S}$.

The next result is a generalization of the case $m = 1$ (cf. [11]).

Theorem 1. Let $n \leq q$. For $q \equiv l \mod.n$, $0 \leq l \leq n - 1$, let $r = \frac{q - l}{n}$. Then $I_S = < I_S(2), I_S(r+1) >$.

Proof: Since $I_{\overline{S}} = \oplus_{d \geq 2} I_{\overline{S}}(d)$, in order to prove the assertion of the theorem it is enough to show that $I_S(d) = \sum_{i=0}^{N} Y_i I_S(d-1)$, for all $d \geq 3$ and $d \neq r + 1$. We consider two cases: a) $3 \leq d \leq r$, and b) $d \geq r+2$. First we observe that if d is any positive integer such that $nd < q+1$, since $I_{\mathbb{P}_m} = < I_{\mathbb{P}_m}(q+1) >$ then $\varphi_f = 0$ for all $f \in I_S(d)$, (where φ_f is as defined in section 3), showing that $I_S(d) \subseteq I_{\overline{S}}(d)$. Since $I_{\overline{S}}(d) = < I_{\overline{S}}(2) >$ then for all $d \geq 3$, we have $I_{\overline{S}}(d) = \sum_{i=1}^{N} Y_i I_{\overline{S}}(d-1)$. In the case a): $3 \leq d \leq r$, i.e., $3n \leq nd \leq q - l < q+1$, let $W = \sum_{i=1}^{N} Y_i I_S(d-1)$. Since $\dim_K(W) = \dim_{\overline{K}}(W \otimes_K \overline{K})$ and $W \otimes_K \overline{K} \simeq W \overline{A}_N = I_{\overline{S}}(d)$, it follows that $\dim_{\overline{K}} I_{\overline{S}}(d) = \dim_K I_S(d)$, and therefore $W = I_S(d)$. In case b): $r + 2 \leq d$, since $\frac{q-l}{n} + 1 \leq d - 1$, we have $q + (n - l) \leq n(d - 1)$, and hence $q+1 \leq n(d-1) < nd$. Thus if $f \in I_S(d)$ then $\varphi_f \in I_{\mathbb{P}_m}(nd) = < X_i^q X_j - X_i X_j^q$, $0 \leq i < j \leq m >$, (cf [10]) and consequently $\varphi_f = \sum_{0 \leq i < j \leq m} g_{ij}(X_i^q X_j - X_i X_j^q)$ where the g_{ij}'s are forms of degree $nd - (q + 1)$. Since $n(d - 1) \geq q + 1$, i.e, $nd - (q+1) \geq n$ the g_{ij}'s can be written as: $g_{ij} = M_{ij}(1)h_{ij}(1) + M_{ij}(l_{ij})h_{ij}(l_{ij})$, where the $M_{ij}(k)$ are monomials of degree n and the $h_{ij}(k)$ are forms of degree $nd - (q+1) - n$. Thus

$$\varphi_f = \sum_{i,j} \sum_{k=1}^{l_{ij}} M_{ij}(k) h_{ij}(k)(X_i^q X_j - X_i X_j^q)$$

Note that for all i, j, k, the form $h_{ij}(k)(X_i^q X_j - X_i X_j^q) \in I_{\mathbb{P}_m}(n(d-1))$. Hence $\varphi_f = M_{i1} H_{i1} + \ldots + M_{is} H_{is}$ with M_{ij} monomials in the variables $X_0, \ldots, X_m$ of degree n and $H_{ij} \in I_{\mathbb{P}_m}(n(d-1))$. Therefore $f = Y_{i1} \lambda_{i1}(H_{i1}) + \ldots + Y_{is} \lambda_{is}(H_{is})$ where each $\lambda_{ij}(H_{ij})$ is in $I_S(d-1)$ and is such that $\varphi_{\lambda_{ij}(H_{ij})} = H_{ij}$. From the above argument we conclude that $f \in \sum_{i=0}^{N} Y_i I_S(d-1)$.

Proposition 1. *Let* $1 \leq d \leq a_S$. *Then the minimal distance of the Reed-Muller type code* $C_S(d)$ *over the Veronese variety is equal to the minimal distance* $\delta_m(nd)$ *of the projective Reed-Muller code of order* nd, *(recall that* $\delta_m(l) = (q-s)q^{m-r-1}$ *where* $l - 1 = r(q-1) + s, 0 \leq s < q - 1$.

Proof: Let $f \in A = K[Y_0, \ldots, Y_N]$ be an homogeneous polynomial of degree d, and let $M_0, \ldots, M_N$ be all the monomials of degree n in the variables $X_0, \ldots, X_m$. Then $f(M_0, \ldots, M_N) \in A[X_0, \ldots, X_m]$ is homogeneous of degree nd. Let $\mathbb{P}_m(K) = \{P_1, \ldots, P_t\}$ and $Q_i = v_n(P_i) = (M_0(P_i), \ldots, M_N(P_i)) \in S$, $i = 1, \ldots, t$. If $W = (f(Q_1), \ldots, f(Q_t)) \in C_S(d)$, then

$$W = (f(M_0, \ldots, M_N)(P_1), \ldots, f(M_0, \ldots, M_N)(P_t))$$

belongs to $C_{\mathbb{P}_m}(nd)$ so $\delta_m(nd) \leq \text{weight}(W)$. The other inequality follows at once from the isomorphism induced by the function φ between the K-vector spaces $A_N(d)/I_S(d)$ and $A_m(nd)/I_{\mathbb{P}_m}(nd)$, as defined in section 3.

5   The case $m = 2$

In the previous section the ideal I_S was described in terms of the homogeneous piece $I_S(2)$, i.e., quadrics and forms of degree $r + 1$, i.e., $I_S(r + 1)$ (with r as defined in the theorem above). In this section we treat the case of the projective plane $\mathbb{P}_2(K)$ closely and, in particular, the dimension of $I_S(2)$ is determined and a set of generators for the homogeneous component $I_S(r + 1)$ is given.

Since for all positive integers d, the K-vector spaces $A_N(d)/I_S(d)$ and $A_m(nd)/I_{\mathbb{P}_m}(nd)$ are isomorphic it follows from relation (1) above that $H_S(n) = H_{\mathbb{P}_2}(2n) = (n+1)(2n+1)$. The number of forms of degree 2 in the variables $Y_0, \ldots, Y_N$ is $\frac{1}{8}[(n+1)(n+2)][(n+1)(n+2)+2]$. Therefore

$$dim_K I_S(2) = \dim_K A_N(2) - H_S(n) = \frac{1}{8}(n+1)n(n-1)(n+6)$$

Let r be as defined in the previous theorem, i.e., $r = \frac{q-l}{n}$ if $q \equiv l \mod.n$. Since $\dim_K I_S(r+1) = \dim_K I_{\mathbb{P}_2}(n(r+1))$, assuming $q+1 \leq nr \leq 2q+1$, it follows from relation (1) of section 3, that $\dim_K I_S(r) = (nr-q+1)^2 - 1$. We recall that if $F_{(0,1)} = X_0^q X_1 - X_0 X_1^q$, $F_{(0,2)} = X_0^q X_2 - X_0 X_2^q$, $F_{(1,2)} = X_1^q X_2 - X_1 X_2^q$, then $I_{\mathbb{P}_2} = < F_{(0,1)}, F_{(0,2)}, F_{(1,2)} >$, (cf.[10]). Let $t = n(r+1) - (q+1)$. Since $F_{(i,j)} \in I_{\mathbb{P}_2}(q+1)$, then $(X_0^a X_1^b X_2^c) F_{(i,j)} \in I_{\mathbb{P}_2}(n(r+1))$ if $a + b + c = t$. The number of monomials $X_0^a X_1^b X_2^c$ with $a + b + c = t$ is $\binom{t+2}{t}$, so there is a total of

$3\binom{t+2}{t}$ elements of the form $(X_0^a X_1^b X_2^c)F_{(i,j)}$, $i,j = 1,2,3$. Since $\dim_K I_{\mathbb{P}_2}(n(r + 1)) = (t + 2)^2 - 1$, it follows that $\dim_K I_{\mathbb{P}_2}(n(r + 1)) \leq 3\binom{t+2}{t}$. Thus there are enough elements in $I_{\mathbb{P}_2}(n(r+1))$ of the form $(X_0^a X_1^b X_2^c)F_{(i,j)}$ to choose from so that a basis for $I_{\mathbb{P}_2}(n(r + 1))$ can be taken.

Since the Veronese mapping is given by the monomials of degree n in the variables X_0, X_1, X_2, we may order them with the lexicographic ordering, so that for instance we can identify the coordinate Y_0 of $\mathbb{P}_N$ with the monomial X_0^n, the coordinate Y_1 with the monomial $X_0^{n-1}X_1$, and so on. Furthermore, we introduce the following notation: $X_{(a,b,c)} = X_0^a X_1^b X_2^c$, where $a + b + c = n$, so that if $\underline{x} = (x_0, x_1, x_2) \in \mathbb{P}_2$, then for instance $X_{(n,0,0)}(\underline{x}) = x_0^n$, $X_{(n-1,1,0)}(\underline{x}) = x_0^{n-1}x_1$, etc.

Now let $q = nr+l$, $0 \leq l \leq n-1$, and let $\widetilde{F}_{(i,j)}=(X_0^a X_1^b X_2^c)F_{(i,j)}$, $i,j = 1,2,3$, $a+b+c = t$, be a basis element of $I_{\mathbb{P}_2}(n(r+1))$ as described above. For instance $\widetilde{F}_{(1,2)} = (X_0^a X_1^b X_2^c)F_{(1,2)} = (X_0^a X_1^b X_2^c)(X_1^q X_2 - X_1 X_2^q) = (X_0^a X_1^{nr+l+b} X_2^{c+1}) - (X_0^a X_1^{b+1} X_2^{nr+l+c}) = X_1^{nr}(X_0^a X_1^{l+b} X_2^{c+1}) - (X_0^a X_1^{b+1} X_2^{c+l})X_2^{nr}$. Observe that $a+(b+l)+(c+1) = n = a+(b+1)+(c+l)$ since $a+b+c = n(r+1)-(q+1)$. With the notation introduced above let, for instance, $G_{(1,2)} = X_{(0,n,0)}^r X_{(a,b+l,c+1)} - X_{(a,b+1,c+l)}X_{(0,0,n)}^r$. Then $G_{(1,2)} \in I_S(r + 1)$ and it is easy to see that $\varphi_{G_{(1,2)}} = \widetilde{F}_{(1,2)}$, where the mapping φ is as defined in section 3. Consequently, a set of generators for the ideal $I_S(r + 1)$ can be obtained as the "pull-back" under the mapping φ of a set of generators of the ideal $I_{\mathbb{P}_2}(n(r + 1))$. We give an example to illustrate the above ideas.

Example. Let $K =\text{GF}(8)$, $v_2 : \mathbb{P}_2(K) \to \mathbb{P}_5(K)$ be the Veronese embedding of degree 2 and let $S = v_2(\mathbb{P}_2(K))$. Since $a_2 + 1 = 2(8 - 1) + 1 \equiv 1 \text{ mod. } 2$ it follows from lemma in section 3 that $a_S = 7$. Since $8 \equiv 0 \text{ mod.} 2$ then $r = 4$ and $I_S = \langle I_S(2), I_S(5) \rangle$. In this case, $\dim_K I_S(2) = \frac{1}{8}(n + 1)n(n - 1)(n + 6) = 6$. If $Y_0, \dots , Y_5$ denote the coordinates of $\mathbb{P}_5(K)$, with the help of Macaulay ([7]), it is easy to see that a set of generators for $I_S(5)$ is:

$$\left\{ \begin{array}{c} Y_3^4 Y_5 + Y_4 Y_5^4, Y_0^4 Y_5 + Y_2 Y_5^4, Y_3^4 Y_4 + Y_3 Y_5^4, Y_1 Y_3^3 Y_4 + Y_1 Y_5^4, Y_0^4 Y_4 + Y_1 Y_5^4, \\ Y_0^4 Y_3 + Y_1 Y_3^4, Y_0^4 Y_2 + Y_0 Y_5^4, Y_0^4 Y_1 + Y_0 Y_3^4 \end{array} \right\}$$

Now we rename the coordinates $Y_0, \dots , Y_5$ as $X_{(2,0,0)}, X_{(1,1,0)}, X_{(1,0,1)}, X_{(0,2,0)}, X_{(0,1,1)}, X_{(0,0,2)}$, respectively by using the lexicographic order. In other words if $\underline{x} = (x_0, x_1, x_2) \in \mathbb{P}_2$, then for instance $X_{(1,1,0)}(\underline{x}) = x_0 x_1$, and $X_{(0,1,1)}(\underline{x}) = x_1 x_2$. Recall that

$$I_{\mathbb{P}_2} = \langle X_0^8 X_1 - X_0 X_1^8, X_0^8 X_2 - X_0 X_2^8, X_1^8 X_2 - X_1 X_2^8 \rangle = \langle F_{(0,1)}, F_{(0,2)}, F_{(1,2)} \rangle$$

We observe that $\dim_K I_{\mathbb{P}_2}(2r) = \dim_K I_S(r) = 8$ and that the elements $\{X_t F_{(i,j)}, t = 0,1,2; 0 \leq i < j \leq 2\}$ are in $I_{\mathbb{P}_2}(10)$, so that a basis for $I_{\mathbb{P}_2}(10)$ can be chosen from these elements. Since $q = 8$, with the notation introduced above, taking $X_2 F_{(1,2)} = X_2(X_1^8 X_2 - X_1 X_2^8) = X_1^8 X_2^2 - X_1 X_2^9$, the element $G_{(1,2)} = X_{(0,2,0)}^4 X_{(0,0,2)} - X_{(0,1,1)} X_{(0,0,2)}^4$ is in $I_S(5)$. In terms of the coordinates Y_i, the element $G_{(1,2)}$ is precisely $Y_3^4 Y_5 + Y_4 Y_5^4$. In this example the corresponding code $C_S(d)$ has dimensions 6, 15, 28, 45, 58, 67 and 72 for $d = 1, 2, \ldots, 7$ respectively, and for instance in the case $d = 2$ it has minimal distance $\delta_S(2) = 40$.

References

1. D. Cox, J. Little, D. O'Shea. *Ideals, Varieties, and Algorithms*. UTM, Springer Verlag, 1992.
2. P. Delsarte, J.M. Goethals, F.J. MacWilliams. On generalized Reed-Muller codes and their relatives. Inform. and Control, vol. 16 (1970), pp. 403-422.
3. I. Duursma, C. Rentería and H. Tapia-Recillas. Reed-Muller codes on Complete Intersections. Submitted for publication.
4. A.V. Geramita, M. Kreuzer and L. Robbiano. Cayley-Bacharach schemes and their canonical modules. Transactions of the AMS, vol. 339, number 1, Sept. (1993), pp. 163-189.
5. J.P. Hansen. Zero-dimensional schemes. Proc. Int. Conf., Ravello, 1992. F. Orrechia and L. Chiantini (eds.), Walter de Gruyter, Berlin, (1994).
6. G. Lachaud. The parameters of the projective Reed-Muller codes. Discrete Mathematics 81 (1990), pp. 217-221.
7. D.R. Grayson, M. Stillman: Macaulay 2 (1998).
8. D.J. Mercier, R. Rolland. Polynômes homogènes à plusieurs variables sur un corps fini $\mathbf{F_q}$ qui s'annulent sur l'espace projectif $\mathbb{P}_m(\mathbf{F_q})$. J. of Pure and Applied Algebra 124 (1998), pp. 227-240
9. C. Rentería, H. Tapia-Recillas. Linear codes associated to the ideal of points in $\mathbb{P}^d$ and its canonical module. Communications in Algebra **24** (3), (1996), pp. 1083-1090.
10. C. Rentería, H. Tapia-Recillas. Reed-Muller Codes: An Ideal Theory Approach. Communications in Algebra, **25**(2), (1997), pp. 401-413.
11. C. Rentería, H. Tapia-Recillas. The a-invariant of some Reed-Muller Codes. To appear in AAECC journal.
12. A.B. Sörensen. Projective Reed-Muller codes. IEEE Trans. on Inf. Theory, vol. 37, No. 6, Nov. (1991), pp. 1567-1576.

Cryptography Primitives Based on a Cellular Automaton

Jesús Urías

Instituto de Investigación en Comunicación Optica
Universidad Autónoma de San Luis Potosí, SLP, 78000 México
jurias@cactus.iico.uaslp.mx,
WWW home page: http://quetzal.iico.uaslp.mx

Abstract. Large families of permutations and a generator of pseudo-random sequences of binary words are implemented in the form of a single square array of XOR gates following the local rule of an expansive cellular automaton.

1 Introduction

We consider a class of block cryptosystems that consist of an indexed family of permutations of binary words and a deterministic generator of pseudorandom sequences of indices that are used to select sequences of permutations. The family of permutations is $\Psi = \{\psi_{\underline{k}} : M \to C \mid \underline{k} \in K\}$, with all three sets M, C, and K the set of binary words of length N, Z_2^N, each. The binary alphabet is $Z_2 = \{0, 1\}$. The words in M are the *clearblocks*; C is the set of enciphered words, or *cipherblocks*; and the words in the set of indices K are the *enciphering keys*. For each $\underline{k} \in K$, every $\underline{m} \in M$ has a cipherblock expression $\psi_{\underline{k}}(\underline{m})$. To encrypt a long plain text it is factored out into a succession of clearblocks $\underline{m}^0, \underline{m}^1, \underline{m}^2, \ldots$ that are transformed sequentially to $\psi_{\underline{k}^0}(\underline{m}^0), \psi_{\underline{k}^1}(\underline{m}^1), \psi_{\underline{k}^2}(\underline{m}^2), \ldots$ by following the sequence of keys $\underline{k}^0, \underline{k}^1, \underline{k}^2, \ldots$ yielded by the pseudorandom generator. To disclose from the sequence of cipherblocks the plain text we need to know the seed that was used to generate the pseudorandom sequence of keys, as to reproduce it, and have access to the family of inverse permutations $\Phi = \{\phi_{\underline{k}} : C \to M \mid \underline{k} \in K\}$ such that for every $\underline{k} \in K$, $\underline{m} = \phi_{\underline{k}}(\psi_{\underline{k}}(\underline{m}))$ for every $\underline{m}$.

The phenomenon of synchronization in coupled cellular automata[1] is used in ref.[2] to implement the families of permutations Ψ and Φ and a pseudorandom generator for cryptography applications. The primitives are implemented in the form of a single bidimensional array of XOR gates, the *unit cipher* (UC). In the next Section the UC is presented in combinatorial terms rather than dynamical ones, as is done in [2]. The necessary definitions concerning cellular automata are given first.

2 The unit cipher

An *elementary cellular automaton* (CA) is a discrete dynamical system with *state space* the set Z_2^Z and evolution operator $\mathcal{A} : Z_2^Z \to Z_2^Z$, with Z the set of integers. An *automaton state* $\underline{x} \in Z_2^Z$ has coordinates $(\underline{x})_i = x_i \in Z_2$ with $i \in Z$. The action of $\mathcal{A}$ on an automaton state $\underline{x}$ is induced by a *local rule* $\mathcal{A}_{\mathcal{L}} : Z_2^3 \to Z_2$, such that $(\mathcal{A}(\underline{x}))_i = \mathcal{A}_{\mathcal{L}}(x_{i-1}, x_i, x_{i+1})$. To introduce dynamics, the automaton state at time $t \geq 0$ is denoted by $\underline{x}^t \in Z_2^Z$ and its evolution is defined iteratively by the rule $\underline{x}^{t+1} = \mathcal{A}(\underline{x}^t)$. Starting from the *initial state* $\underline{x}^0$, the automaton generates the *forward space–time pattern* $\mathbf{x} \in Z_2^{Z \times Z_*}$ with state $(\mathbf{x})^t = \underline{x}^t = \mathcal{A}^t(\underline{x}^0)$ reached at from $\underline{x}^0$ after $t \in Z_*$ time steps. The set of nonnegative integers is denoted by Z_*. The *lattice* that supports the space–time patterns has *nodes* $(i, t) \in Z \times Z_*$, the space index of a node is i and t its time index.

The CA we will consider has local rule $\mathcal{A}_{\mathcal{L}}(x_{i-1}, x_i, x_{i+1}) = x_{i-1} + x_{i+1} \in Z_2$. The equality sign is being used to denote the congruence modulo 2. Figure 1 shows a portion of the $Z \times Z_*$ lattice where the nodes are represented by circles. The value at a node is the sum (modulo 2) of the values presented by the incoming arrows to the node. The outgoing arrows present the value that is taken by the node. The value at node $(i, t+1)$ in Fig. 1 is $x_i^{t+1} = x_{i-1}^t + x_{i+1}^t$.

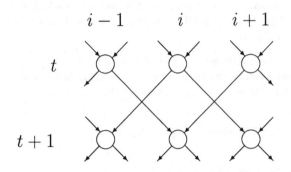

Fig. 1. Detail of the bidimensional array of XOR gates that conform the unit cipher.

The construction of the cryptography primitives is based on the observation that the totality of space–time patterns generated by this CA is closed under the scale transformations $Z \times Z_* \to (2^n \cdot Z + i) \times (2^n \cdot Z_* + t)$, for each $(i, t) \in Z \times Z_*$, in the sense of the following.

Lemma 1 (Scale invariance). *Let* $\mathbf{x} \in Z_2^{Z \times Z_*}$ *be a space–time pattern of the CA with local rule* $x_{i+1}^{t+1} = x_i^t + x_{i+2}^t$, $(i, t) \in Z \times Z_*$. *Then, the **expanded rule*** $x_{i+2^{n-1}}^{t+2^{n-1}} = x_i^t + x_{i+2^n}^t$ *holds for every* $(i, t) \in Z \times Z_*$ *and positive* n.

The proof of Lemma 1 is given in Appendix A.

In view of the expanded rule, the four binary words of length $N = 2^n - 1$, $n > 0$, that are surrounding a square pattern of size $N \times N$ in the lattice are similarly related at every scale N. This is the matter of Theorem 1 below but we have to precise things first. We will consider squares in the lattice that consist of the N time running words $(x_1^1, x_1^2, \ldots x_1^N), \ldots, (x_N^1, x_N^2, \ldots x_N^N)$. The first time running word is given a name, $\underline{k}' = (x_1^1, x_1^2, \ldots x_1^N)$. The surrounding words are $\underline{k} = (x_0^1, x_0^2, \ldots x_0^N)$ on the left side, $\underline{c} = (x_{N+1}^1, x_{N+1}^2, \ldots x_{N+1}^N)$ on the right side, $\underline{t} = (x_1^1, x_2^1, \ldots x_N^1)$ on the top, and $\underline{m} = (x_1^{N+1}, x_2^{N+1}, \ldots x_N^{N+1})$ at the bottom. An example for $N = 3$ follows,

$$
\begin{array}{ccccc}
x_0^1 & x_1^1 & x_2^1 & x_3^1 & x_4^1 \\[4pt]
x_0^2 & x_1^2 & \cdot & \cdot & x_4^2 \\[4pt]
x_0^3 & x_1^3 & \cdot & \cdot & x_4^3 \\[4pt]
 & x_1^4 & x_2^4 & x_3^4 &
\end{array}
$$

Only the five distinguished words are shown explicitly. The rest of entries in the square pattern of the example are entered as dots. Note that $\underline{t} = (x_1^1, x_2^1, x_3^1)$ and $\underline{k}' = (x_1^1, x_1^2, x_1^3)$ share the symbol x_1^1. We call the square pattern together with its bordering words, all of size N, and obeying the local automaton rule a *unit cipher* (UC) of size N. All the entries in an UC follow the local rule of the automaton. In particular, it is sufficient and necessary to know $\underline{t}$, $\underline{k}$, and $\underline{c}$ to fully determine the configuration of a UC, see Fig. 1. The totality of configurations is 2^{3N} for an UC of size N.

Theorem 1 (The primitives). *Let an UC of size $N = 2^n - 1$, $n > 0$, have distinguished words as defined above. Then, there exist functions $\phi, \psi, \rho : Z_2^N \times Z_2^N \to Z_2^N$ such that for every configuration of the UC, $\phi_{\underline{k}}(\underline{c}) = \underline{m}$, $\psi_{\underline{k}}(\underline{m}) = \underline{c}$, and $\rho(\underline{k}, \underline{k}') = \underline{t}$.*

The proof of Theorem 1 is given in Appendix B.

3   Concluding remarks

The UC implements the indexed families of permutations Ψ and Φ. The indices $\underline{k}$ are binary words of length $N = 2^n - 1$, $n > 0$. To compute the permutation of a clearblock $\underline{m}$ under $\psi_{\underline{k}}$ using the UC, the words $\underline{k}$ and $\underline{m}$ are written on it and the automaton is made to run backwards in time. Operatively, to perform the computation we have to provide a value for $\underline{k}'$ but Theorem 1 warranties that the output value $\underline{c} = \psi_{\underline{k}}(\underline{m})$ is independent of $\underline{k}'$. The inverse permutation $\phi_{\underline{k}}$, to bring $\underline{c}$ back to $\underline{m}$, is computed using the UC by first writing on it the words $\underline{k}$ and $\underline{c}$ and then running the automaton forward in time. The value of word $\underline{t}$ is operatively necessary to perform the computation, however, it does not affect the result $\underline{m} = \phi_{\underline{k}}(\underline{c})$. The words $\underline{t}$ and $\underline{k}'$ are not relevant for the permutations but they are not garbage. Actually, $\underline{t}$ and $\underline{k}'$ are part of the pseudorandom generator. Sequences of keys are generated by iterating the function ρ, introduced in Theorem 1, according to the rule $\underline{k}^t = \rho(\underline{k}^{t-1}, \underline{k}^{t-2})^r$, where $\underline{k}^r$ is the word $\underline{k}$ in reversed order. The generated keys are found to distribute uniformly[2] due to the ergodic properties of the expansive automaton we are using.[3][4]

Some of the advantages of the CA primitives for cryptography we have presented are the following. (i) The pseudorandom generator and the families of permutations are contained in a single array of XOR gates, the CU. (ii) The local nature of the automaton rule corresponds to short wiring when the CU is transported to a VLSI device, allowing for a very fast system clocking. (iii) Large size of blocks can be handled easily. (iv) The bidimensional arrays for information processing we are proposing fit naturally in the emerging technology of electronic devices that is being based on quantum dots.[5] (v) Multiple UC's can be interconnected to design involved cryptosystems.

Acknowledgements

This work received partial support from FAI–UASLP and CONACyT.

A   Proof of Lemma 1

The case $n = 1$ is the local rule and for $n = 2$ we apply to x_{i+4}^{t+2} the local rule twice to find that $x_{i+4}^{t+2} = x_{i+1}^{t+1} + x_{i+3}^{t+1} = x_i^t + 2x_{i+2}^t + x_{i+4}^t$. For the induction step, assume the expanded rule holds for every $0 < n < N$, we prove it holds for N. First observe that the identity

$$2x_{i+2^{N-1}}^t + 2x_{i+2^{N-2}}^{t+2^{N-2}} + 2x_{i+3\cdot 2^{N-2}}^{t+2^{N-2}} = 0$$

can be written in the form

$$x_i^t + x_{i+2^N}^t = [x_i^t + x_{i+2^{N-1}}^t + x_{i+2^{N-2}}^{t+2^{N-2}}]$$
$$+x_{i+2^{N-1}}^t + x_{i+2^N}^t + x_{i+3\cdot 2^{N-2}}^{t+2^{N-2}} + x_{i+2^{N-2}}^{t+2^{N-2}} + x_{i+3\cdot 2^{N-2}}^{t+2^{N-2}}.$$

The three terms within brackets add up to zero by the expanded rule for $N-2$. Introducing the integers $j_1 = i + 2^{N-1}$, $j_2 = i + 2^{N-2}$, and $t' = t + 2^{N-2}$ the identity takes the form

$$x_i^t + x_{i+2^N}^t = [x_{j_1}^t + x_{j_1+2^{N-1}}^t + x_{j_1+2^{N-2}}^{t+2^{N-2}}] + x_{j_2}^{t'} + x_{j_2+2^{N-1}}^{t'}.$$

Again, the expanded rule for $N-2$ makes the first three terms add up to zero and applied to the last two terms it yields

$$x_{j_2}^{t'} + x_{j_2+2^{N-1}}^{t'} = x_{j_2+2^{N-2}}^{t'+2^{N-2}} = x_{i+2^{N-1}}^{t+2^{N-1}}.$$

B   Proof of Theorem 1

The existence of function ρ follows immediately from the bipermutative character of the automaton that makes possible to run it backwards in time as to fill the left upper half triangle of the UC configuration beginning from the last symbols in $\underline{k}$ and $\underline{k}'$. To prove the existence of function $\phi_{\underline{k}}$ we only have to prove that *the word $\underline{m}$ that is reached by running the automaton forward in time is fully determined by the distinguished words $\underline{k}$ and $\underline{c}$.* For $n = 1$ it follows trivially as seen in Fig. 1.

Assume it is true for $n < M$, we prove that it is also true for $n = M$. Define the addition (mod 2) of words as $(x_1, x_2, \dots) \oplus (y_1, y_2, \dots) = (x_1+y_1, x_2+y_2, \dots)$. In the UC of size $2^M - 1$ we use the expanded rule of the scale invariance Lemma 1 to compute the word

$$(x_{2^{M-1}}^{2^{M-1}+1}, x_{2^{M-1}}^{2^{M-1}+2}, \dots, x_{2^{M}-1}^{2^M}) = (x_0^1, x_0^2, \dots, x_0^{2^{M-1}}) \oplus (x_{2^M}^1, x_{2^M}^2, \dots, x_{2^M}^{2^{M-1}}).$$

This reduces our problem of computing $\underline{m}$ for the UC of size $2^M - 1$ (without using $\underline{t}$) to the problem of two UC's of size $2^{M-1} - 1$. Very similarly, the existence of function $\psi_{\underline{k}}$ is proved with the help of Lemma 1 and the bipermutative character of the automaton rule.

References

1. Jesús Urías, Gelasio Salazar, and Edgardo Ugalde, Synchronization of cellular automaton pairs, *Chaos*, **8**, 814–818 (1998).
2. Jesús Urías, Edgardo Ugalde and Gelasio Salazar, A cryptosystem based on cellular automata, *Chaos*, **8**, 819–822 (1998).
3. F. Blanchard, Petr Kurka, and A. Maass, Topological and measure–theoretic properties of one–dimensional cellular automata, *Physica* **D 103**, 86–99 (1997).
4. M.A. Shereshevsky, Ergodic properties of certain surjective cellular automata, *Monatshefte fur Mathematik* **114**, 305–316 (1992).
5. C.S. Lent and P.D. Tougwa, A device architecture for computing with quantum dots, *Proceedings of the IEEE*. **85**, 541 (1997).

Factoring the Semigroup Determinant of a Finite Commutative Chain Ring

Jay A. Wood[*]

Department of Mathematics, Computer Science & Statistics
Purdue University Calumet
Hammond, Indiana 46323–2094 USA
wood@calumet.purdue.edu
http://www.calumet.purdue.edu/public/math/wood/

Abstract. The semigroup determinant of a finite commutative chain ring factors completely into linear factors involving the characters of the group of units of the ring. The factorization proves the equivalence of two approaches to the extension theorem for general weight functions on linear codes over finite commutative chain rings.

1 Introduction

In 1896, Frobenius factored the group determinant of a finite group [2] (also see [1] and [3] and the references therein). If G is a finite group and a_g, $g \in G$, are indeterminates, the group determinant of G is $\det(a_{gh^{-1}})$, i.e., the determinant of the $|G| \times |G|$ matrix whose entry in position (g, h) is the indeterminate $a_{gh^{-1}}$ indexed by the product $gh^{-1} \in G$.

The matrix $A = (a_{gh^{-1}})$ represents the left regular representation of G with respect to the basis of the group algebra $\mathbb{C}[G]$ consisting of the elements of G. The factorization of $\det A$ reflects the decomposition of $\mathbb{C}[G]$ into irreducible representations. In fact, Frobenius' work on factoring $\det A$ led directly to his development of the representation theory for finite groups.

If G is abelian, the basis of characters of G will diagonalize A with typical diagonal entry

$$\hat{a}(\pi) = \sum_{g \in G} a_g \pi(g) \ ,$$

the Fourier coefficient of a with respect to π. Thus

$$\det A = \prod_{\pi \in \widehat{G}} \hat{a}(\pi) \ ,$$

in the abelian case.

[*] Partially supported by NSA grant MDA904-96-1-0067, and by Purdue University Calumet Scholarly Research Awards. This paper is an expanded version of results presented at the International Conference on Coding Theory, Cryptography and Related Areas, Guanajuato, Mexico, April 21, 1998. The paper is in final form and no version of it will be submitted for publication elsewhere.

Now suppose that S is a finite semigroup and that a_s, $s \in S$, are indeterminates. As in the group case, form a matrix $\mathcal{A} = (a_{st})$ of size $|S| \times |S|$, where the (s,t) entry is the indeterminate indexed by the product $st \in S$. The main result of this paper is a factorization of $\det \mathcal{A}$ in the case where the semigroup S is the multiplicative semigroup of a finite commutative chain ring, i.e., of a finite, commutative, local ring R with principal maximal ideal. Unlike in the group case, the matrix $\mathcal{A}$ does not represent the regular representation.

The factorization is explicit, and it involves the Fourier coefficients of a (thought of as a function on S) restricted to certain orbits of the group $\mathcal{U}$ of units of R. The key to the matter is to pick a good basis for the reduced semigroup algebra $\mathbb{C}_0[S]$. Multiplication by $\mathcal{U}$ partitions S into $\mathcal{U}$-orbits. Each orbit can be identified with a quotient space of $\mathcal{U}$ which, since $\mathcal{U}$ is abelian, is actually a quotient group of $\mathcal{U}$. The characters of the quotient group provide a good basis along the orbit, and the union of all the characters of all the orbits provides a good basis for $\mathbb{C}_0[S]$. Expressed in terms of this good basis, the matrix $\mathcal{A}$ takes on an especially simple form. The formula for $\det \mathcal{A}$ then follows directly.

In [5], the author discusses the extension theorem for general weight functions on linear codes over finite commutative chain rings. In that paper there are two approaches to the extension problem, one involving $\det \mathcal{A}$ and the other involving the regular representation of $\mathbb{C}_0[S]$. The factorization of $\det \mathcal{A}$ given here shows that those two approaches are equivalent.

Acknowledgments. I thank the referee for several improvements to this paper.

2 Preliminaries

Let S be a multiplicative semigroup with 0 element. The *complex semigroup algebra* $\mathbb{C}[S]$ is a complex vector space with basis e_s, $s \in S$. By defining a multiplication of basis elements by $e_s e_t = e_{st}$ (where the latter is indexed by the semigroup product st) and extending linearly, $\mathbb{C}[S]$ becomes a complex algebra. Observe that the ideal $(e_0) = \mathbb{C}e_0$ generated by e_0 is a two-sided ideal. The quotient algebra $\mathbb{C}_0[S] = \mathbb{C}[S]/(e_0)$ is the *reduced semigroup algebra* of S; it has dimension $|S| - 1$ over $\mathbb{C}$.

Let R be a finite commutative *chain ring*, that is, a finite ring with 1 that is commutative, local, and whose maximal ideal $\mathfrak{m}$ is principal (say, $\mathfrak{m} = (m) = Rm$). It is easy to see ([4, Lemma 13]) that the ideals of R fit into the following chain:

$$R \supset (m) \supset (m^2) \supset \cdots \supset (m^l) \supset (0) \ .$$

Here, l is the smallest nonnegative integer such that $m^{l+1} = 0$.

Let $\mathcal{U}$ be the group of units of R. Then $\mathcal{U}$ acts on R by multiplication, and the orbits of $\mathcal{U}$ are exactly the set-theoretic differences $\mathcal{O}_0 = \mathcal{U} = R \setminus (m)$, $\mathcal{O}_1 = (m) \setminus (m^2)$, $\mathcal{O}_2 = (m^2) \setminus (m^3)$, ..., $\mathcal{O}_l = (m^l) \setminus (0)$, $\mathcal{O}_{l+1} = (0)$. If $x \in \mathcal{O}_i$, the *stabilizer subgroup* of x is $\mathcal{U}_x = \{u \in \mathcal{U} : ux = x\}$. Because $\mathcal{U}$ is abelian, if $x, y \in \mathcal{O}_i$, then $\mathcal{U}_x = \mathcal{U}_y$. We will simply write $\mathcal{U}_i = \mathrm{Stab}(\mathcal{O}_i)$ for this common stabilizer subgroup of the orbit $\mathcal{O}_i$. Note that $\mathcal{O}_i = \{um^i : u \in \mathcal{U}/\mathcal{U}_i\}$ provides a bijection between $\mathcal{O}_i$ and $\mathcal{U}/\mathcal{U}_i$. Observe that $\mathcal{U}_i \subset \mathcal{U}_{i+1}$.

The complex characters of $\mathcal{U}$ form a group denoted by $\widehat{\mathcal{U}}$. Given a subgroup U of $\mathcal{U}$, the *annihilator* of U is $(\widehat{\mathcal{U}} : U) = \{\pi \in \widehat{\mathcal{U}} : \pi(u) = 1, \; u \in U\}$. It follows that $(\widehat{\mathcal{U}} : U) \cong (\mathcal{U}/U)\widehat{}$, so that $\left|(\widehat{\mathcal{U}} : U)\right| = |\mathcal{U}| / |U|$.

A pair $(\pi, \mathcal{O}_i)$ consisting of a character π of $\mathcal{U}$ and a $\mathcal{U}$-orbit $\mathcal{O}_i$ is *admissible* if $\mathcal{U}_i \subset \ker \pi$, i.e., if $\pi \in (\widehat{\mathcal{U}} : \mathcal{U}_i)$. Since $\mathcal{U}_i \subset \mathcal{U}_{i+1}$, every $\pi \in \widehat{\mathcal{U}}$ has a largest non-negative integer i_π such that $(\pi, \mathcal{O}_{i_\pi})$ is admissible. We make one exception. Denote by π_0 the trivial character of $\mathcal{U}$; i.e., $\pi_0(u) = 1$, for all $u \in \mathcal{U}$. The admissible pair $(\pi_0, \mathcal{O}_{l+1} = \{0\})$, is called the *trivial* admissible pair. We exclude this trivial pair, and we set $i_{\pi_0} = l$ for the trivial character π_0.

3   The factorization

Let R be a finite commutative chain ring with group of units $\mathcal{U}$. For every element $r \in R$ assign an indeterminate a_r.

Form a square matrix $\mathcal{A}$ of size $|R|$, the *semigroup matrix*, as follows. The rows and columns of $\mathcal{A}$ are indexed by the elements of R. The entry in row r and column s is the indeterminate a_{rs} associated to the element $rs \in R$.

In a similar fashion, form a square matrix $\mathcal{A}_0$ of size $|R| - 1$, the *reduced semigroup matrix*. Within $\mathcal{A}_0$, set $a_0 = 0$. The rows and columns of $\mathcal{A}_0$ are indexed by the non-zero elements of R. The entry in row r and column s is again a_{rs} (which will equal 0 if $rs = 0$).

We wish to factor $\det \mathcal{A}$ and $\det \mathcal{A}_0$. Because $\mathcal{A}_0$ is more relevant to coding theory (see Remark 1), we will focus our attention on $\det \mathcal{A}_0$, and then describe briefly what happens for $\det \mathcal{A}$.

Given an admissible pair $(\pi, \mathcal{O}_i)$, set

$$\hat{a}(\pi, i) = \sum_{u \in \mathcal{U}/\mathcal{U}_i} \pi(u) a_{um^i} \; .$$

Since $(\pi, \mathcal{O}_i)$ is admissible, this expression is well-defined. In fact, $\hat{a}(\pi, i)$ may be viewed as the Fourier coefficient of the restriction of a to the orbit $\mathcal{O}_i$ (which is bijective to the quotient group $\mathcal{U}/\mathcal{U}_i$; $\pi \in (\widehat{\mathcal{U}} : \mathcal{U}_i) \cong (\mathcal{U}/\mathcal{U}_i)\widehat{}$).

Theorem 1. *For any finite commutative chain ring R with group of units $\mathcal{U}$, $\det \mathcal{A}_0$ factors into linear factors:*

$$\det \mathcal{A}_0 = C \prod_{\pi \in \widehat{\mathcal{U}}} \hat{a}(\pi, i_\pi)^{1+i_\pi} \; ,$$

where C is a non-zero integral constant.

Theorem 2. *For any finite commutative chain ring R with group of units $\mathcal{U}$, $\det \mathcal{A}$ factors into linear factors:*

$$\det \mathcal{A} = C \, a_0 \left(\hat{a}(\pi_0, i_{\pi_0}) - d \, a_0\right)^{1+i_{\pi_0}} \prod_{\pi \in \widehat{\mathcal{U}}, \pi \neq \pi_0} \hat{a}(\pi, i_\pi)^{1+i_\pi} \; ,$$

where C is a non-zero integral constant and $d = [\mathcal{U} : \mathcal{U}_l] = |\mathcal{O}_l|$ is the index of $\mathcal{U}_l$ in $\mathcal{U}$.

These theorems are proved at the end of Sect. 5.

4   A good basis

In order to calculate $\det \mathcal{A}_0$, we actually calculate $P^{-1}\mathcal{A}_0 P$ for a particularly nice choice of matrix P. Factoring $\det \mathcal{A}_0 = \det(P^{-1}\mathcal{A}_0 P)$ is then an easy task. For notational convenience, set $M_i = |\mathcal{O}_i|$.

Lemma 3. *The number of non-trivial admissible pairs $(\pi, \mathcal{O}_i)$, $\pi \in \widehat{\mathcal{U}}$, equals $|R| - 1$.*

Proof. For a fixed i, $(\pi, \mathcal{O}_i)$ is admissible if and only if $\pi \in (\widehat{\mathcal{U} : \mathcal{U}_i})$. There are $|\mathcal{U}| / |\mathcal{U}_i|$ such π's. But since $\mathcal{U}_i$ is the stabilizer subgroup of the orbit $\mathcal{O}_i$, $|\mathcal{U}| / |\mathcal{U}_i|$ is also equal to $M_i = |\mathcal{O}_i|$. Since $|R| = \sum_i M_i$, the result follows once we exclude the trivial admissible pair.   □

We now define a matrix P. The rows of P are indexed by the non-zero elements of R while the columns of P are indexed by the non-trivial admissible pairs $(\pi, \mathcal{O}_i)$, $\pi \in \widehat{\mathcal{U}}$. By Lemma 3, P is a square matrix. The entry $p_{r,(\pi,\mathcal{O}_i)}$ in row r and column $(\pi, \mathcal{O}_i)$, $\pi \in \widehat{\mathcal{U}}$, is

$$p_{r,(\pi,\mathcal{O}_i)} = \begin{cases} 0, & \text{if } r \notin \mathcal{O}_i \ , \\ \frac{1}{M_i}\pi(u), & \text{if } r = um^i \in \mathcal{O}_i \ . \end{cases}$$

Since $(\pi, \mathcal{O}_i)$ is admissible, $\pi \in (\widehat{\mathcal{U} : \mathcal{U}_i})$. This implies that $\pi(u)$ is well-defined, depending only on r.

Let the matrix Q be an "unscaled" version of P:

$$q_{r,(\pi,\mathcal{O}_i)} = \begin{cases} 0, & \text{if } r \notin \mathcal{O}_i \ , \\ \pi(u), & \text{if } r = um^i \in \mathcal{O}_i \ . \end{cases}$$

Lemma 4. $P^{-1} = \overline{Q}^t$, *the conjugate transpose of Q.*

Proof. This follows easily from the orthogonality relations on characters, viewing $\pi \in (\widehat{\mathcal{U} : \mathcal{U}_i})$ as a character of $\mathcal{U}/\mathcal{U}_i$.   □

The columns of P are a good basis for $\mathbb{C}_0[S]$, where S is the multiplicative semigroup of the finite commutative chain ring R. For each orbit $\mathcal{O}_i$, we use the characters of $\mathcal{U}/\mathcal{U}_i$.

5   Calculations

We begin with a technical lemma.

Lemma 5. *Suppose* $\psi \in (\widehat{\mathcal{U} : \mathcal{U}_j}) \cong (\mathcal{U}/\mathcal{U}_j)\widehat{}$. *Then*

$$\sum_{x \in \mathcal{U}_{i+j}/\mathcal{U}_j} \psi(x) = \begin{cases} |\mathcal{U}_{i+j}| / |\mathcal{U}_j|, & \text{if } \psi \in (\widehat{\mathcal{U} : \mathcal{U}_{i+j}}) \ , \\ 0, & \text{if } \psi \notin (\widehat{\mathcal{U} : \mathcal{U}_{i+j}}) \ . \end{cases}$$

Proof. This is an application of the following general result. If H is a subgroup of G and $\pi \in \widehat{G}$, then

$$\sum_{h \in H} \pi(h) = \begin{cases} |H|, & \text{if } \pi \in (\widehat{G} : H) \ , \\ 0, & \text{if } \pi \notin (\widehat{G} : H) \ . \end{cases}$$

The proof is an exercise for the reader.   □

Theorem 6. *Let b be the entry of $P^{-1}A_0 P$ in row $(\pi, \mathcal{O}_i)$ and column $(\psi, \mathcal{O}_j)$, $\pi, \psi \in \widehat{\mathcal{U}}$. Then*

$$b = \begin{cases} 0, & \text{if } \psi \neq \bar{\pi} \ , \\ 0, & \text{if } \psi = \bar{\pi}, \text{ but } (\pi, \mathcal{O}_{i+j}) \text{ is not admissible } , \\ C\,\hat{a}(\bar{\pi}, i+j), & \text{if } \psi = \bar{\pi} \text{ and } (\pi, \mathcal{O}_{i+j}) \text{ is admissible } , \end{cases}$$

where C is a non-zero integral constant (that depends only on i, j).

Proof. The entry b is given by the following summation:

$$b = \sum_{r \neq 0} \bar{q}_{r,(\pi, \mathcal{O}_i)} \sum_{s \neq 0} a_{rs} p_{s,(\psi, \mathcal{O}_j)} \ .$$

The q-term vanishes unless $r \in \mathcal{O}_i$, so that $r = \alpha m^i$, $\alpha \in \mathcal{U}/\mathcal{U}_i$. Similarly, the p-term vanishes unless $s = \beta m^j$, $\beta \in \mathcal{U}/\mathcal{U}_j$. Thus

$$b = \sum_{\alpha \in \mathcal{U}/\mathcal{U}_i} \sum_{\beta \in \mathcal{U}/\mathcal{U}_j} \frac{1}{M_j} \bar{\pi}(\alpha)\psi(\beta) a_{\alpha\beta m^{i+j}} \ .$$

Now we isolate the coefficient of $a_{\gamma m^{i+j}}$, $\gamma \in \mathcal{U}/\mathcal{U}_{i+j}$. This coefficient has the form

$$c = \sum \frac{1}{M_j} \bar{\pi}(\alpha)\psi(\beta) \ ,$$

where the sum runs over pairs of $\alpha \in \mathcal{U}/\mathcal{U}_i$, $\beta \in \mathcal{U}/\mathcal{U}_j$, satisfying $\alpha\beta = \gamma$ in $\mathcal{U}/\mathcal{U}_{i+j}$.

Since $\mathcal{U}_i \subset \mathcal{U}_{i+j}$, we see that $\mathcal{U}_{i+j}/\mathcal{U}_i$ is the kernel of the natural projection $\mathcal{U}/\mathcal{U}_i \twoheadrightarrow \mathcal{U}/\mathcal{U}_{i+j}$. Similarly, $\mathcal{U}_{i+j}/\mathcal{U}_j$ is the kernel of $\mathcal{U}/\mathcal{U}_j \twoheadrightarrow \mathcal{U}/\mathcal{U}_{i+j}$. Fix an $\alpha \in \mathcal{U}/\mathcal{U}_i$, with image $\tilde{\alpha} \in \mathcal{U}/\mathcal{U}_{i+j}$. Since $\mathcal{U}/\mathcal{U}_{i+j}$ is a group, set $\tilde{\beta} = \tilde{\alpha}^{-1}\gamma \in \mathcal{U}/\mathcal{U}_{i+j}$.

Every β in the preimage of $\tilde{\beta}$ contributes a term to the coefficient c. If β_0 is one particular preimage of $\tilde{\beta}$ (β_0 depends on α), then every other preimage has the form $\beta = \beta_0 \delta$, $\delta \in \mathcal{U}_{i+j}/\mathcal{U}_j$. Then

$$c = \sum_{\alpha \in \mathcal{U}/\mathcal{U}_i} \sum_{\delta \in \mathcal{U}_{i+j}/\mathcal{U}_j} \frac{1}{M_j}\bar{\pi}(\alpha)\psi(\beta_0\delta) = \sum_\alpha \frac{1}{M_j}\bar{\pi}(\alpha)\psi(\beta_0)\sum_\delta \psi(\delta) \ .$$

By Lemma 3, $c = 0$ unless $\psi \in (\widehat{\mathcal{U}} : \mathcal{U}_{i+j})$, in which case $\psi(\beta_0)$ is independent of the particular choice of preimage of $\tilde{\beta}$.

Repeating this argument with the roles of α, β reversed, we see that $c = 0$ unless both π and ψ belong to $(\widehat{\mathcal{U}} : \mathcal{U}_{i+j})$. In that case $\pi(\alpha) = \pi(\tilde{\alpha})$ and $\psi(\beta) = \psi(\tilde{\beta})$ are well-defined. Thus

$$c = |\mathcal{U}_{i+j}/\mathcal{U}_i| \, |\mathcal{U}_{i+j}/\mathcal{U}_j| \sum_{\tilde{\alpha}\tilde{\beta}=\gamma} \frac{1}{M_j}\bar{\pi}(\tilde{\alpha})\psi(\tilde{\beta}) \ .$$

But

$$\sum_{\tilde{\alpha}\tilde{\beta}=\gamma} \bar{\pi}(\tilde{\alpha})\psi(\tilde{\beta}) = \sum_{\tilde{\alpha}} \bar{\pi}(\tilde{\alpha})\psi(\tilde{\alpha}^{-1}\gamma) = \left(\sum_{\tilde{\alpha}} \bar{\pi}(\tilde{\alpha})\bar{\psi}(\tilde{\alpha})\right)\psi(\gamma) \ .$$

By the orthogonality relations of characters, this last expression equals 0 unless $\psi = \bar{\pi}$, in which case it equals $|\mathcal{U}/\mathcal{U}_{i+j}|\psi(\gamma) = |\mathcal{U}/\mathcal{U}_{i+j}|\bar{\pi}(\gamma)$.

In conclusion, the entry b equals 0 unless $\psi = \bar{\pi}$ belongs to $(\widehat{\mathcal{U}} : \mathcal{U}_{i+j})$. This means that $(\pi, \mathcal{O}_{i+j})$ must be admissible. In particular, $i + j \leq i_\pi$ is required. If these conditions are satisfied, then

$$b = |\mathcal{U}_{i+j}/\mathcal{U}_i| \, |\mathcal{U}_{i+j}/\mathcal{U}_j| \, |\mathcal{U}/\mathcal{U}_{i+j}| \frac{1}{M_j} \sum_{\gamma \in \mathcal{U}/\mathcal{U}_{i+j}} \psi(\gamma)a_{\gamma m^{i+j}}$$

$$= |\mathcal{U}_{i+j}/\mathcal{U}_i| \sum_{\gamma \in \mathcal{U}/\mathcal{U}_{i+j}} \psi(\gamma)a_{\gamma m^{i+j}} \ ,$$

which has the form $b = C\,\hat{a}(\psi, i + j)$, where C is the non-zero integral constant $|\mathcal{U}_{i+j}/\mathcal{U}_i|$. $\qquad\square$

Proof of Theorem 1. Fix a character $\pi \in \widehat{\mathcal{U}}$. The submatrix of $P^{-1}\mathcal{A}_0 P$ with row indices $(\pi, \mathcal{O}_i)$ and column indices $(\bar{\pi}, \mathcal{O}_j)$, $0 \leq i, j \leq i_\pi$, has the form (neglecting the non-zero constants)

$$\begin{pmatrix} \hat{a}(\bar{\pi},0) & \hat{a}(\bar{\pi},1) & \ldots & \hat{a}(\bar{\pi},i_\pi - 1) & \hat{a}(\bar{\pi},i_\pi) \\ \hat{a}(\bar{\pi},1) & \hat{a}(\bar{\pi},2) & \ldots & \hat{a}(\bar{\pi},i_\pi) & 0 \\ \vdots & & \ldots & & \vdots \\ \hat{a}(\bar{\pi},i_\pi - 1) & \hat{a}(\bar{\pi},i_\pi) & & 0 & 0 \\ \hat{a}(\bar{\pi},i_\pi) & 0 & \ldots & 0 & 0 \end{pmatrix} \ .$$

The determinant of this submatrix is $\hat{a}(\bar{\pi}, i_\pi)^{1+i_\pi}$ times a non-zero integral constant. From this, Theorem 1 follows. $\qquad\square$

Proof of Theorem 2. When dealing with $\mathcal{A}$, one uses a matrix P which is square of size $|R|$, using all elements of R and all admissible pairs, including the trivial pair $(\pi_0, \{0\})$.

In calculating $P^{-1}\mathcal{A}P$, the only difference arises in the submatrix involving the trivial character π_0. Neglecting the non-zero constants, that submatrix is

$$
\begin{pmatrix}
\hat{a}(\pi_0, 0) & \hat{a}(\pi_0, 1) & \cdots & \hat{a}(\pi_0, i_{\pi_0} - 1) & \hat{a}(\pi_0, i_{\pi_0}) & a_0 \\
\hat{a}(\pi_0, 1) & \hat{a}(\pi_0, 2) & \cdots & \hat{a}(\pi_0, i_{\pi_0}) & a_0 & a_0 \\
\vdots & & \cdots & & \vdots & \vdots \\
\hat{a}(\pi_0, i_{\pi_0} - 1) & \hat{a}(\pi_0, i_{\pi_0}) & & a_0 & a_0 & a_0 \\
\hat{a}(\pi_0, i_{\pi_0}) & a_0 & \cdots & a_0 & a_0 & a_0 \\
a_0 & a_0 & \cdots & a_0 & a_0 & a_0
\end{pmatrix} .
$$

In any given row, the (neglected) coefficients for a_0 all equal d times the (neglected) coefficient for $\hat{a}(\pi_0, i_{\pi_0})$. Evaluating the determinant is then an easy exercise involving elementary row and column operations. $\qquad\square$

6   A relative case

We discuss briefly the situation where some symmetry is imposed on the indeterminates a_r. To be precise, suppose U is a subgroup of the group of units $\mathcal{U}$ of a finite commutative chain ring R. Multiplication defines an action of U on R, and we denote the space of orbits by R/U. We impose the conditions $a_{ur} = a_r$, $u \in U$, $r \in R$, on the indeterminates. Notice that a_r now depends only upon the U-orbit $\text{orb}(r)$ of r.

Form a square matrix $\mathcal{A}^U$ of size $|R/U|$, as follows. The rows and columns of $\mathcal{A}^U$ are indexed by the U-orbits. The entry in row $\text{orb}(r)$ and column $\text{orb}(s)$ is the indeterminate a_{rs} associated to $\text{orb}(rs)$. This is well-defined.

Similarly, form a reduced square matrix $\mathcal{A}_0^U$ of size $|R/U| - 1$, by setting $a_0 = 0$ and indexing the rows and columns of $\mathcal{A}_0^U$ by the non-zero U-orbits. The entry in row $\text{orb}(r)$ and column $\text{orb}(s)$ is again the indeterminate a_{rs}. This too is well-defined. When $U = \{1\}$, the matrices $\mathcal{A}^U$ and $\mathcal{A}_0^U$ are just $\mathcal{A}$ and $\mathcal{A}_0$. We now wish to factor $\det \mathcal{A}^U$ and $\det \mathcal{A}_0^U$.

Given an admissible pair $(\pi, \mathcal{O}_i)$ with $\pi \in (\widehat{U} : U)$, set

$$
\breve{a}(\pi, i) = \sum_{u \in \mathcal{U}/U\mathcal{U}_i} \pi(u) a_{um^i} .
$$

Notice that $\hat{a}(\pi, i) = E_i \breve{a}(\pi, i)$, where E_i equals the size of a U-orbit in $\mathcal{O}_i$.

Theorem 7. *For any subgroup U of the group of units $\mathcal{U}$ of a finite commutative chain ring R, $\det \mathcal{A}_0^U$ factors into linear factors:*

$$
\det \mathcal{A}_0^U = C \prod_{\pi \in (\widehat{\mathcal{U}}:U)} \breve{a}(\pi, i_\pi)^{1+i_\pi} ,
$$

where C is a non-zero integral constant.

Theorem 8. *For any subgroup U of the group of units $\mathcal{U}$ of a finite commutative chain ring R, $\det \mathcal{A}^U$ factors into linear factors:*

$$\det \mathcal{A}^U = C\, a_0\, (\check{a}(\pi_0, i_{\pi_0}) - d\, a_0)^{1+i_{\pi_0}} \prod_{\pi \in (\widehat{\mathcal{U}}:U),\, \pi \neq \pi_0} \check{a}(\pi, i_\pi)^{1+i_\pi} \ ,$$

where C is a non-zero integral constant and $d = [\mathcal{U} : U\,\mathcal{U}_l]$ is the index of $U\,\mathcal{U}_l$ in $\mathcal{U}$.

Idea of proofs. View everything as living on the quotient group $\mathcal{U}/U$ and apply Theorems 1 and 2. □

Remark 9. By applying Theorem 4, one obtains the equivalence of the two approaches discussed in [5] to the extension theorem for general weight functions on linear codes over finite commutative chain rings.

7 Examples

Example 10. To illustrate Theorems 1 and 3, consider the finite commutative chain ring $R = \mathbb{Z}/(9)$. The maximal ideal is $\mathfrak{m} = (3)$, and $\mathfrak{m}^2 = (0)$, so that $l = 1$. The group of units is $\mathcal{U} = \{1, 2, 4, 5, 7, 8\}$, which is cyclic of order 6. Either 2 or 5 generates $\mathcal{U}$.

The characters of $\mathcal{U}$ are determined by their values at the generator $2 \in \mathcal{U}$. Let $\xi = \exp(2\pi i/6)$, a primitive 6th root of unity in $\mathbb{C}$. Of course, $\xi^3 = -1$. Then the formulas $\pi_j(2) = \xi^j$, $j = 0, 1, \ldots, 5$, determine the 6 characters of $\mathcal{U}$. Observe that

$$\ker \pi_0 = \{1, 2, 4, 5, 7, 8\} \ ,$$
$$\ker \pi_1 = \ker \pi_5 = \{1\} \ ,$$
$$\ker \pi_2 = \ker \pi_4 = \{1, 8\} \ ,$$
$$\ker \pi_3 = \{1, 4, 7\} \ .$$

The $\mathcal{U}$-orbits $\mathcal{O}_i$ and their stabilizer subgroups $\mathcal{U}_i$ are as follows.

$$
\begin{aligned}
\mathcal{O}_0 &= \{1, 2, 4, 5, 7, 8\} & \mathcal{U}_0 &= \{1\} \\
\mathcal{O}_1 &= \{3, 6\} & \mathcal{U}_1 &= \{1, 4, 7\} \\
\mathcal{O}_2 &= \{0\} & \mathcal{U}_2 &= \{1, 2, 4, 5, 7, 8\}
\end{aligned}
$$

Comparing $\mathcal{U}_i$ to $\ker \pi_j$, we see that $i_{\pi_0} = 1$ (remember that $(\pi_0, \mathcal{O}_2)$ is trivial and thus excluded), $i_{\pi_1} = i_{\pi_2} = i_{\pi_4} = i_{\pi_5} = 0$, and $i_{\pi_3} = 1$.

In writing down the matrix $\mathcal{A}_0$, we will use the order $1, 2, 4, 8, 7, 5, 3, 6$ on the non-zero elements of R. Remember that $a_0 = 0$. Then

$$\mathcal{A}_0 = \begin{pmatrix} a_1 & a_2 & a_4 & a_8 & a_7 & a_5 & a_3 & a_6 \\ a_2 & a_4 & a_8 & a_7 & a_5 & a_1 & a_6 & a_3 \\ a_4 & a_8 & a_7 & a_5 & a_1 & a_2 & a_3 & a_6 \\ a_8 & a_7 & a_5 & a_1 & a_2 & a_4 & a_6 & a_3 \\ a_7 & a_5 & a_1 & a_2 & a_4 & a_8 & a_3 & a_6 \\ a_5 & a_1 & a_2 & a_4 & a_8 & a_7 & a_6 & a_3 \\ a_3 & a_6 & a_3 & a_6 & a_3 & a_6 & 0 & 0 \\ a_6 & a_3 & a_6 & a_3 & a_6 & a_3 & 0 & 0 \end{pmatrix}.$$

To write down the matrix P, we need to order the non-trivial admissible pairs $(\pi_j, \mathcal{O}_i)$. We choose to write the real-valued characters first, followed by the others, grouped in conjugate pairs. Thus the order used is $(\pi_0, \mathcal{O}_0)$, $(\pi_0, \mathcal{O}_1)$, $(\pi_3, \mathcal{O}_0)$, $(\pi_3, \mathcal{O}_1)$, $(\pi_1, \mathcal{O}_0)$, $(\pi_5, \mathcal{O}_0)$, $(\pi_2, \mathcal{O}_0)$, $(\pi_4, \mathcal{O}_0)$. Then

$$P = \frac{1}{6} \begin{pmatrix} 1 & 0 & 1 & 0 & 1 & 1 & 1 & 1 \\ 1 & 0 & -1 & 0 & \xi & -\xi^2 & \xi^2 & -\xi \\ 1 & 0 & 1 & 0 & \xi^2 & -\xi & -\xi & \xi^2 \\ 1 & 0 & -1 & 0 & -1 & -1 & 1 & 1 \\ 1 & 0 & 1 & 0 & -\xi & \xi^2 & \xi^2 & -\xi \\ 1 & 0 & -1 & 0 & -\xi^2 & \xi & -\xi & \xi^2 \\ 0 & 3 & 0 & 3 & 0 & 0 & 0 & 0 \\ 0 & 3 & 0 & -3 & 0 & 0 & 0 & 0 \end{pmatrix}.$$

The reader will verify that $P^{-1}\mathcal{A}_0 P$ has the block diagonal form

$$P^{-1}\mathcal{A}_0 P = \begin{pmatrix} T_0 & 0 & 0 & 0 \\ 0 & T_3 & 0 & 0 \\ 0 & 0 & T_1 & 0 \\ 0 & 0 & 0 & T_2 \end{pmatrix},$$

where

$$T_0 = \begin{pmatrix} \hat{a}(\pi_0, 0) & 3\hat{a}(\pi_0, 1) \\ \hat{a}(\pi_0, 1) & 0 \end{pmatrix}, \quad T_3 = \begin{pmatrix} \hat{a}(\pi_3, 0) & 3\hat{a}(\pi_3, 1) \\ \hat{a}(\pi_3, 1) & 0 \end{pmatrix},$$

$$T_1 = \begin{pmatrix} 0 & \hat{a}(\pi_5, 0) \\ \hat{a}(\pi_1, 0) & 0 \end{pmatrix}, \quad T_2 = \begin{pmatrix} 0 & \hat{a}(\pi_4, 0) \\ \hat{a}(\pi_2, 0) & 0 \end{pmatrix}. \tag{1}$$

Thus $\det \mathcal{A}_0 = 9\hat{a}(\pi_0, 1)^2 \hat{a}(\pi_1, 0)\hat{a}(\pi_2, 0)\hat{a}(\pi_3, 1)^2 \hat{a}(\pi_4, 0)\hat{a}(\pi_5, 0)$.

Example 11. To illustrate Theorem 2, we continue to examine $R = \mathbb{Z}/(9)$. In writing down the matrix $\mathcal{A}$, we will use the order $1, 2, 4, 8, 7, 5, 3, 6, 0$ on the elements of R. In the matrix $\mathcal{A}$, a_0 need no longer be zero. The order on admissible pairs used to write down the matrix P is $(\pi_0, \mathcal{O}_0)$, $(\pi_0, \mathcal{O}_1)$, $(\pi_0, \mathcal{O}_2)$, $(\pi_3, \mathcal{O}_0)$,

$(\pi_3, \mathcal{O}_1)$, $(\pi_1, \mathcal{O}_0)$, $(\pi_5, \mathcal{O}_0)$, $(\pi_2, \mathcal{O}_0)$, $(\pi_4, \mathcal{O}_0)$. The matrices $\mathcal{A}$ and P are then

$$
\begin{pmatrix}
a_1 & a_2 & a_4 & a_8 & a_7 & a_5 & a_3 & a_6 & a_0 \\
a_2 & a_4 & a_8 & a_7 & a_5 & a_1 & a_6 & a_3 & a_0 \\
a_4 & a_8 & a_7 & a_5 & a_1 & a_2 & a_3 & a_6 & a_0 \\
a_8 & a_7 & a_5 & a_1 & a_2 & a_4 & a_6 & a_3 & a_0 \\
a_7 & a_5 & a_1 & a_2 & a_4 & a_8 & a_3 & a_6 & a_0 \\
a_5 & a_1 & a_2 & a_4 & a_8 & a_7 & a_6 & a_3 & a_0 \\
a_3 & a_6 & a_3 & a_6 & a_3 & a_6 & a_0 & a_0 & a_0 \\
a_6 & a_3 & a_6 & a_3 & a_6 & a_3 & a_0 & a_0 & a_0 \\
a_0 & a_0 & a_0 & a_0 & a_0 & a_0 & a_0 & a_0 & a_0
\end{pmatrix}, \quad
\frac{1}{6}
\begin{pmatrix}
1 & 0 & 0 & 1 & 0 & 1 & 1 & 1 & 1 \\
1 & 0 & 0 & -1 & 0 & \xi & -\xi^2 & \xi^2 & -\xi \\
1 & 0 & 0 & 1 & 0 & \xi^2 & -\xi & -\xi & \xi^2 \\
1 & 0 & 0 & -1 & 0 & -1 & -1 & 1 & 1 \\
1 & 0 & 0 & 1 & 0 & -\xi & \xi^2 & \xi^2 & -\xi \\
1 & 0 & 0 & -1 & 0 & -\xi^2 & \xi & -\xi & \xi^2 \\
0 & 3 & 0 & 0 & 3 & 0 & 0 & 0 & 0 \\
0 & 3 & 0 & 0 & -3 & 0 & 0 & 0 & 0 \\
0 & 0 & 6 & 0 & 0 & 0 & 0 & 0 & 0
\end{pmatrix}.
$$

Then the reader will verify that $P^{-1}AP$ has the block diagonal form

$$
P^{-1}AP =
\begin{pmatrix}
T_0 & 0 & 0 & 0 \\
0 & T_3 & 0 & 0 \\
0 & 0 & T_1 & 0 \\
0 & 0 & 0 & T_2
\end{pmatrix},
$$

where

$$
T_0 =
\begin{pmatrix}
\hat{a}(\pi_0, 0) & 3\hat{a}(\pi_0, 1) & 6a_0 \\
\hat{a}(\pi_0, 1) & 2a_0 & 2a_0 \\
a_0 & a_0 & a_0
\end{pmatrix},
$$

and the other T_i are as in (1). Thus

$$
\det \mathcal{A} = 9a_0 \left(\hat{a}(\pi_0, 1) - 2a_0\right)^2 \hat{a}(\pi_1, 0)\hat{a}(\pi_2, 0)\hat{a}(\pi_3, 1)^2 \hat{a}(\pi_4, 0)\hat{a}(\pi_5, 0) .
$$

Example 12. To illustrate Theorem 4, we consider the subgroup $U = \{1, 8\} \subset \mathcal{U}$ inside $R = \mathbb{Z}/(9)$. Only the characters π_0, π_2, π_4 belong to $(\mathcal{U} : U)$.

The U-orbits are $\{1, 8\}$, $\{2, 7\}$, $\{4, 5\}$, $\{3, 6\}$, $\{0\}$. The matrix $\mathcal{A}_0^U$ is

$$
\mathcal{A}_0^U =
\begin{pmatrix}
a_1 & a_2 & a_4 & a_3 \\
a_2 & a_4 & a_1 & a_3 \\
a_4 & a_1 & a_2 & a_3 \\
a_3 & a_3 & a_3 & 0
\end{pmatrix}.
$$

To write down P, we use only those non-trivial admissible orbits with $\pi \in (\mathcal{U} : U)$. They are $(\pi_0, \mathcal{O}_0)$, $(\pi_0, \mathcal{O}_1)$, $(\pi_2, \mathcal{O}_0)$, and $(\pi_4, \mathcal{O}_0)$. We write $\omega = \xi^2$, a primitive 3rd root of unity in $\mathbb{C}$. Then

$$
P = \frac{1}{3}
\begin{pmatrix}
1 & 0 & 1 & 1 \\
1 & 0 & \omega & \omega^2 \\
1 & 0 & \omega^2 & \omega \\
0 & 3 & 0 & 0
\end{pmatrix}.
$$

The reader will verify that

$$
P^{-1}\mathcal{A}_0^U P =
\begin{pmatrix}
\breve{a}(\pi_0, 0) & 3\breve{a}(\pi_0, 1) & 0 & 0 \\
\breve{a}(\pi_0, 1) & 0 & 0 & 0 \\
0 & 0 & 0 & \breve{a}(\pi_4, 0) \\
0 & 0 & \breve{a}(\pi_2, 0) & 0
\end{pmatrix}.
$$

Thus, $\det \mathcal{A}_0^U = 3\breve{a}(\pi_0, 1)^2 \breve{a}(\pi_2, 0)\breve{a}(\pi_4, 0)$.

Example 13. To illustrate Theorem 5, we indicate how to change the appropriate matrices from the previous example.

$$
\mathcal{A}^U = \begin{pmatrix} a_1 & a_2 & a_4 & a_3 & a_0 \\ a_2 & a_4 & a_1 & a_3 & a_0 \\ a_4 & a_1 & a_2 & a_3 & a_0 \\ a_3 & a_3 & a_3 & a_0 & a_0 \\ a_0 & a_0 & a_0 & a_0 & a_0 \end{pmatrix}, \quad P = \frac{1}{3} \begin{pmatrix} 1 & 0 & 0 & 1 & 1 \\ 1 & 0 & 0 & \omega & \omega^2 \\ 1 & 0 & 0 & \omega^2 & \omega \\ 0 & 3 & 0 & 0 & 0 \\ 0 & 0 & 3 & 0 & 0 \end{pmatrix}.
$$

Then

$$
P^{-1}\mathcal{A}^U P = \begin{pmatrix} \breve{a}(\pi_0, 0) & 3\breve{a}(\pi_0, 1) & 3a_0 & 0 & 0 \\ \breve{a}(\pi_0, 1) & a_0 & a_0 & 0 & 0 \\ a_0 & a_0 & a_0 & 0 & 0 \\ 0 & 0 & 0 & 0 & \breve{a}(\pi_4, 0) \\ 0 & 0 & 0 & \breve{a}(\pi_2, 0) & 0 \end{pmatrix}.
$$

Thus, $\det \mathcal{A}^U = 3a_0 \left(\breve{a}(\pi_0, 1) - a_0\right)^2 \breve{a}(\pi_2, 0)\breve{a}(\pi_4, 0)$.

References

1. Conrad, K.: The origin of representation theory. Preprint, 1998
2. Frobenius, F. G.: *Gesammelte Abhandlungen.* Berlin: Springer-Verlag 1968
3. Lam, T. Y.: Representations of finite groups: a hundred years. Notices Amer. Math. Soc. **45** (1998) 361–372, 465–474
4. Wood, J. A.: Extension theorems for linear codes over finite rings. In *Applied Algebra, Algebraic Algorithms and Error-Correcting Codes* (T. Mora, H. Mattson, eds.). LNCS **1255**, Berlin: Springer-Verlag (1997) 329–340
5. Wood, J. A.: Weight functions and the extension theorem for linear codes over finite rings. In *Finite Fields: Theory, Applications and Algorithms* (R. C. Mullin, G. L. Mullen, eds.). Contemp. Math. **225**, Providence: Amer. Math. Soc. (1999) 231–243

Printing: Mercedes-Druck, Berlin
Binding: Stürtz AG, Würzburg